VOLUME FOUR HUNDRED AND FORTY-THREE

METHODS IN
ENZYMOLOGY

Angiogenesis: *In Vitro* Systems

METHODS IN ENZYMOLOGY

Editors-in-Chief

JOHN N. ABELSON AND MELVIN I. SIMON

Division of Biology
California Institute of Technology
Pasadena, California

Founding Editors

SIDNEY P. COLOWICK AND NATHAN O. KAPLAN

VOLUME FOUR HUNDRED AND FORTY-THREE

Methods in

ENZYMOLOGY

Angiogenesis: *In Vitro* Systems

EDITED BY

DAVID A. CHERESH
University of California, San Diego
Moore's Cancer Center
La Jolla, CA, USA

AMSTERDAM • BOSTON • HEIDELBERG • LONDON
NEW YORK • OXFORD • PARIS • SAN DIEGO
SAN FRANCISCO • SINGAPORE • SYDNEY • TOKYO
Academic Press is an imprint of Elsevier

ELSEVIER

Contents

4. An Optimized Three-Dimensional *In Vitro* Model for the Analysis of Angiogenesis

Martin N. Nakatsu and Christopher C. W. Hughes

5. *In Vitro* Three Dimensional Collagen Matrix Models of Endothelial Lumen Formation During Vasculogenesis and Angiogenesis

Wonshill Koh, Amber N. Stratman, Anastasia Sacharidou, and George E. Davis

6. *In Vitro* Differentiation of Mouse Embryonic Stem Cells into Primitive Blood Vessels

Svetlana N. Rylova, Paramjeet K. Randhawa, and Victoria L. Bautch

CONTRIBUTORS

Alfred C. Aplin
Department of Pathology, University of Washington, Seattle, Washington

Victoria L. Bautch
Carolina Cardiovascular Biology Center and Department of Biology, The University of North Carolina at Chapel Hill, Chapel Hill North Carolina

Bradford C. Berk
Aab Cardiovascular Research Institute, University of Rochester School of Medicine and Dentistry, West Henrietta, New York

Diane R. Bielenberg
Vascular Biology Program, Department of Surgery, Children's Hospital, Harvard Medical School, Boston, Massachusetts

Brad A. Bryan
Schepens Eye Research Institute, and Departments of Ophthalmology and Pathology, Harvard Medical School, Boston, Massachusetts

Keith Burridge
Department of Cell and Developmental Biology, Lineberger Comprehensive Cancer Center, University of North Carolina at Chapel Hill, Chapel Hill, North Carolina

Tatiana V. Byzova
Department of Molecular Cardiology, JJ Jacobs Center for Thrombosis and Vascular Biology, The Cleveland Clinic Foundation, Cleveland, Ohio

Lena Claesson-Welsh
Uppsala University, Department of Genetics and Pathology, Rudbeck Laboratory, Uppsala, Sweden

Patricia A. D'Amore
Schepens Eye Research Institute, and Departments of Ophthalmology and Pathology, Harvard Medical School, Boston, Massachusetts

George E. Davis
Department of Pathology and Anatomical Sciences and Department of Medical Pharmacology and Physiology, School of Medicine and Dalton Cardiovascular Center, University of Missouri-Columbia, Columbia, Missouri

Behrad Derakhshan
Vascular Biology and Therapeutics Program and Department of Pharmacology, Yale University School of Medicine, New Haven, Connecticut

Eric Fogel
Division of Pathology and Laboratory Medicine, Veterans Administration Puget Sound Health Care System, Seattle, Washington

Keigi Fujiwara
Aab Cardiovascular Research Institute, University of Rochester School of Medicine and Dentistry, West Henrietta, New York

Kenneth D. Harrison
Vascular Biology and Therapeutics Program and Department of Pharmacology, Yale University School of Medicine, New Haven, Connecticut

Hayden Huang
Brigham and Women's Hospital, Cambridge, Masachusetts

Christopher C. W. Hughes
Department of Molecular Biology and Biochemistry, University of California Irvine, Irvine, California

Donald E. Ingber
Vascular Biology Program, Departments of Pathology & Surgery, Children's Hospital and Harvard Medical School, Boston, Massachusetts

Maxine Jonas
BioTrove, Masachusetts

Michael Klagsbrun
Department of Pathology and Vascular Biology Program, Department of Surgery, Children's Hospital, Harvard Medical School, Boston, Massachusetts

Wonshill Koh
Department of Medical Pharmacology and Physiology, School of Medicine and Dalton Cardiovascular Center, University of Missouri-Columbia, Columbia, Missouri

Xiujuan Li
Uppsala University, Department of Genetics and Pathology, Rudbeck Laboratory, Uppsala, Sweden

F. William Luscinskas
Department of Pathology, Brigham and Women's Hospital and Harvard Medical School, Center for Excellence in Vascular Biology, Boston, Massachusetts

Ganapati H. Mahabeleshwar
Department of Molecular Cardiology, JJ Jacobs Center for Thrombosis and Vascular Biology, The Cleveland Clinic Foundation, Cleveland, Ohio

Akiko Mammoto
Vascular Biology Program, Departments of Pathology & Surgery, Children's Hospital and Harvard Medical School, Boston, Massachusetts

Tadanori Mammoto
Vascular Biology Program, Departments of Pathology & Surgery, Children's Hospital and Harvard Medical School, Boston, Massachusetts

Manuela Martins-Green
Department of Cell Biology and Neuroscience, University of California, Riverside, Riverside, California

Qing Robert Miao
Vascular Biology and Therapeutics Program and Department of Pharmacology, Yale University School of Medicine, New Haven, Connecticut

William A. Muller
Department of Pathology, Northwestern University Feinberg School of Medicine, Chicago, Illinois

Martin N. Nakatsu
Department of Molecular Biology and Biochemistry, University of California Irvine, Irvine, California

Roberto F. Nicosia
Division of Pathology and Laboratory Medicine, Veterans Administration Puget Sound Health Care System, Seattle, Washington, and Department of Pathology, University of Washington, Seattle, Washington

Melissa Petreaca
Graduate Program in Cell, Molecular, and Developmental Biology, University of California, Riverside, Riverside, California

Paramjeet K. Randhawa
Department of Biology, The University of North Carolina at Chapel Hill, Chapel Hill, North Carolina

Cynthia A. Reinhart-King
Department of Biomedical Engineering, Cornell University, Ithaca, New York

Svetlana N. Rylova
Department of Biology, The University of North Carolina at Chapel Hill, Chapel Hill, North Carolina

Anastasia Sacharidou
Department of Medical Pharmacology and Physiology, School of Medicine and Dalton Cardiovascular Center, University of Missouri-Columbia, Columbia, Missouri

Julia E. Sero
Vascular Biology Program, Departments of Pathology & Surgery, Children's
Hospital and Harvard Medical School, Boston, Massachusetts

William C. Sessa
Vascular Biology and Therapeutics Program and Department of Pharmacology,
Yale University School of Medicine, New Haven, Connecticut

Masabumi Shibuya
Department of Molecular Oncology, Graduate School of Medicine and Dentistry,
Tokyo Medical and Dental University, Tokyo, Japan

Akio Shimizu
Vascular Biology Program, Department of Surgery, Children's Hospital, Harvard
Medical School, Boston, Massachusetts

Peter T. C. So
Departments of Mechanical and Biological Engineering, Massachusetts Institute of
Technology, Cambridge, Massachusetts

Amber N. Stratman
Department of Medical Pharmacology and Physiology, School of Medicine and
Dalton Cardiovascular Center, University of Missouri-Columbia, Columbia,
Missouri

Erika S. Wittchen
Department of Cell and Developmental Biology, Lineberger Comprehensive
Cancer Center, University of North Carolina at Chapel Hill, Chapel Hill,
North Carolina

Min Yao
Wellman Center for Photomedicine, Massachusetts General Hospital, Harvard
Medical School, Boston, Massachusetts

Penelope Zorzi
Department of Pathology, University of Washington, Seattle, Washington

Preface: A Tribute to Dr. Judah Folkman

The field of angiogenesis has recently lost its pioneer and leader, Dr. Judah Folkman. This was a tremendous loss to many of us who knew him and the field in general. Dr. Folkman inspired a generation of scientists lead an effort to translate basic discovery toward new therapeutics for a wide range of diseases including: cancer, blinding eye disease and inflammatory disease. Due in large part to Dr. Folkman's efforts and direction we now have the first generation of therapeutics that disrupt angiogenesis in patients suffering from cancer and macular degeneration. While Dr. Folkman clearly passed away before his time, he did live long enough to observe that many thousands of patients who are now better off due to anti-angiogenic therapy.

I had a rather interesting initiation to the field of anti-angiogenesis that was completely inspired by Dr. Folkman. In the mid-1980s as a junior faculty at the Scripps Research Institute I was studying what many of us in the field were beginning to appreciate were a family of cell adhesion receptors, later termed integrins. I had developed a monoclonal antibody (LM609) to the vitronectin receptor later referred to as integrin $\alpha v \beta 3$. During the course of my work, LM609 was used to stain a variety of diseased and normal tissues. To my surprise LM609 reacted strongly with blood vessels in tumors and inflammatory sites but failed to react with blood vessels in normal tissues. After seeing this result I began to read up on the emerging field of angiogenesis research. It was clear that most of the literature in the field came from Dr. Folkman or one of his disciples. I immediately contacted Dr. Folkman. By the time I finished describing our results I realized that he was excited as I was about our studies. In fact before I could ask him any questions he suggested I come to his lab to learn the chick chorioallantoic member (CAM) assay to determine whether LM609 might have an impact on angiogenesis in a quantitative animal model.

Naturally I arranged a trip to the Folkman lab within the next couple of weeks. I had never been to Harvard and was a bit intimidated by the place. As I took the elevator up to the Folkman lab and introduced myself to his administrative assistant who welcomed me and indicated that Dr. Folkman was expecting me. Within minutes Dr. Folkman, clad in a lab coat greated me and suggested we get started. At this point I assumed he was going to introduce me to one of his students or technicians who would then proceed to show me the CAM assay step by step. To my surprise, Dr. Folkman lead me to a hood, sat down and immediately started to instruct me how to induce angiogenesis on the CAM. In fact, the next I knew I was sitting at the hood next to Dr. Folkman going through the procedure in detail. Therefore, I can say I learned the technique from the master. Ultimately, Dr. Folkman introduced me to several members of the Folkman lab including Drs. Donald Inber, Pat D'Amore and Mike Klagsburn. I remember how enthusiastic and interactive each of these folks were. In fact, I am happy to say that I still maintain close contact with these individuals and have had many opportunities over the years to discuss science and reminisce about the past. In fact, Don, Pat and have all kindly contributed chapters to Methods in Enzymology volumes on Angiogenesis.

It was on my airplane ride home that I began to realize my career was about to make a change in course. From that point forward I began to focus on the role of adhesion receptors in angiogenesis and began to realize that blocking angiogenesis with integrin antagonists could have a very impressive impact on the growth of tumors in mice. Importantly, two of the agents we developed, including humanized LM609 have shown clinical activity in patients with late stage cancer.

Since my initiation to field I have since followed Dr. Folkman's work and have attended dozens of his lectures. Listening to a Folkman lecture is like watching one of your favorite movies, you can watch it over and over again and still find something interesting to think about. It was difficult for anyone to attend his lecture and not come away excited about science in general and angiogenesis in particular. The field of angiogenesis has matured over the past 25 years due in large part to Dr. Folkman's drive, enthusiasm, perseverance, and kindness. Dr. Folkman's leadership has helped to recruit many scientists and physicians from the academic and private sectors to focus on new approaches to develop inhibitors of angiogenesis.

In the early days there were a limited number of technological approaches to measure and study angiogenesis. The CAM assay was among the first quantitative approaches to measure the growth of newly forming blood vessels. From this humble beginning the field has exploded and as a result we now have a wide range of techniques approaches and animal models designed to monitor and study the growth of new blood vessels in development, tissue remodeling and disease. These methods are described in detail in this volume by many of the current leaders of the field.

David A. Cheresh

METHODS IN ENZYMOLOGY

CHAPTER ONE

APPROACHES FOR STUDYING ANGIOGENESIS-RELATED SIGNAL TRANSDUCTION

Behrad Derakhshan,* Kenneth D. Harrison,* Qing Robert Miao, *and* William C. Sessa

Contents

Vascular Biology and Therapeutics Program and Department of Pharmacology, Yale University School of Medicine, New Haven, Connecticut
* Equal contributors

Methods in Enzymology, Volume 443
ISSN 0076-6879, DOI: 10.1016/S0076-6879(08)02001-6

Abstract

Understanding how extracellular growth factors activate intracellular pathways that promote angiogenesis is a broad area of research. In this chapter, we outline the systematic dissection of vascular endothelial growth factor (VEGF)–mediated activation of endothelial nitric oxide synthase and other downstream targets that are relevant to the angiogenic response. These approaches may also be applied to most other angiogenic-factor signaling cascades.

1. INTRODUCTION

Cells are able to recognize and respond to a variety of extracellular stimuli. Signal transduction, the process of converting extracellular signals into intracellular responses, involves three basic components: (1) an extracellular stimulus, (2) a transducer, and (3) an intracellular signal. External signals may be composed of a variety of inputs; growth factors, hormones, and neurotransmitters are a few of the factors that serve to activate signal transduction cascades. This diverse array of extracellular inputs leads to an equally varied number of possible internal signaling mediators. Most often, the transducer of the signal is a cell surface receptor (e.g., a G protein–coupled receptor, ion channel, or receptor tyrosine kinase) that binds the extracellular factor and initiates the intracellular signaling pathway. This usually results in the generation of a second messenger (e.g., cAMP, Ca^{2+}, or phosphoinositides), which serves to propagate the initial signaling event and thus leads to an amplified intracellular response. Phosphorylation and dephosphorylation of proteins also play an important role in signal propagation, especially with regard to the control of signal amplification.

The kinetics underlying these events, as well as the degree of crosstalk between initially distinct pathways in the cell, leads to a large amount of complexity when trying to understand the physiologic consequence of a given extracellular stimulus. Because of the presence of multiple inputs at any given time, a static understanding of one signaling pathway is not sufficient to describe the information being processed in the cell. Moreover, some receptors and pathway intermediates are cell specific, leading to difficulties in forecasting physiologic outcome after activation of a particular signaling pathway.

2. ANGIOGENESIS

Angiogenesis, the sprouting of newly formed blood vessels from the preexisting vasculature, is a key process occurring in normal growth and development as well as in disease states, such as that observed during tumor progression from a dormant to a malignant state. During the angiogenic cascade, blood vessels undergo permeability changes that lead to leaking of key plasma proteins into the extracellular space (Ferrara, 1999; 2005). Ensuing degradation of the extracellular matrix and detachment of pericytes from the *lamina densa,* results in the release of pro-angiogenic growth factors that promote endothelial pruning and reorganization into intact, newly synthesized and fully functional blood vessels (Bryan and D'Amore, 2007).

Extensive studies of the vascular endothelial growth factor (VEGFA) have revealed VEGFA as a potent angiogenic mitogen and survival factor. VEGFA belongs to the gene family that additionally encodes placental growth factor (*PLGF*), *VEGFB, VEGFC,* and *VEGFD* (Ferrara, 1999). For the purposes of this chapter, whenever the term VEGF is used, we refer to the VEGFA family member.

2.1. VEGF signaling

VEGF binds and activates two distinct receptors with tyrosine kinase activity, VEGFR1 and VEGFR2, leading to the recruitment and subsequent phosphorylation of adaptor proteins and other downstream effectors that bind to the receptor tyrosine kinase, such as Grb2 and phospholipase Cγ (PLCγ) (Kowanetz and Ferrara, 2006). These receptors also serve to recruit and activate phosphatidylinositol 3'-kinase (PI3K). PI3K activation leads to phosphorylation of PtdIns(4,5)P (PIP$_2$), generating PtdIns(3,4,5)P (PIP$_3$), a binding site for the pleckstrin homology (PH) domain of Akt. Akt binds PIP$_3$ and is subsequently phosphorylated by phosphoinositide-dependent kinase-1 (PDK1) and target of rapamycin complex-2 (TORC2) on Thr308 and Ser473, respectively (Alessi *et al.,* 1997; Sarbassov *et al.,* 2005; Stephens *et al.,* 1998). Once activated, Akt phosphorylates eNOS on Ser1179, leading to maximal stimulation of eNOS-derived NO. This NO serves to regulate several aspects of vascular function by activating soluble guanylate cyclase (sGC) to produce cGMP, through nitrosation of iron-sulfur–containing proteins and by various nitro-sylation reactions involving proteins with reactive thiols (see Derakhshan *et al.,* 2007; Stamler *et al.,* 1997; 2001). VEGF can also activate PLCγ, resulting in the generation of inositol-1,4,5-trisphosphate (InsP$_3$) and dia-cylglycerol (DAG), which, in the case of the former, leads to the release of intracellular calcium stores and the subsequent binding of activated calmodulin to eNOS (thereby increasing NO release). This increase in intracellular calcium also activates Cam kinase II (CaMKII), which also serves to phosphorylate Ser1179 of eNOS and increase NO production (Fleming *et al.,* 2001). Conversely, DAG activates protein kinase C (PKC) isoforms, leading to phosphorylation of eNOS Thr497, an event that negatively regulates NO release (Fleming *et al.,* 2001). The ultimate differences in NO production downstream of VEGF-induced PLCγ activation

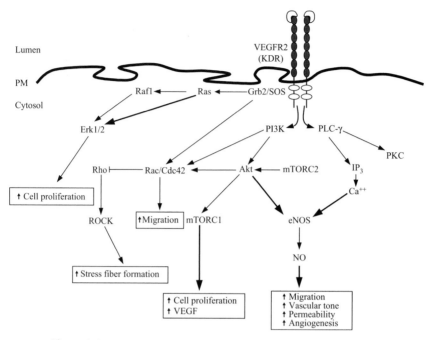

Figure 1.1   VEGF mediated signal transduction in endothelial cells.

highlight the importance of weighing positive and negative phosphorylation events on eNOS function.

The past few years have seen great leaps in our understanding of the mechanisms by which cells integrate and process a complex array of information. We will review several approaches used to study cell-signaling pathways, some of which are summarized in Fig. 1.1. As a template for these approaches, our focus will be VEGF-induced signal transduction pathways in endothelial cells that lead to the generation of nitric oxide (NO) by endothelial nitric oxide synthase (eNOS).

2.2. Stimulation of Akt phosphorylation in cell lysates

Akt is a central regulator of cell survival, migration, and metabolism (Manning and Cantley, 2007). Akt also regulates levels of eNOS-dependent NO release through phosphorylation of Ser1179 (Dimmeler *et al.*, 1999; Fulton *et al.*, 1999). Thus, an understanding of Akt activity in endothelial cells serves as an important step in assessing the potential for endothelial survival and migration. PI3K-generated phosphoinositides recruit Akt to the plasma membrane through its pleckstrin homology domain. Once at the PM, Akt is dually phosphorylated at Thr308 and Ser473 by two distinct kinases, PDK1 and TORC2, respectively (Alessi *et al.*, 1997; Sarbassov *et al.*, 2005; Stephens *et al.*, 1998). Because of the requirement of these phospho-sites for full activation of Akt, one can measure Akt activity not only by *in vitro* kinase assays but also by the simpler approach of phosphorylation state of the kinase.

Several endothelial cell lines have been used to study Akt signaling. We have used VEGF-stimulated bovine aortic endothelial cells (BAECs) because of the relatively robust nature of Akt signaling in these cells. When performing these experiments, one must maintain a high signal-to-noise ratio with regard to the Akt signaling pathway. One can ensure optimal conditions by focusing on: (1) the use of high-quality phospho-specific antibodies; (2) the use of phosphatase inhibitors to maintain the endogenously generated level of Akt phosphorylation; and (3) loading an optimal amount of protein in the gel to allow for efficient detection of phospho-Akt.

1. Culture cells in 6-well plates to 90% confluency. Wash 2× with sterile PBS and replace media with media containing reduced serum (0.1 to 0.5% heat-inactivated FBS) overnight (12 to 16 h).
2. Preincubate select wells with LY294002 (10 μM, 30 min) to assess VEGF-dependent PI3K activation.
3. Add VEGF (10 to 50 ng/ml, from 50 μg/ml stock) to each well for 5 to 10 min.
4. Aspirate media, wash 2× with ice-cold PBS, and lyse cells with RIPA buffer containing protease and phosphatase inhibitors. Rotate lysates at 4 degrees for 30 min.
5. Centrifuge lysates at 13,000 rpm to clear insoluble material.
6. Determine protein concentrations and prepare lysates for Western blotting.

2.3. Detection of Akt phosphorylation

We routinely use the LiCor Odyssey Imaging system for dual-color imaging of protein phosphorylation. Through the use of monoclonal and polyclonal antibodies followed by incubating the blot with secondaries from different species each coupled to fluorophores of different emission wavelengths (e.g., 680 nm and 800 nm), this system allows for the simultaneous detection of phosphorylated and total species of protein on the same blot. In this manner we quantitatively determine the relative degree of phospho-Akt/total Akt in a given experimental condition.

1. Perform SDS-PAGE.
2. Transfer proteins to nitrocellulose.
3. Block nitrocellulose in 0.1% casein/Tris-buffered saline (TBS) (Bio-Rad).

4. Incubate with primary antibody overnight at 4 °C in blocking solution—examples of primary antibodies used in our laboratory include phospho-S473 Akt (Cell Signaling)/total Akt (Cell Signaling) and phospho-T308 Akt (Cell Signaling).
5. Wash the blot 3 × 10 min with TBS + 0.1% Tween.
6. Incubate with secondary antibodies (680 nm and 800 nm emission) 45 min at RT.
7. Wash 6× 5 min with TBST. Rinse with TBS.
8. Detect phospho-Akt relative to total Akt levels with LiCor scanner.

2.4. Akt *in vitro* kinase assay

2.4.1. Preparation of BAEC for Akt *in vitro* kinase assay

BAEC are grown to 90% confluency in 6-well plates, deprived of serum for 36 to 48 h, and subsequently stimulated with VEGF (50 ng/ml). *In vitro* assays are carried by use of the Akt Kinase Assay Kit (Nonradioactive) from Cell Signaling Technologies, essentially as described in the accompanying product insert. A brief description of the assay steps is provided in the following.

1. After VEGF stimulation, aspirate the media and wash BAEC twice in ice-cold sterile PBS. Add lysis buffer to the cells, incubate on ice for 5 min and scrape the cells with a rubber policeman.
2. Dounce homogenize samples with 40 strokes and incubate cell lysates at 4 °C for 45 min with rotation. Centrifuge lysates to remove cellular debris.
3. Measure and adjust the protein concentration to 1 mg/ml for each assay sample.
4. Wash the immobilized Akt beads twice with PBS, once in lysis buffer and resuspend the beads in an equal volume of lysis buffer (1:1). Add beads to the protein lysates and carry out the kinase assay as described by the manufacturer, ensuring efficient incubation at 30 °C with gentle agitation to ensure the reaction is homogeneous. Perform SDS–PAGE for assay samples as described earlier by use of appropriate antibodies.

2.5. Use of dominant negative mutants to dissect signaling pathways

Dominant negative mutants interfere with the endogenous, wild–type protein. This could occur, for instance, through a deleterious mutation in the activation domain of a transcription factor while the DNA binding capacity remains intact (thereby preventing the wild-type transcription factor from binding DNA and effecting transactivation). Another mechanism by which dominant negatives might function is through proteins, which require dimerization for optimal activity.

Our laboratory has recently used the dominant negative approach centered on the latter mechanism to dissect the intricacies of the Akt-eNOS-Hsp90 axis (Miao *et al.*, 2008). Existing primarily as a homodimer (Minami *et al.*, 1991; Nemoto and Sato, 1998), Hsp90 is a molecular chaperone that is important for efficient phosphorylation of eNOS by Akt. An ATP-binding pocket in Hsp90 is critical for its chaperone function, and inhibition of ATP binding by use of inhibitors such as geldanamycin or 17-AAG locks the chaperone in an "open" conformation and results in degradation of its client proteins (Basso *et al.*, 2002; Whitesell *et al.*, 1994). Structure and function analysis of Hsp90 reveals Asp88 as the conserved ATP binding site (Miao *et al.*, 2008). This site thus functions as a template for the design of mutants deficient in ATP binding and, therefore, defective in chaperone activity.

To ascertain the effects of Hsp90 chaperone function on eNOS activity, endothelial cells must be transduced with a virus expressing the mutant Hsp90 construct. We routinely use the AdEasy system for generation of adenovirus (He *et al.*, 1998). With the preceding approaches we showed that adenoviral transduction of endothelial cells with an Hsp90D88N mutant reveals defects in VEGF-stimulated eNOS phosphorylation and NO release. Moreover, Hsp90D88N transduction results in decreased VEGF-induced Rac activation and stress fiber formation (Miao *et al.*, 2008).

2.6. Use of RNA interference to interrogate signaling pathways

The advent of RNA interference technology has considerably enhanced the repertoire of tools used to understand signaling pathways. We have used this approach in several recent studies to gain insights into endothelial cell signaling pathways (Bernatchez *et al.*, 2007; Fernandez-Hernando *et al.*, 2006; Miao *et al.*, 2006; Schleicher *et al.*, 2008; Suarez *et al.*, 2007). The reader is referred to excellent reviews on the proper design of siRNAs and important considerations that should be taken into account when interpreting data with this approach (Moffat and Sabatini, 2006; Pei and Tuschl, 2006). A good starting point for siRNA design might use the neural network-trained algorithm available at www.biopredsi.org (Huesken *et al.*, 2005). This algorithm provides 21 nucleotide siRNA sequences; these siRNA sequences are also available from Qiagen (www.qiagen. com). An absolutely crucial aspect of any RNAi experiment is the inclusion of proper controls. These controls include scrambled siRNA sequences, siRNA sequences denoted as "nonsilencing" (usually defined as such once a BLAST has been performed against the genome of interest), and the use of multiple sequences that target the transcript of interest and produce similar experimental outcomes. For the sake of brevity, a concise sample RNAi approach follows.

1. One must define the concentration of siRNA that produces knockdown of the gene of interest. Dose response and time dependence of knockdown are important constraints to be experimentally determined.
2. BAECs are readily transfectable with siRNA oligonucleotides. We routinely use oligofectamine as a transfection reagent to efficiently transfect BAECs with siRNA oligos.
3. Growth medium is removed from the cells. Cells are washed twice with PBS, and the siRNA transfection medium is added to the cells for 10 to 12 h. Transfection medium includes OPTI-MEM + siRNA of interest + oligofectamine. The ratio of oligofectamine (or transfection reagent of choice) to siRNA must be experimentally determined.
4. After 10 to 12 h of incubation with transfection medium, 2× growth medium is added to the transfection medium. Cells are incubated for 48 to 72 h after initial transfection before th experiment.

3. Phosphorylation of eNOS on Ser1179

Among the numerous potential phosphorylation sites in eNOS, perhaps the best studied is Ser1179 (S1179 in bovine eNOS, equivalent to S1177 in human eNOS and S1176 in murine eNOS) in the reductase domain.

Unstimulated cultured endothelial cells reveal a lack of significant phosphorylation at Ser1179 of eNOS. Phosphorylation and subsequent activation occur rapidly in response to stimuli such as fluid shear-stress (Dimmeler *et al.*, 1999; Gallis *et al.*, 1999), estrogen (Lantin-Hermoso *et al.*, 1997), insulin (Kim *et al.*, 2001), bradykinin (Fleming *et al.*, 2001), and VEGF (Dimmeler *et al.*, 1999; Michell *et al.*, 2001). The phosphorylation of eNOS at Ser1179 is primarily, but not solely, mediated by the serine/threonine kinase Akt/PKB *in vivo*. Stimulation of BAEC with VEGF results in phospho-activation of Akt by means of the phosphatidylinositol-3-OH kinase (PI(3)K) pathway (Fig. 1.1.). Activated Akt can further phosphorylate several cellular targets, including Ser1179 on eNOS. The net effect of Ser1179 phosphorylation is a twofold to threefold increase in NO production compared with basal eNOS activity (Dimmeler *et al.*, 1999; Fulton *et al.*, 1999). The PI(3)K inhibitors, LY294002 and Wortmannin, attenuate VEGF-induced phosphorylation and activation of eNOS (Dimmeler *et al.*, 1999; Fulton *et al.*, 1999). On phosphorylation of Ser1179, the observed increase in eNOS activity results as a consequence of accelerated electron flux from the reductase to the oxygenase domain of eNOS, because of enhanced calcium/CaM sensitization and depressing of the carboxy tail autoinhibitory control element, a 45-amino acid insertion in the FMN-binding module in eNOS (Lane and Gross, 2000; McCabe *et al.*, 2000).

Phosphorylation of eNOS is complemented by multiple protein–protein interactions providing spatial and temporal control of eNOS activity. For example, in endothelial cells, VEGF promotes association of eNOS and Hsp90, contributing to activation of eNOS (Garcia-Cardena *et al.*, 1998). The eNOS–Hsp90 association is an absolute requirement for Akt-mediated stimulation of eNOS (Brouet *et al.*, 2001). The net effect of Akt phosphorylation is an enhanced electron flux from the reductase domain to the oxygenase domain of eNOS and thus accelerated production of NO. In the resting state, eNOS is localized to plasmalemmal caveolae through direct tethering to the scaffolding domain of caveolin-1, leading to tonic inhibition of NO synthesis (Feron *et al.*, 1996; Garcia-Cardena *et al.*, 1996; 1997; Michel *et al.*, 1997; Venema *et al.*, 1997). The current dogma is that in response to VEGF, Hsp90 along with Ca^{++}-bound calmodulin disrupt the inhibitory eNOS-caveolin-1 complex, allowing Ca^{++}/CaM to bind eNOS, revealing a docking site for Akt-dependent phosphorylation and supporting eNOS catalysis—the conversion of L-arginine to NO.

3.1. Detection of eNOS phosphorylation by use of phospho-specific antibodies

The method described can be used both with a LiCor Odyssey Imaging system for dual-color imaging (as described later) or with enhanced chemiluminescence. Before commencing the experiment, serum-starve confluent BAEC, grown in a 6-well plate, for 48 h. Stimulate eNOS S1179 phosphorylation by adding VEGF for 0, 1, 3, 5, 10 and 30 min and stop the reaction immediately by aspirating off the media and adding ice-cold lysis buffer directly to the cells. To ensure maximal signal, add fresh phosphatase inhibitors to the lysis buffer.

1. Resolve the isolated proteins by SDS-PAGE and perform the Western blot by transferring onto nitrocellulose for 2 h at 300 mA at 4 °C.
2. Block the membranes for 90 min in 5% milk made up in TBST.
3. Incubate with primary antibody overnight at 4 °C, Zymed pS1179 (1:1000)/BD Biosciences total eNOS (1:2500).
4. Wash 3× for 10 min with TBST and incubate with secondary antibodies for 45 min at RT.
5. Wash in TBST for 5 min. Repeat the TBST washes 6 times and rinse with TBS before detection.
6. Detect eNOS phosphorylation on Ser1179 relative to total eNOS levels by use of the LiCor scanner. A typical blot is shown in Fig. 1.2. Maximal eNOS phosphorylation is observed at 5 min on stimulation with 50 ng/ml

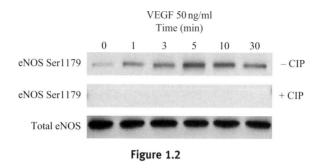

Figure 1.2

VEGF. Treatment with calf intestinal phosphatase (CIP) results in complete ablation of the phosphorylation signal.

3.2. Detection of NO from endothelial cells in culture

3.2.1. RFL-6 reporter bioassay

NO-mediated vasodilation depends principally on nitrosylation of the heme–iron in soluble guanylate cyclase (sGC), resulting in enzyme activation and intracellular cGMP accumulation (Forstermann *et al.*, 1994; Waldman and Murad, 1987). By use of a reporter cell line with robust levels of sGC, one can assay the short half-life of NO by quantifying levels of intracellular cGMP by ELISA (Forstermann *et al.*, 1994).

3.2.2. Sample preparation from BAEC in culture

1. BAECs are cultured as described earlier. Rat lung fibroblast cells (RFL-6; ATCC number: CCL-192) are grown in Ham's F-12 media supplemented with 10% (v/v) FBS, penicillin (100 units/ml)/streptomycin (100 μg/ml), L-glutamine (2 mM), and Fungizone (2.5 μg/ml) (complete HAM) at 37 °C in a humidified atmosphere of 5% CO_2. RFL-6 cells express high levels of sGC but are devoid of any NOS activity, making them an ideal cell type to assay conditioned media from stimulated endothelial cells.
2. BAEC-derived NO bioactivity was measured indirectly on the basis of the increase in cGMP elicited in RFL-6 reporter cells, after exposure to BAEC preconditioned media (Forstermann *et al.*, 1994). Wash BAEC grown in 6-well plates twice in Locke's buffer (10 mM HEPES, pH 7.4, 154 mM NaCl, 5.6 mM KCl, 2 mM $CaCl_2$, 1 mM $MgCl_2$, 3.6 mM $NaHCO_3$, 5.6 mM glucose, 500 μM arginine) and subsequently equilibrate in Locke's buffer in the presence or absence of 3 mM N^{ω}-methyl-L-arginine (NMA) for 30 min.

3. Incubate BAEC in Locke's buffer supplemented with superoxide dis-
   mutase (SOD, 100 U/ml) ± 3 mM NMA for 5 min. Subsequently,
   incubate relevant wells with fresh media in the presence or absence of
   VEGF (50 ng/ml) for 5 min.
4. Wash RFL-6 cells in Locke's buffer containing 300 μM 3-isobutyl
   1-methylxanthine (IBMX) and equilibrate in the same buffer for 20
   min. Two minutes before transferring the BAEC-conditioned media,
   incubate RFL-6 cells in Locke's buffer supplemented with 300 μM
   IBMX and 100 U/ml SOD. Aspirate the RFL-6 media and transfer
   BAEC conditioned media to the RFL-6 monolayer and incubated for 4
   min at RT. Rapidly aspirate off the media and quench cGMP production within seconds by addition of 109 μl of 0.1 N ice-cold HCl.
5. Incubate RFL-6 cells at RT for 20 min and lyse BAEC on ice, by the
   addition of RIPA buffer, and quantify protein to determine eNOS
   expression levels by Western blotting.
6. Harvest RFL-6 cells with a rubber policeman and clear lysates of insoluble material by centrifugation at 3500 rpm. Transfer supernatants into
   fresh tubes and lyophilize in a speed-vac for 4 to 6 h. Resuspend the
   lyophilized samples in 109 μl of enzyme immunoassay buffer (EIA)
   (Cayman Chemical, MI) and store at $-80\,°C$ until day of cGMP analysis.

3.2.3. cGMP quantification by enzyme-linked immunosorbent assay (ELISA)

Samples are assayed by use of a Cyclic GMP EIA Kit (Cayman), essentially
as described by the manufacturer.

1. Load 50 μl of samples and standards into 96-well plates. Each plate
   should contain two blank wells, two nonspecific binding wells, two
   maximum binding wells, and a total activity well.
2. Assay standards in duplicate and test samples in triplicate. Prepare plates
   by the addition of cGMP acetylcholine esterase tracer and cGMP antise-
   rum to the appropriate wells and incubated at 4 °C for 18 h in the dark.
3. Wash 5× with wash buffer and develop plates by addition of Ellman's
   Reagent for 90 min in the dark with shaking at RT.
4. Read plates by use of a Spectrophotometer equipped with a 405-nm
   filter. cGMP amounts are expressed as pmol of cGMP per mg of protein.

3.2.4. Citrulline assay for eNOS activity

The enzymatic, five-electron oxidation of the equivalent guanido nitrogen
of L-arginine to citrulline and NO by the NOS family of enzymes, has been
very well documented (see Stuehr and Griffith, 1992). The reaction requires
NADPH, molecular oxygen, calcium/calmodulin, and the cofactor tetra-
hydrobiopterin. The citrulline assay described herein monitors the conversion
of radiolabelled [³H]L-arginine to [³H]L-citrulline, because the stoichiometry

of this reaction is equivalent to the enzymatic release of NO (Bredt and Snyder, 1989; 1990).

3.2.4.1. Preparation of the cation-exchange column

1. Prepare the resin (Dowex AG 50WX-8 (H^+ *or* Na^+)—100 to 200 mesh) by placing 25 g in a 500-ml beaker and adding 250 ml of 0.5 N HCl.
2. Stir for 2 h at RT and subsequently allow resin beads to settle; decant HCl and unsettled light particles.
3. Wash the resin several times with 100 ml of dH_2O, stir for 3 min, allow particles to settle, and decant the supernatant; check the pH with pH paper and stop washes when pH is between 5 and 6.
4. Add 250 ml of 0.5 N NaOH and stir for 2 h at RT (allow the beads to settle and remove the supernatant).
5. Wash the settled resin as in step 3, until the pH is at or below 8.
6. Equilibrate the resin with three successive 5-ml washes with stop buffer (50 mM HEPES, pH 5.5, 5 mM EDTA, 5 mM EGTA, 2 mM L-citrulline).

3.2.4.2. Purification of radiolabeled arginine (${_L}$-[^{3}H]arginine or ${_L}$-[^{14}C] arginine)

1. Equilibrate Dowex resin with 5 ml of ice-cold stop buffer.
2. Apply the radioactive arginine to 1.0 ml of equilibrated resin in a disposable flow column.
3. Wash the column with 6 ml of ice-cold distilled water.
4. Elute the arginine with 6 ml (in consecutive 2 ml washes) of 0.5 M NH_4OH into a glass test tube; wash Dowex once with 6 ml of ice-cold water and twice with ice-cold stop buffer.
5. Transfer the eluted arginine to microcentrifuge tubes and lyophilize in a speed-vac.
6. Resuspend the arginine in 2% ethanol (same volume as was originally loaded on the Dowex).

3.2.4.3. Preparation of BAEC for detection of NO

1. Wash cells in PSS (22 mM HEPES, pH 7.4, 125 mM NaCl, 5 mM KCl, 1 mM $MgCl_2$, 1.5 mM $CaCl_2$, 0.16 mM Na_2PO_4, 0.4 mM NaH_2PO_4, 5 mM $NaHCO_3$, 5.6 mM D-glucose) and equilibrate cells for 60 min; at this point add 3 mM NMA to appropriate wells.
2. Wash cells in PSS twice to remove excess NMA and add cold arginine (9.8 μM) and 3.3 μCi of ${_L}$-[^{3}H]arginine for 15 min in the presence or absence of VEGF (50 ng/ml).
3. Rinse cells quickly with ice-cold Stop-PBS (1× PBS, 5 mM arginine, 4 mM EDTA) and fix cells in ice-cold absolute ethanol for 10 min on ice.

4. Place in a 37 °C incubator until the ethanol evaporates and stop the reaction with of stop buffer for 5 min (keep a sample of the lysate for protein analysis or add RIPA to the cells and lyse to measure protein content).
5. After the 5 min incubation, transfer samples to 2-ml Eppendorf tubes with equilibrated beads and add the equilibrated Dowex resin. Rotate the tubes at 4 °C for 10 min.
6. Spin the resin down and remove the supernatant for counting by use of a conventional scintillation counter.

4. MEASURING RELEASED NO BY NO-SPECIFIC CHEMILUMINESCENCE

After stimulation with VEGF, incubate BAEC in serum-free media and analyze samplings of conditioned media to detect nitrite levels, a stable breakdown product of NO in aqueous solution.

4.1. Basal NO (nitrite) accumulation

1. Plate BAEC in 6-well plates and allow them to reach 90% confluency.
2. The day before the experiment, add 2 ml of serum-containing media to the cells.
3. After 16 to 24 h, collect 500-μl samples of conditioned media into Eppendorf tubes and centrifuge to sediment any floating cells (these samples are used to determine the basal accumulated NO release).
4. Deprive BAEC of serum to quiesce the cells for 4 h before stimulus.
5. Transfer 200 μl of the supernatant to fresh tubes and add two volumes of chilled EtOH (-20 °C) and incubate on ice for 1 h to precipitate serum proteins.
6. After precipitation, centrifuge at 14,000 rpm for 10 min and transfer all of the supernatant to new tubes.
7. Dry supernatants to remove EtOH by use of a speed-vac and resuspend samples in 200 μl of dH$_2$O. Store the samples at -20 °C until NO release measurements.

4.2. NO (nitrite) accumulation before and after agonist stimulation

1. Subsequent to incubation of BAEC in serum-free media, remove the media and replace with an additional 2 ml of serum-free media to allow NO accumulation for 30 min at 37 °C (the amount of accumulated NO is used to determine NO release prestimulation).

2. Sample the media and centrifuge to sediment floating cells and store the soup at $-20\,°C$.
3. Stimulate BAEC by adding VEGF directly to the media and allow NO to accumulate for further 30 min at $37\,°C$ in the presence or absence of NMA (the amount of accumulated NO is that induced on VEGF-stimulation).
4. Sample the media and centrifuge to sediment floating cells and store the soup at $-20\,°C$.
5. Wash the cells with cold PBS twice and lyse cells in RIPA for 30 min on ice to determine the protein concentration (express as NO release per milligram of protein).

Nitrite and nitrate can be measured by use of a NO chemiluminescence detector (Sievers NO analyzer). Cytosolic nitrite and nitrate is reduced to NO by incubation with glacial acetic acid containing $NaIO_4$. The resulting NO is measured by the chemiluminescence derived from its reaction with ozone, according to the following two reactions (where $h\nu$ is direct chemiluminescence):

$$NO + O_3 \rightarrow NO_2^- + O_2 \qquad (1.1)$$

$$NO_2^- \rightarrow NO_2 + h\nu \qquad (1.2)$$

4.3. VEGF-induced stress fiber formation

Migration of endothelial cells is critical for the formation of new blood vessels during angiogenesis. At the cellular level, an event indicative of the process of migration is the formation of stress fibers. Stress fibers are a specific cytoskeletal organization of actin monomers. Actin filament elongation and actin-myosin filament contraction are key regulatory events that provide the primary driving forces for cell movement (Hall, 2005). VEGF can stimulate the activity of small GTPases (such as Rho, Rac, and Cdc42) leading to cytoskeletal rearrangements that regulate cell contractility through the formation of lamellipodia and filopodia (Soga *et al.*, 2001b; Wojciak-Stothard *et al.*, 2001; Wojciak-Stothard and Ridley, 2002). Rho promotes actin polymerization through activation of mDia, a formin-containing protein that binds to filament barbed ends and promotes their linear elongation. Rho also activates p160Rho kinase (ROCK), which in turn phosphorylates and thereby inactivates myosin light-chain phosphatase. Inactivation of myosin light chain phosphatase leads to an increase in myosin II activity, resulting in the cross-linking of actin filaments and the generation of contractile force. The other two VEGF-stimulated small GTPases mentioned previously, Rac and Cdc42, both activate the Arp2/3 complex, which interacts with preexisting filaments to generate a branched cytoskeletal network. Rac and Cdc42 mediate their effects through interactions with complexes containing the effectors Sra-1 or WASP, respectively. In this manner, Rac and Cdc42 function to regulate the formation of lamellipodia and filopodia, respectively. These processes are essential for directed cell movement (Hall, 2005).

Accumulating evidence indicates that VEGF-directed cell migration is associated with regulation of the actin cytoskeleton through activation of Rac GTPase (Soga *et al.*, 2001a,b). Chemotaxis induced by VEGF on a specific extracellular matrix is a balance of VEGF receptor activation and integrin-stimulation (Breier, 2000; Soldi *et al.*, 1999). We will focus here on cytoskeletal signaling regulated by VEGF. Previous studies have shown that VEGF induces edge ruffling (Waltenberger *et al.*, 1994) and stress fiber formation in cultured endothelial cells (Morales-Ruiz *et al.*, 2000; Rousseau *et al.*, 1997). Recent studies that used Rac-1-null cells have demonstrated that Rac1 plays a key role in actin lamellipodia induction, cell-matrix adhesion, and in the regulation of cell migration by stimulating lamellipodial extension (Guo *et al.*, 2006).

4.4. VEGF-induced Rac activation

Rho, Rac, and Cdc42 are members of the Ras GTPase superfamily and thus share sequence homology and enzymatic function (Fryer and Field, 2005). The Ras GTPases cycle between GTP-bound active states and GDP-bound inactive states. In endothelial cells, Rho, Rac, and Cdc42 regulate stress fiber formation and cell–cell junctions in response to VEGF (Soga *et al.*, 2001b) (Wojciak-Stothard *et al.*, 2001; Wojciak-Stothard and Ridley, 2002). A large body of evidence indicates that Rac is a key regulator of cell migration and endothelial morphogenesis because of its ability to stimulate lamellipodial extension (Connolly *et al.*, 2002; Soga *et al.*, 2001a,b). Current studies regarding the precise mechanisms of Rac activation are controversial. Most studies support the notion that Rac is activated by both tyrosine kinase and G-protein–coupled receptors. Rac activation is often dependent on PI3K activity, because PI3K is thought promote the activation of guanine nucleotide exchange factors (GEFs, Vav, and Tiam1) (Das *et al.*, 2000; Fleming *et al.*, 2000; 2004; Quilliam *et al.*, 2002; Welch *et al.*, 2003) that stimulate exchange of GDP for GTP, thereby activating Rac. Moreover, inhibitors of PI3K block Rac activation (Rickert *et al.*, 2000; Royal *et al.*, 2000; Sander *et al.*, 1998). GTPases are inactivated by GTPase activating protein (GAPs; e.g., Bcr), which stimulate the intrinsic GTPase activity of the small GTPase (Bernards, 2003; Bernards and Settleman, 2004). Therefore, mutant forms of Rac GTPases that favor an activated GTP-bound state (V12) or inactive GDP-bound state (N17) are valuable tools to for the explanation of Rac GTPase function *in vitro*. Soga *et al.* demonstrated that Rac is an important regulator of stress fiber formation in endothelial cells (Soga *et al.*, 2001a,b) with Rac mutants. They found constitutively active V12Rac to induce the haptotaxis of endothelial cells on collagen- and vitronectin-coated supports, an activity that was further stimulated by treatment with VEGF. In agreement with these results, the dominant-negative Rac mutant N17Rac inhibited the basal and VEGF-stimulated chemotaxis of endothelial cells on collagen- and vitronectin-coated supports.

A novel technique for the detection of Rac activation in living cells has been developed. Results from these studies show that active Rac is localized primarily toward the leading edge of migrating cells (Kraynov *et al.*, 2000). This technique has been discussed in a previous volume of *Methods in Enzymology* (Chamberlain *et al.*, 2000).

In quiescent BAEC, F-actin was found mostly in peripheral membrane structures and unorganized fibers throughout the cell. As expected, VEGF induced the formation of long, condensed stress fibers, and VEGF treatment resulted in Rac activation in BAEC. Treatment of BAEC with the PI3K inhibitor LY294002 blocked VEGF-induced Rac activation, whereas a Rac1 inhibitor (Gao *et al.*, 2004) did not suppress VEGF-stimulated phosphorylation of Akt in BAEC, indicating that PI-3 kinase, but not Rac, is likely upstream of Akt. Consequently, both LY294002 and the Rac1 inhibitor suppressed VEGF-stimulated BAEC cell migration. Our previous results also show that VEGF-induced cell migration and F-actin rearrangement are dependent on Akt activity (Ackah *et al.*, 2005; Morales-Ruiz *et al.*, 2000) and that constitutively activated Akt is sufficient to cause cellular chemokinesis, most likely because of its effects on stress fiber formation (Morales-Ruiz *et al.*, 2000).

NO can influence the tractional forces in activated endothelial cells and influence remodeling of focal adhesions, perhaps by influencing tyrosine phosphorylation of focal adhesion kinase (Goligorsky *et al.*, 1999). However, whether Rac activation and stress fiber formation are modulated by NO production is still unclear and is the focus of current studies.

4.5. Stress fiber formation assay

Stress fiber formation can be determined by immunofluorescent staining of F-actin by use of fluorescent phallotoxins (Molecular Probes, Invitrogen). The vial contents should be dissolved in 1.5 ml methanol to yield a final concentration of 6.6 μM.

1. Coat glass coverslips with 0.1% gelatin in PBS, pH 7.4, for 30 min.
2. Aspirate gelatin and let air-dry for 30 min.
3. Plate cells on gelatin-coated coverslips and let cells grow to desired confluency.
4. Arrest cell growth with serum-free medium.

5. Stimulate cells with growth factors or cytokines for desired time.
6. Wash cells with PBS twice.
7. Fix cells in 2% paraformaldehyde in PBS for 15min on ice.
8. Wash cells with PBS twice.
9. Permeabilize cells with 0.1% Triton-X 100 in PBS for 5 min.
10. Wash cells with PBS twice.
11. Block with 1% bovine serum albumin (BSA) in PBS for 30 min.
12. Prepare staining solution at 1:50 dilution (dilute 5 μl methanolic stock solution into 250 μl PBS containing 1%BSA).
13. Add 250 μl staining solution to each coverslip.
14. Incubate at room temperature for 30 min.
15. Wash cells with PBS three times.
16. Mount coverslips in permanent anti-fade mounting medium such as ProLong® Gold reagent or Cytoseal.
17. Store in the dark at 4 °C.

4.6. Approaches to dissect Rac1 function in cells

To examine the role of Rac1 in VEGF signaling pathways, several tools are available such as constitutively active or dominant-negative Rac1, Rac1 inhibitors, small interference RNA (siRNA) directed against Rac1 expression, and Rac1 knockout cells. The most common approach used thus far has been the application of constitutively active or dominant-negative Rac1 developed by Ridley et al. (1992). Constitutively active Rac1 (V12 Rac1) consists of a mutation of amino acid 12 from Gly (G) to Val (V), thereby decreasing the intrinsic GTPase activity of Rac1, making the protein less responsive to GAPs (Ridley et al., 1992). Dominant-negative Rac1 (N17 Rac1), consisting of a mutation of amino acid 17 from Ser to Asn, has preferential affinity for GDP and thus primarily remains in a GDP-bound inactive state. The eukaryotic expression vector encoding V12 Rac1 or N17 Rac1 is introduced into cells by several established approaches, including microinjection into the nuclei of cells (Ridley et al., 1992) or by the previously mentioned approach of adenoviral transduction (Wojciak-Stothard et al., 2001). A cell-permeable inhibitor that specifically and reversibly inhibits Rac1 GDP/GTP exchange activity by interfering with the interaction between Rac1 and Rac-specific GEFs (guanine nucleotide exchange factors) Trio and Tiam1 is also widely applied (Gao et al., 2004). This inhibitor can effectively inhibit Rac1-mediated cellular functions in NIH3T3 and PC-3 cells at 50 to 100 μM and exhibits no effect on Cdc42 or RhoA activation (Desire et al., 2005; Gao et al., 2004). Recently, siRNA for Rac1 was applied (Xue et al., 2004), and several companies provide validated siRNA for Rac1 and other small GTPases. In addition, Rac-1-null mouse embryonic fibroblasts (MEFs) have been generated by infecting Rac$^{loxp/loxp}$ MEF cells with adenovirus expressing the Cre recombinase.

This approach further demonstrated that Rac1 plays a key role in actin lamellipodia induction, cell–matrix adhesion, and in the regulation of cell migration by stimulating lamellipodial extension (Guo *et al.*, 2006).

4.7. Rac activity assay

The Rac activity assay has been described in a previous volume of *Methods in Enzymology* (Benard and Bokoch, 2002). At present, several commercial kits for Rac activity measurements are available. The EZ-Detect Rac1 Activation kit from Pierce (Product#89856) has provided superior results. The activation of Rac1 is determined by isolating Rac-GTP by way of its specific downstream effector PAK1. The p21-binding domain of PAK1 is expressed as a GST-fusion protein (GST-PBD) and immobilized on a glutathione resin (Benard *et al.*, 1999; Benard and Bokoch, 2002). The isolated active Rac1 is then detected by a SDS-PAGE with a specific mouse monoclonal antibody.

1. Cells are grown to 90 to 100% confluency in 100-mm culture dishes.
2. After treatments, cell are lysed in 500 μl lysis/binding/washing buffer (100 ml, contains 25 mM Tris-HCl, pH 7.5, 150 mM NaCl, 5 mM MgCl$_2$, 1% NP-40, 1 mM DTT, and 5% glycerol) at 4 °C.
3. For preparation of positive or negative controls, the clarified cell lysate (500 μg) is treated with 0.1 mM GTPγS or 1.0 mM GDP in the presence of 10 mM EDTA, pH 8.0 at 30 °C for 15 min (to activate or inactivate Rac1). The nucleotide exchange reaction is terminated by adding MgCl$_2$.
4. The treated cell lysates (500 μg) are incubated with GST-Pak1-PBD in the presence of SwellGel Immobilized Glutathione resin (Pierce) at 4 °C for 1 h.
5. After incubation, the mixture is centrifuged at 8000g to remove the unbound proteins.
6. The resins are washed three times with lysis/binding/wash buffer and the sample is eluted by adding 50 μl of 2× SDS sample buffer and boiling at 95 °C for 5 min.
7. Half (25 μl) of the sample volumes are analyzed by SDS-PAGE (4 to 20% polyacrylamide mini-gel) and transferred to a nitrocellulose membrane.
8. The active Rac1 is detected by Western blotting with a specific mouse monoclonal antibody.

5. Concluding Remarks

Despite the complexity underlying angiogenic signaling cascades, great progress has been made in recent years in our understanding of how particular agonists activate the components of specific pathways. However, an entirely linear signaling pathway does not exist. Major challenges remain regarding our ability to relate these reductionist approaches to a systematic understanding of cell signaling networks and, in particular, how activation of these networks relate to a given phenotype *in vivo*. It is our expectation that novel insights gained through continued use of the approaches outlined in this review will provide the necessary foundation to address the future challenges that lie ahead.

REFERENCES

Alessi, D. R., James, S. R., Downes, C. P., Holmes, A. B., Gaffney, P. R. J., Reese, C. B., and Cohen, P. (1997). Characterization of a 3-phosphoinositide–dependent protein kinase which phosphorylates and activates protein kinase B[alpha]. *Curr. Biol.* **7,** 261–269.

Ackah, E., Yu, J., Zoellner, S., Iwakiri, Y., Skurk, C., Shibata, R., Ouchi, N., Easton, R. M., Galasso, G., Birnbaum, M. J., Walsh, K., and Sessa, W. C. (2005). Akt1/protein kinase Balpha is critical for ischemic and VEGF-mediated angiogenesis. *J. Clin. Invest.* **115,** 2119–2127.

Alessi, D. R., James, S. R., Downes, C. P., Holmes, A. B., Gaffney, P. R. J., Reese, C. B., and Cohen, P. (1997). Characterization of a 3-phosphoinositide–dependent protein kinase which phosphorylates and activates protein kinase B[alpha]. *Curr. Biol.* **7,** 261–269.

Basso, A. D., Solit, D. B., Chiosis, G., Giri, B., Tsichlis, P., and Rosen, N. (2002). Akt Forms an intracellular complex with heat shock protein 90 (Hsp90) and Cdc37 and is destabilized by inhibitors of Hsp90 function. *J. Biol. Chem.* **277,** 39858–39866.

Benard, V., Bohl, B. P., and Bokoch, G. M. (1999). Characterization of rac and cdc42 activation in chemoattractant-stimulated human neutrophils using a novel assay for active GTPases. *J. Biol. Chem.* **274,** 13198–13204.

Benard, V., and Bokoch, G. M. (2002). Assay of Cdc42, Rac, and Rho GTPase activation by affinity methods. *Methods Enzymol.* **345,** 349–359.

Bernards, A. (2003). GAPs galore! A survey of putative Ras superfamily GTPase activating proteins in man and *Drosophila*. *Biochim. Biophys. Acta* **1603,** 47–82.

Bernards, A., and Settleman, J. (2004). GAP control: Regulating the regulators of small GTPases. *Trends Cell. Biol.* **14,** 377–385.

Bernatchez, P. N., Acevedo, L., Fernandez-Hernando, C., Murata, T., Chalouni, C., Kim, J., Erdjument-Bromage, H., Shah, V., Gratton, J.-P., McNally, E. M., Tempst, P., and Sessa, W. C. (2007). Myoferlin regulates vascular endothelial growth factor receptor-2 stability and function. *J. Biol. Chem.* **282,** 30745–30753.

Bredt, D. S., and Snyder, S. H. (1989). Nitric oxide mediates glutamate-linked enhancement of cGMP levels in the cerebellum. *Proc. Natl. Acad. Sci. USA* **86,** 9030–9033.

Bredt, D. S., and Snyder, S. H. (1990). Isolation of nitric oxide synthetase, a calmodulin-requiring enzyme. *Proc. Natl. Acad. Sci. USA* **87,** 682–685.

Breier, G. (2000). Functions of the VEGF/VEGF receptor system in the vascular system. *Semin. Thromb. Hemost.* **26,** 553–559.

Brouet, A., Sonveaux, P., Dessy, C., Balligand, J. L., and Feron, O. (2001). Hsp90 ensures the transition from the early Ca^{2+}-dependent to the late phosphorylation-dependent activation of the endothelial nitric-oxide synthase in vascular endothelial growth factor-exposed endothelial cells. *J. Biol. Chem.* **276,** 32663–32669.

Bryan, B. A., and D'Amore, P. A. (2007). What tangled webs they weave: Rho-GTPase control of angiogenesis. *Cell. Mol. Life Sci.* **64,** 2053–2065.

Chamberlain, C. E., Kraynov, V. S., and Hahn, K. M. (2000). Imaging spatiotemporal dynamics of Rac activation *in vivo* with FLAIR. *Methods Enzymol.* **325**, 389–400.

Connolly, J. O., Simpson, N., Hewlett, L., and Hall, A. (2002). Rac regulates endothelial morphogenesis and capillary assembly. *Mol. Biol. Cell* **13**, 2474–2485.

Das, B., Shu, X., Day, G. J., Han, J., Krishna, U. M., Falck, J. R., and Broek, D. (2000). Control of intramolecular interactions between the pleckstrin homology and Dbl homology domains of Vav and Sos1 regulates Rac binding. *J. Biol. Chem.* **275**, 15074–15081.

Derakhshan, B., Hao, G., and Gross, S. S. (2007). Balancing reactivity against selectivity: The evolution of protein S-nitrosylation as an effector of cell signaling by nitric oxide. *Cardiovasc. Res.* **75**, 210–219.

Desire, L., Bourdin, J., Loiseau, N., Peillon, H., Picard, V., De Oliveira, C., Bachelot, F., Leblond, B., Taverne, T., Beausoleil, E., Lacombe, S., Drouin, D., *et al.* (2005). RAC1 inhibition targets amyloid precursor protein processing by gamma-secretase and decreases Abeta production *in vitro* and *in vivo*. *J. Biol. Chem.* **280**, 37516–37525.

Dimmeler, S., Fleming, I., Fisslthaler, B., Hermann, C., Busse, R., and Zeiher, A. M. (1999). Activation of nitric oxide synthase in endothelial cells by Akt-dependent phosphorylation. *Nature* **399**, 601–605.

Fernandez-Hernando, C., Fukata, M., Bernatchez, P. N., Fukata, Y., Lin, M. I., Bredt, D. S., and Sessa, W. C. (2006). Identification of Golgi-localized acyl transferases that palmitoylate and regulate endothelial nitric oxide synthase. *J. Cell. Biol.* **174**, 369–377.

Feron, O., Belhassen, L., Kobzik, L., Smith, T. W., Kelly, R. A., and Michel, T. (1996). Endothelial nitric oxide synthase targeting to caveolae. Specific interactions with caveolin isoforms in cardiac myocytes and endothelial cells. *J. Biol. Chem.* **271**, 22810–22814.

Ferrara, N. (1999). Role of vascular endothelial growth factor in the regulation of angiogenesis. *Kidney Int.* **56**, 794–814.

Ferrara, N. (2005). The role of VEGF in the regulation of physiological and pathological angiogenesis. *Exs.* **94**, 209–231.

Fleming, I., Fisslthaler, B., Dimmeler, S., Kemp, B. E., and Busse, R. (2001). Phosphorylation of Thr(495) regulates Ca(2$^+$)/calmodulin-dependent endothelial nitric oxide synthase activity. *Circ. Res.* **88**, E68–E75.

Fleming, I. N., Batty, I. H., Prescott, A. R., Gray, A., Kular, G. S., Stewart, H., and Downes, C. P. (2004). Inositol phospholipids regulate the guanine-nucleotide-exchange factor Tiam1 by facilitating its binding to the plasma membrane and regulating GDP/GTP exchange on Rac1. *Biochem. J.* **382**, 857–865.

Fleming, I. N., Gray, A., and Downes, C. P. (2000). Regulation of the Rac1-specific exchange factor Tiam1 involves both phosphoinositide 3-kinase-dependent and -independent components. *Biochem. J.* **351**, 173–182.

Forstermann, U., Closs, E. I., Pollock, J. S., Nakane, M., Schwarz, P., Gath, I., and Kleinert, H. (1994). Nitric oxide synthase isozymes. Characterization, purification, molecular cloning, and functions. *Hypertension* **23**, 1121–1131.

Fryer, B. H., and Field, J. (2005). Rho, Rac, Pak and angiogenesis: Old roles and newly identified responsibilities in endothelial cells. *Cancer Lett.* **229**, 13–23.

Fulton, D., Gratton, J.-P., McCabe, T. J., Fontana, J., Fujio, Y., Walsh, K., Franke, T. F., Papapetropoulos, A., and Sessa, W. C. (1999). Regulation of endothelium-derived nitric oxide production by the protein kinase Akt. *Nature* **399**, 597–601.

Gallis, B., Corthals, G. L., Goodlett, D. R., Ueba, H., Kim, F., Presnell, S. R., Figeys, D., Harrison, D. G., Berk, B. C., Aebersold, R., and Corson, M. A. (1999). Identification of flow-dependent endothelial nitric-oxide synthase phosphorylation sites by mass spectrometry and regulation of phosphorylation and nitric oxide production by the phosphatidylinositol 3-kinase inhibitor LY294002. *J. Biol. Chem.* **274**, 30101–30108.

Gao, Y., Dickerson, J. B., Guo, F., Zheng, J., and Zheng, Y. (2004). Rational design and characterization of a Rac GTPase-specific small molecule inhibitor. *Proc. Natl. Acad. Sci. USA* **101**, 7618–7623.

Garcia-Cardena, G., Fan, R., Shah, V., Sorrentino, R., Cirino, G., Papapetropoulos, A., and Sessa, W. C. (1998). Dynamic activation of endothelial nitric oxide synthase by Hsp90. *Nature* **392**, 821–824.

Garcia-Cardena, G., Fan, R., Stern, D. F., Liu, J., and Sessa, W. C. (1996). Endothelial nitric oxide synthase is regulated by tyrosine phosphorylation and interacts with caveolin-1. *J. Biol. Chem.* **271**, 27237–27240.

Garcia-Cardena, G., Martasek, P., Masters, B. S., Skidd, P. M., Couet, J., Li, S., Lisanti, M. P., and Sessa, W. C. (1997). Dissecting the interaction between nitric oxide synthase (NOS) and caveolin. Functional significance of the nos caveolin binding domain *in vivo*. *J. Biol. Chem.* **272**, 25437–25440.

Goligorsky, M. S., Abedi, H., Noiri, E., Takhtajan, A., Lense, S., Romanov, V., and Zachary, I. (1999). Nitric oxide modulation of focal adhesions in endothelial cells. *Am. J. Physiol.* **276**, C1271–C1281.

Guo, F., Debidda, M., Yang, L., Williams, D. A., and Zheng, Y. (2006). Genetic deletion of Rac1 GTPase reveals its critical role in actin stress fiber formation and focal adhesion complex assembly. *J. Biol. Chem.* **281**, 18652–18659.

Hall, A. (2005). Rho GTPases and the control of cell behaviour. *Biochem. Soc. Trans.* **33**, 891–895.

He, T.-C., Zhou, S., da Costa, L. T., Yu, J., Kinzler, K. W., and Vogelstein, B. (1998). A simplified system for generating recombinant adenoviruses. *Proc. Natl. Acad. Sci. USA* **95**, 2509–2514.

Huesken, D., Lange, J., Mickanin, C., Weiler, J., Asselbergs, F., Warner, J., Meloon, B., Engel, S., Rosenberg, A., Cohen, D., Labow, M., Reinhardt, M., *et al.* (2005). Design of a genome-wide siRNA library using an artificial neural network. *Nature Biotech.* **23**, 995–1001.

Kim, F., Gallis, B., and Corson, M. A. (2001). TNF-alpha inhibits flow and insulin signaling leading to NO production in aortic endothelial cells. *Am. J. Physiol. Cell. Physiol.* **280**, C1057–C1065.

Kowanetz, M., and Ferrara, N. (2006). Vascular endothelial growth factor signaling pathways: Therapeutic perspective. *Clin Cancer Res.* **12**, 5018–5022.

Kraynov, V. S., Chamberlain, C., Bokoch, G. M., Schwartz, M. A., Slabaugh, S., and Hahn, K. M. (2000). Localized Rac activation dynamics visualized in living cells. *Science* **290**, 333–337.

Lane, P., and Gross, S. S. (2000). The autoinhibitory control element and calmodulin conspire to provide physiological modulation of endothelial and neuronal nitric oxide synthase activity. *Acta Physiol. Scand.* **168**, 53–63.

Lantin-Hermoso, R. L., Rosenfeld, C. R., Yuhanna, I. S., German, Z., Chen, Z., and Shaul, P. W. (1997). Estrogen acutely stimulates nitric oxide synthase activity in fetal pulmonary artery endothelium. *Am. J. Physiol.* **273**, L119–L126.

Manning, B. D., and Cantley, L. C. (2007). AKT/PKB signaling: Navigating downstream. *Cell* **129**, 1261–1274.

McCabe, T. J., Fulton, D., Roman, L. J., and Sessa, W. C. (2000). Enhanced electron flux and reduced calmodulin dissociation may explain "calcium-independent" eNOS activation by phosphorylation. *J. Biol. Chem.* **275**, 6123–6128.

Miao, R. Q., Fontana, J., Fulton, D., Lin, M. I., Harrison, K. D., and Sessa, W. C. (2008). Dominant-negative Hsp90 reduces VEGF-stimulated nitric oxide release and migration in endothelial cells. *Arterioscler. Thromb. Vasc. Biol.* **28**, 105–111.

Miao, R. Q., Gao, Y., Harrison, K. D., Prendergast, J., Acevedo, L. M., Yu, J., Hu, F., Strittmatter, S. M., and Sessa, W. C. (2006). Identification of a receptor necessary for

Nogo-B stimulated chemotaxis and morphogenesis of endothelial cells. *Proc. Natl. Acad. Sci. USA* **103,** 10997–11002.

Michel, J. B., Feron, O., Sacks, D., and Michel, T. (1997). Reciprocal regulation of endothelial nitric-oxide synthase by Ca2$^+$-calmodulin and caveolin. *J. Biol. Chem.* **272,** 15583–15586.

Michell, B. J., Chen, Z., Tiganis, T., Stapleton, D., Katsis, F., Power, D. A., Sim, A. T., and Kemp, B. E. (2001). Coordinated control of endothelial nitric-oxide synthase phosphorylation by protein kinase C and the cAMP-dependent protein kinase. *J. Biol. Chem.* **276,** 17625–17628.

Minami, Y., Kawasaki, H., Miyata, Y., Suzuki, K., and Yahara, I. (1991). Analysis of native forms and isoform compositions of the mouse 90-kDa heat shock protein, HSP90. *J. Biol. Chem.* **266,** 10099–10103.

Moffat, J., and Sabatini, D. M. (2006). Building mammalian signalling pathways with RNAi screens. *Nature Rev. Cell. Mol. Biol.* **7,** 177–187.

Morales-Ruiz, M., Fulton, D., Sowa, G., Languino, L. R., Fujio, Y., Walsh, K., and Sessa, W. C. (2000). Vascular endothelial growth factor-stimulated actin reorganization and migration of endothelial cells is regulated via the serine/threonine kinase Akt. *Circ. Res.* **86,** 892–896.

Nemoto, T., and Sato, N. (1998). Oligomeric forms of the 90-kDa heat shock protein. *Biochem. J.* **330,** 989–995.

Pei, Y., and Tuschl, T. (2006). On the art of identifying effective and specific siRNAs. *Nature Methods* **3,** 670–676.

Quilliam, L. A., Rebhun, J. F., and Castro, A. F. (2002). A growing family of guanine nucleotide exchange factors is responsible for activation of Ras-family GTPases. *Progr. Nucl. Acid Res. Mol. Biol.* **71,** 391–444.

Rickert, P., Weiner, O. D., Wang, F., Bourne, H. R., and Servant, G. (2000). Leukocytes navigate by compass: Roles of PI3Kgamma and its lipid products. *Trends Cell. Biol.* **10,** 466–473.

Ridley, A. J., Paterson, H. F., Johnston, C. L., Diekmann, D., and Hall, A. (1992). The small GTP-binding protein rac regulates growth factor-induced membrane ruffling. *Cell* **70,** 401–410.

Rousseau, S., Houle, F., Landry, J., and Huot, J. (1997). p38 MAP kinase activation by vascular endothelial growth factor mediates actin reorganization and cell migration in human endothelial cells. *Oncogene* **15,** 2169–2177.

Royal, I., Lamarche-Vane, N., Lamorte, L., Kaibuchi, K., and Park, M. (2000). Activation of cdc42, rac, PAK, and rho-kinase in response to hepatocyte growth factor differentially regulates epithelial cell colony spreading and dissociation. *Mol. Biol. Cell* **11,** 1709–1725.

Sander, E. E., van Delft, S., ten Klooster, J. P., Reid, T., van der Kammen, R. A., Michiels, F., and Collard, J. G. (1998). Matrix-dependent Tiam1/Rac signaling in epithelial cells promotes either cell–cell adhesion or cell migration and is regulated by phosphatidylinositol 3-kinase. *J. Cell. Biol.* **143,** 1385–1398.

Sarbassov, D. D., Guertin, D. A., Ali, S. M., and Sabatini, D. M. (2005). Phosphorylation and regulation of Akt/PKB by the Rictor-mTOR complex. *Science* **307,** 1098–1101.

Schleicher, M., Shepherd, B. R., Suarez, Y., Fernandez-Hernando, C., Yu, J., Pan, Y., Acevedo, L. M., Shadel, G. S., and Sessa, W. C. (2008). Prohibitin-1 maintains the angiogenic capacity of endothelial cells by regulating mitochondrial function and senescence. *J. Cell Biol.* **180,** 101–112.

Soga, N., Connolly, J. O., Chellaiah, M., Kawamura, J., and Hruska, K. A. (2001a). Rac regulates vascular endothelial growth factor stimulated motility. *Cell. Commun. Adhes.* **8,** 1–13.

Soga, N., Namba, N., McAllister, S., Cornelius, L., Teitelbaum, S. L., Dowdy, S. F., Kawamura, J., and Hruska, K. A. (2001b). Rho family GTPases regulate VEGF-stimulated endothelial cell motility. *Exp. Cell. Res.* **269,** 73–87.

Soldi, R., Mitola, S., Strasly, M., Defilippi, P., Tarone, G., and Bussolino, F. (1999). Role of alphavbeta3 integrin in the activation of vascular endothelial growth factor receptor-2. *EMBO J.* **18,** 882–892.

Stamler, J. S., Lamas, S., and Fang, F. C. (2001). Nitrosylation. The prototypic redox-based signaling mechanism. *Cell* **106,** 675–683.

Stamler, J. S., Toone, E. J., Lipton, S. A., and Sucher, N. J. (1997). (S)NO signals: Translocation, regulation, and a consensus motif. *Neuron* **18,** 691–696.

Stephens, L., Anderson, K., Stokoe, D., Erdjument-Bromage, H., Painter, G. F., Holmes, A. B., Gaffney, P. R. N. J., Reese, C. B., McCormick, F., Tempst, P., Coadwell, J., and Hawkins, P. T. (1998). Protein kinase B kinases that mediate phosphatidylinositol 3,4,5-trisphosphate-dependent activation of protein kinase B. *Science* **279,** 710–714.

Stuehr, D. J., and Griffith, O. W. (1992). Mammalian nitric oxide synthases. *Adv. Enzymol. Relat. Areas Mol. Biol.* **65,** 287–346.

Suarez, Y., Fernandez-Hernando, C., Pober, J. S., and Sessa, W. C. (2007). Dicer dependent microRNAs regulate gene expression and functions in human endothelial cells. *Circ. Res.* **100,** 1164–1173.

Venema, V. J., Zou, R., Ju, H., Marrero, M. B., and Venema, R. C. (1997). Caveolin-1 detergent solubility and association with endothelial nitric oxide synthase is modulated by tyrosine phosphorylation. *Biochem. Biophys. Res. Commun.* **236,** 155–161.

Waldman, S. A., and Murad, F. (1987). Cyclic GMP synthesis and function. *Pharmacol. Rev.* **39,** 163–196.

Waltenberger, J., Claesson-Welsh, L., Siegbahn, A., Shibuya, M., and Heldin, C. H. (1994). Different signal transduction properties of KDR and Flt1, two receptors for vascular endothelial growth factor. *J. Biol. Chem.* **269,** 26988–26995.

Welch, H. C., Coadwell, W. J., Stephens, L. R., and Hawkins, P. T. (2003). Phosphoinositide 3-kinase-dependent activation of Rac. *FEBS Lett.* **546,** 93–97.

Whitesell, L., Mimnaugh, E., Costa, B., Myers, C., and Neckers, L. (1994). Inhibition of heat shock protein HSP90-pp60v-src heteroprotein complex formation by benzoquinone ansamycins: Essential role for stress proteins in oncogenic transformation. *Proc. Natl. Acad. Sci. USA* **91,** 8324–8328.

Wojciak-Stothard, B., Potempa, S., Eichholtz, T., and Ridley, A. J. (2001). Rho and Rac but not Cdc42 regulate endothelial cell permeability. *J. Cell Sci.* **114,** 1343–1355.

Wojciak-Stothard, B., and Ridley, A. J. (2002). Rho GTPases and the regulation of endothelial permeability. *Vascul. Pharmacol.* **39,** 187–199.

Xue, Y., Bi, F., Zhang, X., Pan, Y., Liu, N., Zheng, Y., and Fan, D. (2004). Inhibition of endothelial cell proliferation by targeting Rac1 GTPase with small interference RNA in tumor cells. *Biochem. Biophys. Res. Commun.* **320,** 1309–1315.

CHAPTER TWO

PHYSIOLOGIC STRESS-MEDIATED SIGNALING IN THE ENDOTHELIUM

Cynthia A. Reinhart-King,[†] Keigi Fujiwara,* *and* Bradford C. Berk*

Contents

Abstract

Although the vasculature was once thought to be a passive conduit for blood, it is now known that the endothelium is responsible for healthy vascular homeostasis and the progression of many cardiovascular-related diseases. Because the endothelium lines blood vessels, it is subjected to the mechanical forces due to of blood flow. It is now well established that endothelial cells transduce these mechanical signals into chemical signals that are evident in the

* Aab Cardiovascular Research Institute, University of Rochester School of Medicine and Dentistry, West Henrietta, New York
† Department of Biomedical Engineering, Cornell University, Ithaca, New York

Methods in Enzymology, Volume 443
ISSN 0076-6879, DOI: 10.1016/S0076-6879(08)02002-8

mechanoregulation of a number of signal transduction pathways and endothe-lial cell phenotype. Despite the significant volume of work in the field of endothelial cell mechanotransduction, the exact mechanism by which mechani-cal forces are sensed and transduced into chemical signals is not yet well established. In this chapter, we focus on the specific role of fluid shear stress, the frictional drag force caused by blood flow, and cyclic stretch caused by the pumping action of the heart, in regulating vascular homeostasis and vascular signaling. The regulation of flow-mediated signaling in the endothelium is typically studied with well-characterized *in vitro* flow and stretch devices. Here, we examine various platforms used to analyze flow-mediated and stretch-mediated signals and describe the method for the implementation of these techniques.

1. Introduction

The endothelial cell layer of the vasculature directly contacts the circulating blood, providing a dynamic barrier to the surrounding tissue. Because of its unique position in the body, it is exposed to three primary mechanical forces caused by blood flow: pressure caused by the hydrostatic force within the vessel, hoop stresses caused by the balance between the cell–cell contacts and vasomotion of the vessel, and the shear stresses caused by the friction of the blood flow against the vessel wall. In addition, the endothelium modulates a number of biologic processes within the vessel wall, including active regulation of vascular tone and blood pressure through the production of nitric oxide, control over the coagulation cas-cade and fibrinolytic processes through the production of prothrombotic and antithrombotic factors, regulation of vascular remodeling through the production of growth factors and vasoactive substances, and control over the inflammatory response by providing a platform by which leukocytes can home to inflamed tissue. As such, the endothelium mediates the progression of several pathologic conditions, including chronic inflammation, wound healing, and the development of various cardiovascular diseases including atherosclerosis.

In healthy blood vessels, endothelial cells are typically quiescent, pre-senting an antithrombotic nonadhesive surface to the passing blood (Garin and Berk, 2006; Pearson, 2000). However, on activation by cytokines, such as tumor necrosis factor (TNF-α) or interleukin-1β (IL-1β), or other stimuli, endothelial cells upregulate a number of genes, including E-selectin, intercellular adhesion molecule (ICAM)-1, vascular cell adhe-sion molecule (VCAM)-1, monocyte chemoattractant protein-1 (MCP-1), interleukin-1 and 8, plasminogen activator inhibitor-1 (PAI-1), and tissue factor (TF) (Cybulsky and Gimbrone, 1991; Diamond *et al.*, 1989; Malek *et al.*, 1999). (Fig. 2.1) In general, endothelial cell turnover increases, and

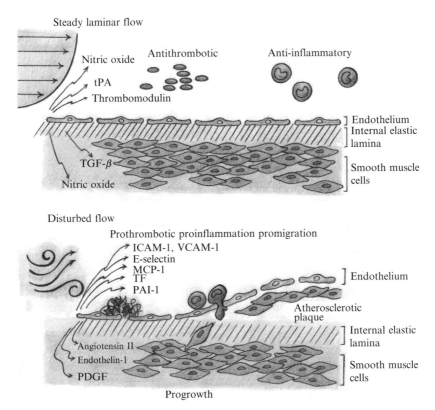

Figure 2.1 Flow–mediated effects on atherosclerotic plaque formation. Steady laminar flow promotes endothelial cell survival and is antithrombotic and antiinflammatory. Disturbed flow elicits the opposite response, stimulating endothelial apoptosis, vascular smooth muscle growth, thrombosis, and monocyte infiltration, resulting in a proatherogenic phenotype. Image drawn by Ms. Becky Zhao.

the production of antithrombotic mediators decreases. Endothelial cell activation is required for a number of adaptive processes, including clotting and leukocyte adhesion during wound healing (Hajjar and Deora, 2000). However, chronic endothelial cell activation is linked to chronic inflammatory diseases, including atherosclerosis (Davignon and Ganz, 2004).

Chronic endothelial cell activation and increased incidence of atherosclerotic lesion formation is often detected in branch points or bifurcations within the vasculature (Gimbrone *et al.*, 2000; Malek *et al.*, 1999; World *et al.*, 2006). Notably, the blood flow patterns, fluid shear stresses, and stretch patterns imposed on the endothelium are also strikingly different in these regions than those measured in the straight portions of the vessel (Chien, 2007). The nature and magnitude of blood flow, shear, and vessel wall stretch is largely determined by the shape and structure of the blood vessel and the cardiac cycle. The shear stresses found in most major human arteries have been found to be 2 to 20 dyn/cm^2, with localized increases to 30 to 100 dyn/cm^2 near branches and areas of sharp wall curvature (Dewey *et al.*, 1981). Typically, in the straight portions of a blood vessel, the flow is laminar. In these regions, endothelial cells align and elongate parallel to the direction of flow (Fig. 2.2). This realignment corresponds with a streamlining of the cell that reduces resistance to flow and is speculated to mediate the subsequent signaling response. In contrast, flow within abrupt curvatures, such as occurs at bifurcations, is typically disturbed, exhibiting flow reversal, separation, and low velocity. As a result, the endothelial cells do not reorient like those located in the straight portions of the vessel. Because the cells do not align with the flow, their topology exposes them to greater shear stress gradients across the length of the cell, and these areas are also more prone to atherosclerosis (Barbee *et al.*, 1994). For instance, within the carotid bifurcation, where atherosclerosis often develops, the flow separates, disrupting the laminar profile and producing disturbed streamlines (Ku *et al.*, 1985). The lateral wall experiences areas of flow reversal and recirculation varying with

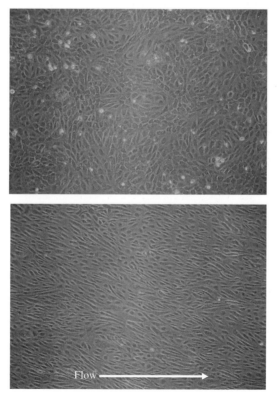

Figure 2.2   Phase image of HUVECs before and after 24 h of flow at 20 dyn/cm^2 with a cone and plate flow device as described in the text. Image taken at 10×.

the cardiac cycle, resulting in a time-averaged shear stress close to zero. Because areas subjected to laminar shear stresses are generally free from plaque formation and lesions correlate with areas of disturbed flow, it is believed that laminar shear stresses impose an atheroprotective force on the vasculature and help maintain healthy vascular homeostasis (Berk *et al.*, 2002; Traub and Berk, 1998). A similar general theme based on endothelial cell response to stretch has been advanced by Chien (Chien, 2007). However, the mechanism by which laminar flow and unidirectional stretch are detected and translated into an atheroprotective force by the endothelium remains unclear.

Significant evidence exists that laminar flow at physiologic shear stresses inhibits platelet aggregation and can enhance endothelial cell survival by preventing apoptosis (Dimmeler *et al.*, 1996; Garin *et al.*, 2007; Yoshizumi *et al.*, 2003). Physiologic levels of cyclic stretch also seem to provide beneficial effects on endothelial barrier function (Fujiwara, 2003). In addition, significant evidence exists that disturbed flow that exerts time-averaged low shear stresses on the endothelium induces endothelial expression of proapoptotic, proinflammatory, and procoagulant genes (Berk, 2008; Davies, 2007; Garcia-Cardena and Gimbrone, 2006). However, the upstream events that initiate the signaling cascade in response to shear stress are still unclear.

Because many questions regarding how endothelial cells mechanotranduce fluid shear stress and stretch into intracellular responses, the atheroprotective effects of these forces and the subsequent flow-mediated and stretch-mediated signaling continue to be an active area of research. Although there have been significant advances in this field with animal models, *in vivo* work presents a number of challenges. In addition to the difficulties associated with isolating the effects of the mechanical signals caused by blood flow from the chemical humoral effectors, it is also difficult to visualize various force fields within the vasculature and to relate those stresses to specific endothelial cell phenotype in real time. Therefore, *in vitro* methods have been designed to mimic *in vivo* forces for the studies of flow-mediated and stretch-mediated signaling in the endothelium. Here, we will discuss several methods used to subject cells to flow, and we will also describe methods we have used to stretch cells to compare these mechanosignaling responses to the response elicited by flow.

2. Parallel Plate Flow System

The parallel plate flow chamber is one of the most commonly used platforms for subjecting monolayers of endothelial cells to uniform laminar shear stresses. Its advantages include the following (Kandlikar *et al.*, 2005):

- The fluid shear stress is relatively uniform within the chamber.
- It can be used on a temperature-controlled microscope stage, allowing for real-time visualization of the cells and/or flow patterns within the chamber.
- It is capable of applying a wide range of flow rates and fluid shear stresses.
- It allows for continuous sampling of the conditioned medium.
- It can be scaled down to micron-sized channels, with smaller volumes to conserve the use of rare or expensive molecules.

2.1. Design of a parallel plate flow chamber

Most parallel plate flow chambers contain several common components, depicted in Fig. 2.3. Typically, a reusable plastic upper surface is used to sandwich a gasket onto a tissue culture dish or glass plate such as a micro-slide. The gasket contains a rectangular cutout, creating the side walls of the flow domain. The upper surface contains ports at either end of the channel for the inlet and outlet of flow. Some devices also use a vacuum line to keep the apparatus pressed together without the use of anchors or clips. This helps to ensure that the height of the flow path created by the opening in the

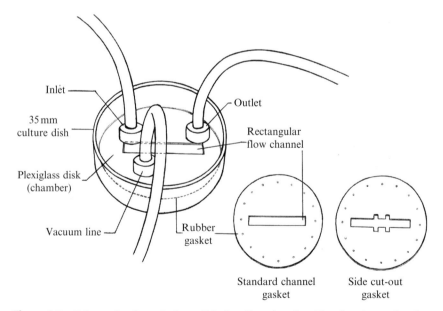

Figure 2.3 Schematic of a typical parallel plate flow chamber. The chamber and gasket are designed to fit within a standard 35-mm tissue culture dish. Insets indicate gasket designs for uniform laminar flow and disturbed flow containing spatial gradients of shear. Image drawn by Ms. Becky Zhao.

Table 2.1 Commercially available parallel plate flow chambers with relevant dimensions

Model	Manufacturer	Channel height (μm)
Circular flow chamber	Glycotech Corporation	125–150
ECIS flow system	Applied Biophysics	1000
Stovall flow cell	Stovall Life Science, Inc.	1000
Flow chamber system	Oligene	300–500
Adhesion flow chamber	Immunogenics, Inc.	250
Vacucell	C&L Instruments, Inc.	250

gasket is uniform. Alternately, the chamber may be secured by screws set at the perimeter, which must be tightened equally to ensure that the flow path is level.

Many parallel plate flow systems are commercially available with varying channel dimensions, dictating a range of fluid shear stresses. Several commercially available systems are listed in Table 2.1 (Kandlikar *et al.*, 2005), with their manufacturer and reported range of channel dimensions. Most commercially available flow chambers are designed to fit in commonly used 100-mm or 35-mm tissue culture–treated dishes, and many of these devices contain an optically transparent flow path, allowing for direct visualization of cell behavior under flow.

2.2. Calculation of the imposed fluid shear stress in a parallel plate flow chamber

Given a flow channel, where the length and width of the channel is much greater than the height, the flow profile can be approximated as flow between infinite parallel plates, often referred to as plane Poiseuille flow. For a Newtonian fluid, the shear stress at the surface of the plate is:

$$\tau = \frac{6Q\mu}{wh^2} \qquad (2.1)$$

where Q is the flow rate, μ is the fluid viscosity, w is the width of the channel, and h is the height of the channel (i.e., the thickness of the gasket). To achieve a specific shear stress, the corresponding flow rate should be used. Flow rate can be controlled in several ways. For short-term studies of flow-mediated response, syringe pumps in withdraw-mode can pull media from a reservoir through the channel. However, depending on the capacity of the syringe and the number of syringes connected in parallel, the maximum duration for the application of flow ranges from 1 to 3 h. To apply flow for longer times, a flow loop is needed, where the culture medium is moved by a peristaltic pump. However, peristaltic pumps do create some pulsation within the flow. There are several methods for minimizing this pulsation: (1) install a commercially available pulse dampener in the loop; (2) use a peristaltic pump with a larger number of rollers (>8 to 10) to smooth the flow; and (3) add two reservoirs into the flow loop, up and downstream of the flow chamber. The flow through the chamber is pressure driven on the basis of the height of the upper reservoir, and media is returned to the upper reservoir from the lower reservoir through a peristaltic pump (Go *et al.*, 1999).

2.3. Protocol for subjecting endothelial cells to laminar fluid shear stress with a parallel-plate flow chamber

2.3.1. Materials

- Glycotech Flow Chamber (cat #. 31-001): The flow chamber kit includes gaskets containing well-characterized channels of defined width, height, and length and quantitative characterization of the flow rates necessary to achieve a given wall shear stress.
- Peristaltic pump (Rainin, Cat #7103-058) or
- Syringe Pump with 60-ml syringes, depending on the length of experimentation.
- Silastic or Tygon Tubing (1/16 in ID).
- Corning 35-mm tissue culture dishes.
- Human umbilical vein endothelial cells (HUVEC) (available through Cascade Biologics or can be isolated from a primary source).
- HUVEC Media: 500 ml DMEM (Cascade Biologics) 10 ml LSGS (Cascade Biologics), 5% FBS (Gibco).
- Vacuum line.

2.3.2. Methods

Plate cells on 35-mm dishes in culture media several days before the experiment and allow the cells to grow to confluence. NOTE: if cells are grown significantly past confluence, cells may release from the dish in cell sheets upon the initiation of flow.

   If the experiments do not involve any biochemistry and instead only require observation and immunocytochemistry, cell can be plated over the entire dish.

   If the desired end experiments involve biochemistry (e.g., Western blotting of cell lysates after flow), then cells should be plated only within the flow channel. To position cells within the flow channel, a template is cut from a sheet of silicon (Scientific Instrument Services, #S11) to match the gasket and channel dimensions, and the silicon is pressed to seal to the culture dish. Cells are cultured within the silicon template, and the template is removed before the flow chamber is assembled.

Sterilize the chamber and gasket by rinsing in 200 proof ethanol and leaving under UV light for >20 min. Lay the gasket out onto a dry Kim wipe and seal the gasket to the chamber by carefully aligning the holes in the chamber with the flow channel in the gasket and pressing down. Push air bubbles out from beneath the gasket by smoothing.

2.3.3. Assembling the tubing and pump

The experiment is performed at 37° and 5% CO_2/95% air atmosphere. To prevent rusting and malfunction of the peristaltic pump, the pump remains outside of the incubator, and tubing is run in and out of the side port of the incubator. It is important to keep the flow path located outside of the incubator as short as possible to minimize the cooling of the media. Likewise, this apparatus can be assembled and placed on a microscope for real-time visualization, given that the microscope is equipped with a temperature/gas controlled incubator.

Assemble the flow path by running the tubing through the flow pump and placing the upstream end of the tubing into a 15-ml conical tube containing cell culture media. Prime the pump, filling the tubing with media. Place a second tube between the outlet port of the chamber and the media–containing conical tube.

Attach the vacuum line to the vacuum port of the chamber.

Aspirate the media from the 35-mm dish containing the cells.

To prevent a bubble from forming within the flow path, the upstream tubing should be attached to the chamber and flow should be started, allowing the port in the upper surface to overflow before the chamber is sealed to the 35-mm dish.

Once the flow is started and the media have exited the inlet port, turn the chamber upside down and invert the 35-mm dish onto the chamber. Once assembled, the vacuum will hold the chamber together and the entire assembly can be turned right side up.

Continue flow for desired time. We have successfully run this system continuously for up to 48 h.

At the completion of the experiment, the chamber can be disassembled by stopping the flow, turning off the vacuum, and lifting the chamber from the 35-mm dish. The cells can then either be fixed for visualization or lysed for biochemical assaying with standard methods.

2.4. Disturbed flow chamber

Because regions of disturbed flow correlate well with regions prone to atherosclerotic lesions, it is equally important to understand endothelial cell behavior in areas containing shear stress gradients. A number of investigators have proposed methods to create these gradients, including uniquely shaped flow paths within the parallel plate design (Usami *et al.*, 1993), introduction of an obstacle into the flow path (Tardy *et al.*, 1997), and use of a backward facing step (Skilbeck *et al.*, 2001). In our own laboratory, we have modified the parallel plate flow chamber to create sharp spatial gradients within the flow path. The chamber contains the standard flow path within the gasket, with rectangular cutouts within the sides of the channel (see Fig. 2.3). The cutouts create shear gradients and vortices within the flow path, dependent on flow rate (Fig. 2.4). This chamber, unlike the backward facing step, allows the decoupling of the effects of spatial shear stress gradients from those caused by flow reattachment at low flow rates, and its flow profile is easy to visualize. Because the design is based on the conventional parallel plate flow chamber, it permits the direct microscopic observation of the cells within the flow path.

The chamber is made using customized rectangular hole punch on a standard parallel plate flow chamber gasket. The flow loop is assembled in the same way the standard parallel plate is assembled. The flow profile is characterized as described below.

2.5. Experimental characterization and visualization of the flow field

2.5.1. Materials

- Parallel plate flow chamber assembly as described previously.
- Fluorescent tracer beads, 2 μm (Invitrogen).

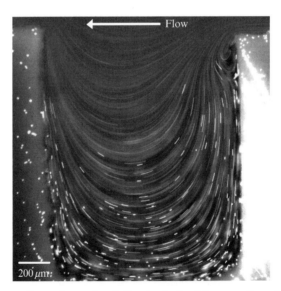

Figure 2.4 Streamlines of tracer beads flowing in the side cutout flow chamber, created as described in the text. Note the formation of a stable flow vortex at the inlet of the cutout.

- Cell culture medium.
- Inverted fluorescent microscope with CCD camera.

2.5.2. Methods

Sonicate beads to disperse aggregates.
Create a dilute suspension of tracer beads in cell culture medium: 10 μl of bead suspension into 20 ml of deionized water.
Assemble the flow chamber apparatus as described previously.
Pump the bead solution through the chamber at known flow rates.

After allowing the flow to equilibrate (>5 sec), collect fluorescent images at known exposure times. The exposure time is chosen on the basis of the flow rate to achieve clear streaks of the beads through the flow path. In the gasket shown, we use exposure times ranging from 0.2 to 1.0 sec for flow rates ranging from 0.1 to 9.0 ml/min.

The streaks created by the tracer beads outline the flow profile. See Fig. 2.4 for an example.

3. Cone and Plate Flow System

The cone and plate flow system works almost identically to a cone and plate viscometer. Actually, many laboratories, including our own, have adapted cone and plate viscometers to subject uniform shear stress distribution within the fluid environment of a cell culture dish. A number of systems are commercially available, including those from Wells-Brookfield, Thermo-Scientific, and Research Equipment Ltd., which can be adapted for use in cell culture. The cone and plate system consists of a cone with a well-defined angle that can rotate about its center axis. It is placed into a tissue culture dish and rotated, such that at sufficiently small Reynolds numbers, it produces a stable laminar flow.

The cone and plate system offers a number of advantages (Papadaki and McIntire, 1999), including the following:

- It produces a relatively stable flow such that small changes in the dimensions do not have significant effects on the wall shear stresses.
- The system requires very little medium because it requires a fixed volume, without a flow loop.
- The fluid flow is not dependent on a hydrostatic pressure gradient.
- Compared with the parallel plate chamber, there is a larger area of cell coverage, increasing the area for sampling and lysate recovery.
- The plates can be tissue culture dishes that are easily replaceable and are excellent for cell growth.

The cone and plate flow chamber does have some disadvantages. It is not easily adaptable for use on a microscope stage and real-time observation, and it does not allow for continuous sampling of the media. However, for studies of flow-mediated signaling, these disadvantages have not limited our experiments. An additional disadvantage of most commercially available viscometers is that they are typically open to the environment, permitting media evaporation. This prevents long-term studies of cell response to flow. We have addressed this problem by implementing a device that uses the cone and plate design but also covers the dish to prevent media loss.

3.1. Design of a cone and plate chamber

We have adopted use of a cone and plate design first described by Jo and colleagues (Boo *et al.*, 2002; Jo *et al.*, 2006). The cone is designed to fit into either a 35-mm or 100-cm tissue culture dish (Fig. 2.5). Unlike a typical cone and plate viscometer and many other commonly used cone and plate devices, this design includes an upper lid that rests on the top of the dish, minimizing evaporation of the cell culture media and permitting long-term flow conditions. This lid also supports the free rotation of the cone on an

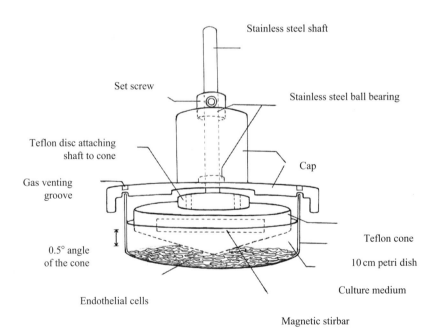

Figure 2.5 Schematic of a cone and plate design, initially described by Jo *et al.* (2006). The device is machined from Teflon and stainless steel, as described in the text. Image drawn by Ms. Becky Zhao.

axle secured by ball bearings and a setscrew. The setscrew controls the height of the cone from this dish. The cone is machined from Teflon to contain a 0.5-degree angle from its peak to its edge and is free from defects and machining marks so at to not affect the flow field. Teflon does not typically bind proteins and additives from the media, and it does not interfere with cell viability. To rotate the cone, a magnetic stir bar is embedded within the Teflon that can be turned with a laboratory magnetic stirrer. The entire system can be assembled in a temperature and CO_2 controlled chamber.

3.2. Calculation of the imposed fluid shear stress in a cone and plate flow chamber

The shear stress profile is uniform across the plate of a cone and plate flow chamber. This uniformity in shear stress is due to the increase of the gap height from the center of the cone countered by the linear increase in fluid velocity as a function of the radius. Because both the gap height and fluid velocity vary linearly with the radius, the shear stress is uniform throughout the flow field (Dewey *et al.*, 1981).

The shear stress imposed on the surface of the plate is calculated on the basis of the rotational speed of the cone:

$$\tau_w = \frac{\omega\mu}{\theta},$$ (2.2)

where ω is the rotation speed, μ is the fluid viscosity, and θ is the angle of the cone. The rotation of the cone in our system is determined by the speed set on the stir plate.

3.3. Protocol for subjecting endothelial cell to laminar fluid shear stress with a cone and plate flow chamber

3.3.1. Materials

- Human umbilical vein endothelial cells (HUVEC) are available through VEC Technologies (Rensselaer, NY) or can be isolated from a primary source.
- HUVEC media: 500 ml DMEM (Cascade Biologics) 10 ml LSGS (Cascade Biologics). 5% FBS (Gibco) carriage return.
- Cone and plate apparatus: needs to be machined to the preceding specifications. We have used the machine shop at the University of Alabama at Birmingham.
- Stir plate (Wheaton Microstir).

- 100-mm or 60-mm tissue culture–treated dishes: Size is chosen on the basis of the size of the cone.

3.3.2. Methods

Plate HUVEC on the tissue culture–treated dishes and allowed to grow to confluence.

Adjust the height of the cone by placing the cone in a dry tissue culture dish containing a #1 coverslip placed at its center. Align the coverslip with the peak of the cone and adjust the setscrew such that the peak of the cone rests on the coverslip

Add fresh media to the dish. The volume of media depends on the volume of the cone. Once the cone is added to the dish, the media should exceed the height of the cone, but should not come as high as the top edge of the dish. If media overflows once the cone is placed on the dish, the dish should be removed, and enough media should be removed to prevent the media from moving up the wall and overflowing. If the media are pulled to the top of the dish edge, capillary action will continue to draw media out of the dish causing the dish to dry out.

Position the cone into the dish, being careful not to trap air bubbles beneath the cone.

Place the cone and plate on the stir plate and initiate flow by turning on the stirplate to the desired speed.

4. Cyclic Stretching of the Endothelium

In addition to fluid shear stress, endothelial cells are exposed to cyclic stretch caused by the pumping action of the heart. For large human arteries, the vessel wall expands circumferentially by 5 to 12% under the normal physiologic condition (Birukov *et al.*, 2003; Nagel *et al.*, 1999). Effects of mechanical stretch on endothelial cells and other cell types have been studied with various types of cell stretch apparatuses. These *in vitro* studies have shown that stretch activates specific signaling pathways, causing cells to exhibit stretch-dependent phenotypes (Shi *et al.*, 2007).

To exert mechanical stretch onto cultured cells, several commercial devices are available, including those from Flexcell International and B-Bridge. Several different models of cell stretch devices are designed for different experimental purposes. In addition, investigators have developed their own apparatuses to meet the needs of specific studies (Arold *et al.*, 2007; Joung *et al.*, 2006; Lee *et al.*, 1996; Liu *et al.*, 2006; Takemasa *et al.*, 1997; Tschumperlin and Margulies, 1998; Yost *et al.*, 2000). Although these designs vary in their specific details, they all implement the same basic technique: cells are grown on thin, stretchable, silicon membranes and mechanically stretched. For cells to be stretched, it is important that they must be firmly anchored to the membrane so that there is no slip between the cells and the membrane and also cells on the membrane should be stretched to the same extent as the membrane. To ensure firm cell attachment, silicon membranes are typically coated with an extracellular matrix protein such as collagen or fibronectin.

Here we describe a protocol for stretching a confluent culture of bovine aortic endothelial cells (BAECs) with a uniaxial cell stretch apparatus (Model NS-500) obtained from K. K. Scholar Tec (Osaka, Japan). The apparatus was designed by Naruse *et al.* (1998). Cells are grown in a stretch chamber (30 × 30 × 10 mm) made of silicon. (Fig. 2.6). The bottom of this chamber is a 100-μm-thick silicon membrane on which cells are plated. The chamber is mounted onto a mechanical stretch apparatus that is controlled by a computer to stretch the entire chamber by 5 to 25% at 1 to 60 cycles/sec. Experiments are performed in a tissue culture incubator (humidified and 5% CO_2 atmosphere). With this setup, cells can be kept under a cyclic stretch condition for several days.

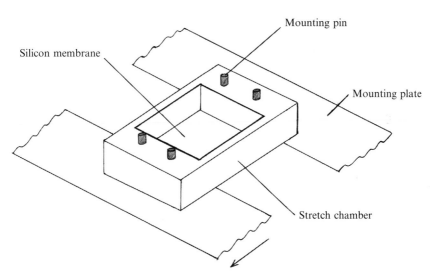

Figure 2.6   Schematic of a cell stretch chamber mounted on the metal plates of a stretch apparatus (Model NS-500). The frame of the chamber is made of 10-mm-thick silicon with a thin silicon membrane glued at the edges to create the bottom of the chamber. Cells are plated on this membrane. The chamber has four pin holes used to secure it to the mounting plates of the stretch apparatus. Initially, the mounting plates are separated such that when a chamber is mounted, it is at its rest length. One of the plates is fixed in position, whereas the other plate is connected to a motor and moved back and forth (arrow), cyclically stretching the chamber. Each mounting plate has 12 pins so that up to six chambers can be stretched simultaneously. Image drawn by Ms. Becky Zhao.

4.1. Protocol for subjecting endothelial cells to cyclic stress

4.1.1. Materials

- Rat Tail Type I Collagen (BD Biosciences).
- 0.02 N Acetic acid.
- Stretch apparatus and chambers as described previously.
- 10-cm Petri dishes.

4.1.2. Methods

- Each stretch chamber is autoclaved, air-dried, and placed in a sterile 10-cm petri dish in a laminar flow hood.
- Coat the bottom of the stretch chambers with rat tail type I collagen by adding 2 ml of 0.1 mg/ml of collagen in 0.02 N acetic acid to each chamber. Allow the solution to evaporate completely in the hood overnight.
- Wash chambers thoroughly with sterile phosphate-buffered saline (PBS: 137 mM NaCl, 8 mM Na$_2$HPO$_4$, 2.7 mM KCl, 1.6 mM KH$_2$PO$_4$, pH 7.4) to remove the acetic acid and sterilize them under UV light for at least 20 min.
- Plate BAECs, between passages 6 and 10, at the desired cell density in 2 ml of media. In our experiments, we plate 4 to 6 × 10^5 cells in 2 ml of DMEM supplemented with 10% FBS, 10 mM HEPES, and penicillin (50 IU/ml)/streptomycin (50 μg/ml), and allow them to form a tight monolayer over 2 to 3 days of culture. Other types of endothelial cells may require different culture media and plating conditions, which should be determined by each investigator.
- Set stretch conditions through the control box positioned outside of the tissue culture incubator. It provides various preset levels of amplitude (5, 6, 8, 10, 12, 15, 20, and 25% uniaxial stretch) and frequency (1, 5, 10, 20, 30, 40, 50, and 60 stretches/min). These values are for the cell stretch apparatus (Model NS-500). Other models may provide different sets of conditions. Our model is capable of stretching six chambers at a time, so that six stretch samples under the same condition or samples from different time points may be obtained.
- The duration of stretch is controlled through the main switch of the apparatus. It can be as short as a second or as long as days. We have grown cells under cyclic stretch for several days. Because cyclic stretch at high amplitude and frequency will cause the medium to splash, the culture medium should be periodically replenished.
- Control cells should be plated in stretch chambers as described previously and mounted in the stretch apparatus without stretching for the same length of time as experimental conditions. To account for the vibrations caused by the stretch apparatus, a second control is placed in a petri dish and on the lid of the stretch apparatus.

NOTES: When cells are cyclically stretched, they also experience shear stress caused by movements of culture medium. To avoid this problem, we have used a single sustained stretch (Chiu *et al.*, 2007). We operate the stretch machine manually such that the chamber is gently stretched to a desired length and kept at that length for the duration of experiments. A single sustained stretch is only feasible for short-term experiments, because it is anticipated that cells may adjust their adhesion points in time. When this "reattachment" takes place, cells would no longer be in a stretched state.

Stretched endothelial cultures may be analyzed for signaling molecule and enzyme activities, protein modifications, morphologic changes, gene and protein expression, and secretion of bioactive substances into the media. Here we describe the protocol for morphologic studies of cells, specifically immunocytochemistry, and harvesting cells for immunoblotting studies.

4.1.2.1. Immunofluorescence

Remove a stretch chamber from the stretch apparatus and place it in a 10-cm petri dish so that the bottom of the chamber (i.e., the thin silicon membrane) is supported. Slightly tilt the dish and remove culture medium by aspiration. Wash the inside of the stretch chamber with warm (room temperature $-37°$) PBS. Fix cells for 5 to 10 min at room temperature by directly adding 3 ml of 3.7% formaldehyde in PBS and wash thoroughly with PBS. If permeabilization is required, treat the fixed cells with 0.1% Triton X-100 in PBS for 1 to 2 min at room temperature and wash with PBS. Cold methanol may be used to fix cells also. After fixation, one may treat the specimen with a blocking solution. When an aspirator is used, one must be careful not to touch the thin membrane, which is easily damaged.

Stain cells with primary and secondary antibodies following a routine immunofluorescence protocol specific for the antibodies used.

To observe immunofluorescently stained cells in the chamber, it is most convenient to use a water immersion objective lens, which can be dipped directly into a PBS-filled stretch chamber. It should also be possible to place a coverslip on the membrane and use a dry or oil immersion lens, although we have not used this approach.

Fixed cells in stretch chambers stained for immunofluorescence should be removed when microscopy is done. Add 2% SDS solution in deionized water to the chamber and scrape cells off with a rubber policeman. Store chambers in a clean beaker with tap water until further and thorough cleaning (see following).

4.1.2.2. Western blotting

Remove a stretch chamber from the stretch apparatus and place it in a 10-cm petri dish so that the bottom of the chamber (i.e., the thin silicon membrane) is supported. Place the dish on ice and by slightly tilting the dish, remove culture medium by aspiration.

Wash the inside of the stretch chamber with cold PBS and add 0.1 ml SDS-PAGE sample buffer. Gently scrape off cells with a rubber policeman and collect the sample into a small vial with a pipette.

The sample is ready for a routine SDS-PAGE, followed by immunoblotting.

With appropriate solutions, one may fractionate stretched cells for biochemical analyses or isolating mRNA.

Chambers may be reused by washing them thoroughly (see later).

4.1.2.3. Cleaning chambers for reuse

Used chambers should be kept in a clean beaker with tap water until further cleaning.

To begin cleaning, place used chambers in a beaker containing hot detergent and sonicate for 15 min.

Rinse thoroughly (>five times) with hot tap water and then with RO (reverse osmosis) water five times. Sonicate in RO water for 15 min and rinse five times with RO water.

Rinse thoroughly with deionized water, air-dry, and keep them in a clean box.

Before reusing the chambers, autoclave.

ACKNOWLEDGMENTS

This work was supported by NHLBI grants HL 64839 and HL 77789 to B. C. B. and an NIH NRSA to C. R. K. (HL 84961).

REFERENCES

Arold, S. P., Wong, J. Y., and Suki, B. (2007). Design of a new stretching apparatus and the effects of cyclic strain and substratum on mouse lung epithelial-12 cells. *Ann. Biomed. Eng.* **35,** 1156–1164.

Barbee, K. A., Davies, P. F., and Lal, R. (1994). Shear stress-induced reorganization of the surface topography of living endothelial cells imaged by atomic force microscopy. *Circ. Res.* **74,** 163–171.

Berk, B. C. (2008). Atheroprotective signaling mechanisms activated by steady laminar flow in endothelial cells. *Circulation.* **117,** 1082–1089.

Berk, B. C., Min, W., Yan, C., Surapisitchat, J., Liu, Y., and Hoefen, R. (2002). Atheroprotective mechanisms activated by fluid shear stress in endothelial cells. *Drug News Perspect* **15,** 133–139.

Birukov, K. G., Jacobson, J. R., Flores, A. A., Ye, S. Q., Birukova, A. A., Verin, A. D., and Garcia, J. G. (2003). Magnitude-dependent regulation of pulmonary endothelial cell barrier function by cyclic stretch. *Am. J. Physiol. Lung Cell Mol. Physiol.* **285,** L785–L797.

Boo, Y. C., Sorescu, G., Boyd, N., Shiojima, I., Walsh, K., Du, J., and Jo, H. (2002). Shear stress stimulates phosphorylation of endothelial nitric-oxide synthase at Ser1179 by Akt-independent mechanisms: Role of protein kinase A. *J. Biol. Chem.* **277,** 3388–3396.

Chien, S. (2007). Mechanotransduction and endothelial cell homeostasis: The wisdom of the cell. *Am. J. Physiol. Heart Circ. Physiol.* **292**, H1209–H1224.

Chiu, Y., McBeath, E., and Fujiwara, K. (2007). Fyn- and stretch-dependent PECAM-1 tyrosine phosphorylation in detergent extracted endothelial cell model. *Mol. Biol. Cell* **18** (Supp), 425.

Cybulsky, M. I., and Gimbrone, M. A., Jr. (1991). Endothelial expression of a mononuclear leukocyte adhesion molecule during atherogenesis. *Science* **251**, 788–791.

Davies, P. F. (2007). Endothelial mechanisms of flow-mediated athero-protection and susceptibility. *Circ. Res.* **101**, 10–12.

Davignon, J., and Ganz, P. (2004). Role of endothelial dysfunction in atherosclerosis. *Circulation* **109**, III27–III32.

Dewey, C. F., Jr., Bussolari, S. R., Gimbrone, M. A., Jr., and Davies, P. F. (1981). The dynamic response of vascular endothelial cells to fluid shear stress. *J. Biomech. Eng.* **103**, 177–185.

Diamond, S. L., Eskin, S. G., and McIntire, L. V. (1989). Fluid flow stimulates tissue plasminogen activator secretion by cultured human endothelial cells. *Science* **243**, 1483–1485.

Dimmeler, S., Haendeler, J., Rippmann, V., Nehls, M., and Zeiher, A. M. (1996). Shear stress inhibits apoptosis of human endothelial cells. *FEBS Lett.* **399**, 71–74.

Fujiwara, K. (2003). Mechanical stresses keep endothelial cells healthy: Beneficial effects of a physiological level of cyclic stretch on endothelial barrier function. *Am. J. Physiol. Lung Cell Mol. Physiol.* **285**, L782–L784.

Garcia-Cardena, G., and Gimbrone, M. A., Jr. (2006). Biomechanical modulation of endothelial phenotype: Implications for health and disease. *Handb. Exp. Pharmacol.* **176**, 79–95.

Garin, G., Abe, J., Mohan, A., Lu, W., Yan, C., Newby, A. C., Rhaman, A., and Berk, B. C. (2007). Flow antagonizes TNF-alpha signaling in endothelial cells by inhibiting caspase-dependent PKC zeta processing. *Circ. Res.* **101**, 97–105.

Garin, G., and Berk, B. C. (2006). Flow-mediated signaling modulates endothelial cell phenotype. *Endothelium* **13**, 375–384.

Gimbrone, M. A., Jr., Topper, J. N., Nagel, T., Anderson, K. R., and Garcia-Cardena, G. (2000). Endothelial dysfunction, hemodynamic forces, and atherogenesis. *Ann. N. Y. Acad. Sci.* **902**, 230–239; discussion 239–240.

Go, Y. M., Park, H., Maland, M. C., and Jo, H. (1999). *In vitro* system to study role of blood flow on nitric oxide production and cell signaling in endothelial cells. *Methods Enzymol.* **301**, 513–522.

Hajjar, K. A., and Deora, A. (2000). New concepts in fibrinolysis and angiogenesis. *Curr. Atheroscler. Rep.* **2**, 417–421.

Jo, H., Song, H., and Mowbray, A. (2006). Role of NADPH oxidases in disturbed flow- and BMP4- induced inflammation and atherosclerosis. *Antioxid. Redox Signal* **8**, 1609–1619.

Joung, I. S., Iwamoto, M. N., Shiu, Y. T., and Quam, C. T. (2006). Cyclic strain modulates tubulogenesis of endothelial cells in a 3D tissue culture model. *Microvasc. Res.* **71**, 1–11.

Kandlikar, S., Garimella, S., Li, D., Colin, S., and King, MR. (2005). "Heat Transfer and Fluid Flow in Minichannels and Microchannels." New York: Elsevier Science, New York.

Ku, D. N., Giddens, D. P., Zarins, C. K., and Glagov, S. (1985). Pulsatile flow and atherosclerosis in the human carotid bifurcation. Positive correlation between plaque location and low oscillating shear stress. *Arteriosclerosis* **5**, 293–302.

Lee, A. A., Delhaas, T., Waldman, L. K., MacKenna, D. A., Villarreal, F. J., and McCulloch, A. D. (1996). An equibiaxial strain system for cultured cells. *Am. J. Physiol.* **271**, C1400–C1408.

Liu, J., Liu, T., Zheng, Y., Zhao, Z., Liu, Y., Cheng, H., Luo, S., and Chen, Y. (2006). Early responses of osteoblast-like cells to different mechanical signals through various signaling pathways. *Biochem. Biophys. Res. Commun.* **348**, 1167–1173.

Malek, A. M., Alper, S. L., and Izumo, S. (1999). Hemodynamic shear stress and its role in atherosclerosis. *JAMA* **282,** 2035–2042.

Nagel, T., Resnick, N., Dewey, C. F., Jr., and Gimbrone, M. A., Jr. (1999). Vascular endothelial cells respond to spatial gradients in fluid shear stress by enhanced activation of transcription factors. *Arterioscler. Thromb. Vasc. Biol.* **19,** 1825–1834.

Naruse, K., Yamada, T., and Sokabe, M. (1998). Involvement of SA channels in orienting response of cultured endothelial cells to cyclic stretch. *Am. J. Physiol.* **274,** H1532–H1538.

Papadaki, M., and McIntire, L. V. (1999). Quantitative measurement of shear-stress effects on endothelial cells. *In* "Tissue Engineering Methods and Protocols" (J. R. M. M. L. Yarmush, ed.), Humana Press, Totowa, NJ.

Pearson, J. D. (2000). Normal endothelial cell function. *Lupus* **9,** 183–188.

Shi, F., Chiu, Y. J., Cho, Y., Bullard, T. A., Sokabe, M., and Fujiwara, K. (2007). Down-regulation of ERK but not MEK phosphorylation in cultured endothelial cells by repeated changes in cyclic stretch. *Cardiovasc. Res.* **73,** 813–822.

Skilbeck, C., Westwood, S. M., Walker, P. G., David, T., and Nash, G. B. (2001). Population of the vessel wall by leukocytes binding to P-selectin in a model of disturbed arterial flow. *Arterioscler. Thromb. Vasc. Biol.* **21,** 1294–1300.

Takemasa, T., Sugimoto, K., and Yamashita, K. (1997). Amplitude-dependent stress fiber reorientation in early response to cyclic strain. *Exp. Cell Res.* **230,** 407–410.

Tardy, Y., Resnick, N., Nagel, T., Gimbrone, M. A., Jr., and Dewey, C. F., Jr. (1997). Shear stress gradients remodel endothelial monolayers *in vitro* via a cell proliferation-migration-loss cycle. *Arterioscler. Thromb. Vasc. Biol.* **17,** 3102–3106.

Traub, O., and Berk, B. C. (1998). Laminar shear stress: Mechanisms by which endothelial cells transduce an atheroprotective force. *Arterioscler. Thromb. Vasc. Biol.* **18,** 677–685.

Tschumperlin, D. J., and Margulies, S. S. (1998). Equibiaxial deformation-induced injury of alveolar epithelial cells *in vitro. Am. J. Physiol.* **275,** L1173–L1183.

Usami, S., Chen, H. H., Zhao, Y., Chien, S., and Skalak, R. (1993). Design and construction of a linear shear stress flow chamber. *Ann. Biomed. Eng.* **21,** 77–83.

World, C. J., Garin, G., and Berk, B. (2006). Vascular shear stress and activation of inflammatory genes. *Curr. Atheroscler. Rep.* **8,** 240–244.

Yoshizumi, M., Abe, J., Tsuchiya, K., Berk, B. C., and Tamaki, T. (2003). Stress and vascular responses: Atheroprotective effect of laminar fluid shear stress in endothelial cells: Possible role of mitogen-activated protein kinases. *J. Pharmacol. Sci.* **91,** 172–176.

Yost, M. J., Simpson, D., Wrona, K., Ridley, S., Ploehn, H. J., Borg, T. K., and Terracio, L. (2000). Design and construction of a uniaxial cell stretcher. *Am. J. Physiol. Heart Circ. Physiol.* **279,** H3124–H3130.

CHAPTER THREE

ENDOTHELIAL CELL ADHESION AND MIGRATION

Cynthia A. Reinhart-King

Contents

Abstract

Endothelial cell adhesion and migration is fundamental to a number of physiologic processes, including vascular development and angiogenesis. It has been investigated in a variety of contexts, including tumorigenesis, wound healing, tissue engineering, and biomaterial design. The chemical and mechanical extracellular environments are critical regulators of these processes, affecting integrin-matrix binding, cell adhesion strength, and cell migration. Understanding the synergy between matrix chemistry and mechanics will ultimately lead to precise control over adhesion and migration. Moreover, a better understanding of endothelial cell adhesion is critical for development of therapeutics and biomaterials for the treatment of endothelial cell dysfunction and the progression of vascular disease. This chapter will focus on the specific interactions between endothelial cells and the extracellular matrix that mediate adhesion and migration. Several engineering methods used to probe and quantify endothelial cell adhesion and migration will be discussed.

Department of Biomedical Engineering, Cornell University, Ithaca, New York

Methods in Enzymology, Volume 443

ISSN 0076-6879, DOI: 10.1016/S0076-6879(08)02003-X

1. INTRODUCTION

The endothelium is the single cell layer lining blood vessels, establishing a semipermeable barrier between blood and surrounding tissue. The formation of blood vessels, from larger arteries to micron-sized capillaries, depends on endothelial cell adhesion to the extracellular matrix and well-organized cellular movements. During the closure of a wound, for example, angiogenesis is essential to revascularization within the newly healed tissue. Endothelial cells migrate from preexisting blood vessels, proliferate, and reorganize with vascular smooth muscle cells and pericytes to form a new capillary network (Davis and Senger, 2005; Lamalice *et al.*, 2007). Similarly, angiogenesis occurs in response to tumor-secreted angiogenic signals that trigger the formation of a vascular network to provide a blood supply to growing tumors (Eliceiri and Cheresh, 2001). Endothelial cell adhesion and migration is also critical in the formation of larger vessels and has been the subject of intense investigation for the optimization of vascular stents. It is now well established that without proper endothelialization, stents tend to occlude because of thrombosis and intimal hyperplasia (Sayers *et al.*, 1998). Therefore, a greater understanding of cell adhesion and migration could be exploited for the design of stents that support rapid endothelialization.

Endothelial cell adhesion and migration is primarily mediated through integrin binding to the extracellular matrix. Integrins are $\alpha\beta$ heterodimeric cell surface receptors that recognize specific extracellular matrix ligands. Integrins not only serve to anchor the cells to their matrix, but they also function as transducers of chemical and mechanical signals between the extracellular and intracellular environments (Miranti, 2002). Because of their critical role in mediating cell adhesion, integrins are important regulators of several endothelial cell–related processes, including vasculogenesis and angiogenesis. Endothelial cells express several integrins associated with several different ECM ligands (Table 3.1). Evidence indicates that specific integrins have specific functions in various vascular processes. For example, α_v is believed to be the subunit primarily responsible for angiogenesis (Friedlander *et al.*, 1995). Additional studies with α_v knockout mice indicate that other integrin subunits may also have a role in angiogenesis and can compensate without the expression of α_v (Bader *et al.*, 1998). Consistent with this finding, α_5 knockout mice also displayed impaired vasculogenesis during development (Yang *et al.*, 1993). The importance of integrin-ECM connections has been confirmed through the use of ECM fragments as anti-angiogenic agents (Eliceiri and Cheresh, 2001). Although there is still much to be learned about the individual contributions of individual integrin subtypes, it is clear that integrin–matrix interactions are requisite for endothelial cell adhesion and migration associated with formation of the vasculature.

Table 3.1 Endothelial cell integrins and ligands

Integrin Subunits	ECM Ligands
$\alpha1\beta1$	collagen/laminin
$\alpha2\beta1$	collagen/laminin
$\alpha3\beta1$	laminin
$\alpha5\beta1$	fibronectin
$\alpha6\beta1$	laminin
$\alpha v\beta3$	vitronectin/fibronectin
$\alpha v\beta5$	vitronectin

In vitro assays of endothelial cell adhesion and migration have led to critical insights into the mechanisms of angiogenesis and vasculogenesis. By investigating adhesion and migration *in vitro*, specific extracellular matrix signals to the cells can be precisely controlled. Most studies involving adhesion-mediated signals have paid special attention to the chemical nature of adhesion-related signals. That is to say, many studies have focused on the specific interaction between certain endothelial integrins with particular extracellular matrix proteins, including fibronectin, laminin, and collagen. Of interest is how the matrix type or density affects adhesion-related cell response. These studies have been useful in dissecting the relative roles of various integrins on endothelial cell adhesion, migration, and tube formation.

More recently, the role of matrix mechanics has emerged as an area of intense interest. It is becomingly increasingly evident that matrix stiffness (or modulus) can alter cell adhesion and subsequent cell signaling responses in a variety of cell types, including vascular smooth muscle cells (Peyton and Putnam, 2005; Wong, 2003), fibroblasts (Lo *et al.*, 2000), mammary epithelial cells (Paszek *et al.*, 2005), neurons (Georges *et al.*, 2006), and endothelial cells (Reinhart-King *et al.*, 2003). Substrate stiffness has been shown to alter cell-substrate adhesive strength, cell contractility, focal adhesion formation, migration speed, cell–cell interactions, and cell assembly. Because *in vivo* compliance varies within tissues and changes in pathologic conditions (Guo *et al.*, 2006; Paszek *et al.*, 2005), the role of matrix mechanics in cell regulation is gaining increasing attention.

Our laboratory, in particular, is interested in how both the chemical (i.e., matrix type, density, and presentation) and the mechanical (i.e., substrate stiffness and external applied forces) environments affect endothelial cell adhesion, migration, and subsequent adhesion-related signaling. In this chapter, methods to prepare cells for the study of adhesion and adhesion-related signals and protocols to control the chemical and mechanical cellular microenvironment will be described, as well as methods to quantify cell adhesion and migration.

2. Cell Preparation

In cell adhesion and migration experiments, it is important that the cells are synchronized and adhesion-related background signals are at a minimum, particularly before assaying for adhesion-related signal transduction. To synchronize the cells and minimize adhesion-related signals, the following protocol can be followed:

1. Culture endothelial cells until confluence.
   NOTE: The cells should not be overgrown, because they can begin to peel up as cell sheets rather than remaining adhered to the dish.
2. Serum-starve the cells for 16 h. Serum starvation is most important for signaling assays to minimize all background signaling; however, for measurements of adhesion strength and migration, it is often sufficient to synchronize the cells using only step 1.
   Media used for serum starvation varies with the type of endothelial cells and culture conditions before starvation. For example, we have found bovine aortic endothelial cells can be serum starved for 16 h by completely removing the serum; however, human umbilical vein endothelial cells can require 2% serum to remain viable over the 16 h window.
3. Wash the cells twice with ample PBS to remove all residual media and then trypsinize. Because the length of time the cells are exposed to trypsin can affect their ability to re-adhere (Brown et al., 2007), it is important to keep this time constant throughout all adhesion experiments. In addition, to help maintain consistency, freshly thawed aliquots of trypsin should be used for each experiment. To reduce trypsin activity after the cells have released from the dish, trypsin inhibitor should be added to the trypsinized cells at an equal or greater volume to the amount of trypsin used to detach the cells.
4. The cells are then centrifuged to a pellet for 5 min at 100g and resuspended in serum-free media.
5. To further minimize adhesion-related signaling events, cells can be placed in suspension in serum-free media for 30 min in conical tubes coated with 1% heat-inactivated BSA. The BSA coating minimizes cell adhesion to the walls of the tube.

The cells can now be plated on the substrate of choice at the desired density.

3. Preparation of Well-Defined Surfaces

To study endothelial cell adhesion and migration quantitatively, a well-defined adhesive substrate must be prepared. This chapter will focus on primarily on 2D adhesion and migration. Advances in the design and use of 3D scaffolds for cell adhesion and migration significantly lags behind the widespread use of 2D supports. Matrigel is perhaps the most prevalently used 3D matrix for the study of endothelial cell adhesion and migration; however, its chemical composition is not well defined. A number of investigators are creating tailored 3D matrices for tissue engineering scaffolds and for use as platform for basic science studies of endothelial cell adhesion and migration (Jun and West, 2005; Raeber *et al.*, 2005), but these systems are not at the "off-the-shelf" stage and are, therefore, not yet widely used. As the 3D materials platform continues to mature, we expect that methods to measure cell adhesion and migration will also evolve. A number of differences exist between 2D and 3D adhesion. Most importantly, in 2D versus 3D, integrins bind the ECM on all sides, thereby altering cell adhesion, morphology, migration, and signaling (Berrier and Yamada, 2007). Quantitative methods to study adhesion strength and migration behavior in 3D are requiring a significant shift in approach to address these fundamental differences.

Preparing a 2D adhesive substrate typically involves immobilizing specific extracellular matrix proteins onto glass or plastic. In general, multiwell plates are used, because the format allows for simultaneous replication of adhesive conditions and minimal use of protein and cells. The regions not occupied by specific cell adhesion proteins are typically backfilled with a blocking agent, most often 10 mg/ml heat-denatured bovine serum albumin (BSA). This is prepared by dissolving the appropriate amount of BSA in calcium and magnesium-free Dulbecco's PBS, sterile-filtered, and heated to 85 °C for 10 min in small volumes to ensure equal heating. Proper heat-inactivated BSA should appear homogenous and slightly hazy. Aggregates indicate that the BSA has been overheated, whereas a clear solution indicates that the BSA is not denatured. To uniformly coat a glass or plastic dish with a specific protein, the following short protocol can be followed.

1. Dissolve the ECM protein of interest in the appropriate buffer. This is typically PBS, but may vary depending on the protein. Salt can be added to maintain the conditions that are oftentimes specified by the supplier of the specific protein. In the case of full-length fibronectin, for example, dissolving the lyophilized protein requires the addition of water and a 30-min incubation time before being mixed or aliquoted for use or storage. The 30-min incubation period helps minimize aggregation.
2. Add the dissolved protein to the well to be coated at the desired concentration, typically ranging from 1 μg/ml to 50 μg/ml.

3. Incubate overnight at 4 °C or 37 °C overnight on a level surface.
4. Aspirate off the remaining solution and wash twice with PBS. Add 1% BSA solution (described previously) and incubate for 1 h at 37 °C.

> NOTE: Because time, temperature, and concentration can alter the amount of protein bound to the culture dish, it is best to choose one time and temperature for all experiments.

One of the drawbacks of this particular method for creating an adhesive surface is that cells have the ability to remodel the protein on the surface of the glass or plastic by secreting additional matrix or degrading the protein already on the surface. Therefore, coating glass or plastic is only typically used if the study to be performed is short term, over just a few hours. To perform longer term adhesion studies, we have adopted the use of polyacrylamide sheets (Wang and Pelham, 1998).

4. PREPARATION OF POLYACRYLAMIDE SUBSTRATES FOR CELL ADHESION STUDIES

The use of polyacrylamide as a cell attachment substrate was first proposed by Pelham and Wang, in 1998 (Wang and Pelham, 1998). Their goal in this study was to create a deformable substrate to assay cell response to changes in substrate stiffness. However, because polyacrylamide is typically inert to cell adhesion and protein adsorption, it also provides an easy-to-characterize, stable platform for controlling protein presentation to cells. That is to say, it is relatively easy to chemically conjugate specific proteins to the polyacrylamide gel, and because polyacrylamide is not easily remodeled by cells, the desired protein presentation changes relatively little over the duration of the experiment compared with protein on glass or plastic (Nelson, 2003). The compliance of the polyacrylamide is varied by controlling the relative ratio of acrylamide to cross linker.

Since the original publication describing polyacrylamide as a cell substrate, the method has been implemented and published by a number of groups (Johnson et al., 2007; Kandow et al., 2007; Klein et al., 2007). In our laboratory, we have adapted the use of polyacrylamide gels to study endothelial response to covalently conjugated ECM proteins at varying densities and changes in mechanical stiffness (Reinhart-King et al., 2003; 2005). Although the specific interactions between endothelial integrins and ECM ligands have received much attention over the past several decades, the effect of ECM mechanics on endothelial cell adhesion and migration has received relatively little attention. By use of polyacrylamide matrices of varying compliance and matrix chemistry, we have shown that both matrix chemistry and mechanics can alter cell spreading, cell migration, and the formation of stable cell–cell contacts (Reinhart-King *et al.*, 2003; 2005). Polyacrylamide substrates have another advantage in addition to their mechanical and chemical tunability; they can be used as substrates for use in traction force microscopy—a method by which the contractile forces exerted by cells on its substrate are measured (Fig. 3.1) (Dembo and Wang, 1999). As cells exert force on the polyacrylamide substrate, the polyacrylamide deforms. These deformations are detected on the basis of fluorescent markers embedded within the substrate. The bead movements are translated into a strain map, and these strains are used to calculate the forces exerted by the cell. The method to perform traction force microscopy will not be presented here, because it typically requires a computationally intensive, custom algorithm to translate substrate strains to traction stresses and is described in detail elsewhere (Dembo and Wang, 1999). In addition,

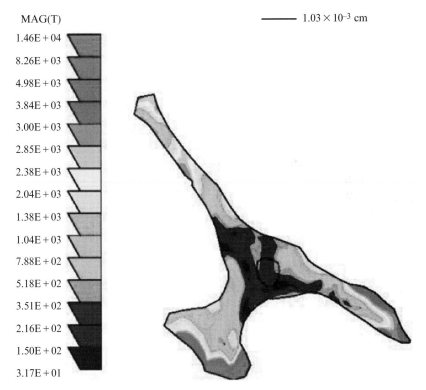

Figure 3.1 Color contour map depicting the traction forces exerted by a bovine aortic endothelial cell on a deformable polyacrylamide substrate derivatized with an RGD-containing peptide. The image was obtained by use of the LITRC traction algorithm written by Micah Dembo at Boston University; he is also the inventor of the basic theory that underlies traction force microscopy. (See color insert.)

there are other methods available to calculate cellular traction forces that are based on the movement of microfabricated cantilevers (Galbraith and Sheetz, 1997) or posts (Tan et al., 2003) by cellular forces.

It is becoming increasingly obvious that the role of mechanics cannot be ignored in the study of endothelial cell adhesion and migration. To control chemistry and mechanics, we use the following protocol to synthesize polyacrylamide substrates of well-defined compliance, presenting well-defined ECM protein chemistries, which has been adapted from Wang and Pelham (Wang and Pelham, 1998).

1. The first part of the protocol involves activating glass coverslips used as a support for the gels. The coverslip size is chosen on the basis of the experiment. For migration/adhesion assays, 22-mm square, No 1 coverslips fit into 6-well plates, allowing for easy manipulation.

2. Briefly pass the glass coverslip through a flame. Heating the glass helps the NaOH in step 3 to spread easily; however, if the glass gets too hot, it will break.

3. With a clean cotton swab, immediately apply 0.1 N NaOH to the flamed side.

4. Repeat steps 1 and 2 until the required number of slides have been coated. Allow the coverslips to dry, producing a thin film of NaOH over the surface.

5. In fume hood: Add $\sim$30 μl of 3-aminopropyl-trimethyoxysilane (APTMS) to each coverslip and spread quickly by use of a glass Pasteur pipette. This amount varies on the basis of the size of the coverslip—the amount should be just enough that it covers the surface of the coverslip.

6. Allow coverslips to dry for approximately 10 min inside the chemical fume hood.

7. Rinse the coverslips thoroughly with DI water.

8. Incubate the coverslips in a solution of glutaraldehyde in PBS (1:140, v/v) for 30 min.

9. Wash each coverslip three times with distilled water. Incubate for 5 min between each rinse.

10. Allow the coverslips to dry.

11. Coat Corning No. 1½, 18-mm circular glass coverslip with Rainex. (The size of this coverslip is based on size of glass used in step 1). The two coverslips act as a sandwich used to caste the gel. This coverslip should be smaller than the one used above. Apply Rainex with a clean cotton swab and allow to dry for at least 5 min. Buff off the excess with a Kimwipe.

12. Combine the following components in a 50-ml centrifuge tube to prepare a gel with stiffness of 2500 Pa: 2.5 ml 40% (w/v) acrylamide (BioRad, Hercules, CA); 1.0 ml 2%(w/v) n,n' methylene-Bis-acrylamide (BioRad,

Hercules, CA); 2.6 ml 0.25 M HEPES, pH 6.0; 12.39 ml ddH$_2$O; and 10 μl TEMED (BioRad, Hercules, CA).

Compliance is based on the amount of acrylamide and bisacrylamide. To vary the compliance, the amounts of acrylamide and bisacrylamide should be adjusted (Yeung et al., 2005).

13. pH solution to 6.0 by adding ~4 to 5 drops of 2 M HCl.

14. Remove 925 ml of acrylamide mixture and degas solution for 30 min under vacuum.

15. Weigh out 5.6 mg of N-succinimidyl ester of acrylamidohexanoic acid (N6). The linker allows for the covalent conjugation of proteins through linkage through primary amines. This particular linker is synthesized by use of the protocol from Pless et al. (Pless et al., 1983); however, a number of other linkers are commercially available (Kandow et al., 2007).

16. Add 70 μl of 200 proof ethyl alcohol (molecular biology grade) to the N-succinimidyl. Pipette the solution up and down until well mixed and add it to the 925 μl of acrylamide mixture.

17. To initiate polymerization, add 5 μl freshly prepared 10% ammonium persulfate (APS). Mix gently by pipetting up and down, being careful not to introduce bubbles.

18. Add 20 μl of gel solution to the activated coverslips from step 1.

19. Gently press the drop of gel solution with Rainex-coated coverslip by carefully touching the coverslip to the edge of the drop and then lowering it slowly with forceps.

20. Allow polymerization to occur for 30 min—*not longer*. The edges of the gel should recede beneath the top coverslip.

21. A few minutes before the gel is done polymerizing, dilute the ECM protein of choice in 50 mM HEPES buffer (pH 8.0) to the desired concentration (typically 1 μg/ml to 1 mg/ml).

22. Peel off coverslip from each gel with a clean razor blade.

23. Add 200 μl of the dissolved protein to the gels; 200 μl is chosen, because it seems to be the minimum amount needed to cover the gel. Spread the protein across the gel by gently pipetting the liquid over the gel's surface, completely covering the gel.

24. Incubate at 4 °C for 2 h.

25. Mix a 1:1000 volume of ethanolamine with 50 mM HEPES, pH 8.0. Each gel requires 200 μl of solution.

26. Dispense 200 μl of the ethanolamine/HEPES solution directly onto the gels. There is no need to rinse off the protein first.

27. Incubate gels at room temperature for 30 min.

28. Place gels in cold ddH$_2$O water and store at 4 °C for up to 2 weeks. Cells can be plated on the substrates just as they would be plated onto glass or plastic.

5. Quantifying Cell Adhesion

Cell adhesion is a complex process, involving the initial contact of a cell to a surface, coordinated receptor-ligand binding and actin polymerization, and finally, establishment of a well-spread state. Several methods have been tailored to understand and quantify each step in this process (Dobereiner *et al.*, 2004; Dubin-Thaler *et al.*, 2004; Giannone *et al.*, 2004; Reinhart-King *et al.*, 2005). Protocols to observe and measure cell adhesion during initial cell–substrate contact, spreading, and complete adhesion will be presented here.

5.1. Observation of endothelial cell spreading dynamics

In recent years, there has been significant interest in the observation of cells during initial adhesion and spreading to understand how intracellular and extracellular forces drive spreading (Cuvelier *et al.*, 2007; Dubin-Thaler *et al.*, 2004; Reinhart-King *et al.*, 2005). We have found that the extracellular matrix composition can alter endothelial cell–spreading dynamics, namely the rate of spreading and the shape changes cells undergo during spreading (Fig. 3.2) (Reinhart-King *et al.*, 2005). We have used measurements of area and perimeter to show that cells on high densities of ECM protein tend to spread isotropically, whereas cells on less ligand tend to spread anisotropically, through extension of thin membrane protrusions. This work has been accomplished by extensive time-lapse studies. Unlike conventional time-lapse studies of cell migration, we have found that endothelial cells are particularly sensitive to light before they have become well attached. To observe cells from their rounded state through to a fully

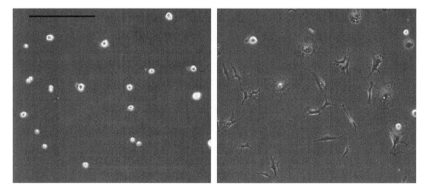

Figure 3.2 Images of bovine aortic endothelial cells plated on an RGD-coated polyacrylamide gel of 2500 Pa stiffness, taken at the time of plating and 4 h later. Scale bar is 200 μm.

spread state, the following considerations increase cell viability on the microscope stage.

1. Bright-field light should be minimized through the use of a shutter. In addition, the bulb intensity should be decreased and the exposure time on the camera should be increased accordingly. In our setup, the light is decreased such that it is virtually undetectable through the eyepieces and requires a 12-sec exposure by use of a Spot RT CCD camera.

2. A green filter is placed into the light path (Olympus, IF550).

3. A temperature and CO_2-controlled chamber is used. One hour before the experiment, the controls are turned on to allow the chamber to equilibrate. In previous experiments, we used a homemade custom-designed chamber that enclosed the microscope, from beneath the objectives to the area above the condenser. We have recently installed an enclosure from Precision Plastics (Beltsville, MD) that is sealed around the microscope with a gasket and is equipped with temperature, CO_2, and humidity control. This also works well for this purpose.

4. The experiments are performed in a dark room to minimize light. In addition, there is minimal traffic in the room to minimize air disturbances and temperature fluctuations.

5. If fluorescence microscopy is used, neutral density filters should be installed in the light path to minimize light on the sample.

If these guidelines are followed, cell spreading can be tracked through the entire process: from touchdown of the cell to a fully extended state.

5.2. Centrifugation assay

Because ECM chemistry and mechanics can alter the strength of endothelial cell adhesion and rate of spreading, it is of interest to measure the affinity between the cell and substrate. The centrifugation assay was developed as a method to quantify receptor-ligand affinity during early adhesion events (McClay et al., 1981). It can be used to measure changes in cell adhesion strength as a function of changes in the ECM mechanics or chemistry. Cells are plated in 96-well plates coated with ECM proteins. After several minutes of adhesion, the plate is inverted and centrifuged to detach cells from the substrate. The data can be reported in terms of either the amount of force needed to detach a certain fraction of cells or as the fraction of cells at each condition detached by a given force (Asthagiri et al., 1999; Guo et al., 2006). In either case, the amount of force applied should be experimentally determined for a given set of conditions.

1. Typically, multiwell plates are used for this assay and are selected so that the wells are as small as possible for a given experiment. This helps in the process of removing air from the wells as explained in step 3. Most often,

a 96-well flat bottom plate is used and coated with the desired ECM proteins and blocked with BSA as described previously. A positive control well should be plated with poly-L-lysine. It is preferable to use the inner wells of the plate for the assay to prevent media from leaking out of the plate when it is inverted for centrifugation. If the assay is to be performed on cells plated on polyacrylamide gels, then 6-well plates can be used, where the polyacrylamide glass support is fixed to the bottom of the wells.

2. Cells are plated in the wells in serum-free media. For a 96-well plate, we use 1×10^5 cells/ml in 100 μl of media. The plates are then incubated at 37 °C, 5% CO_2 for 15 min.

3. At this time, the wells should be filled to the top with serum-free media and sealed. To seal the wells before centrifugation, packing tape should be pressed down over the wells. Air and excess media should be pushed out as the tape is pulled across the wells. It is critical that no air is trapped in the well before being inverted in the next step.

4. The plate is then inverted onto a plate carrier and centrifuged for 10 min at room temperature. Typical speeds might range from 1000 to 2000g. The speed of centrifugation should be determined for the conditions being tested and will vary with the substrate, time of incubation before centrifugation, and time of centrifugation.

5. Plates should be removed from the centrifuge. At this point, cells remaining attached to the plate should be counted. The most straight-forward way to do this is to count the cells with an inverted microscope and compare this number to the number of the cells still adherent in the positive control. If there are many conditions being tested, it can be easier to take images of each well and count them later with automated imaging software like ImageJ. Alternately, the nonadherent cell can be aspirated off and the wells can be fixed with 3.7% paraformaldehyde for 10 min, washed with PBS, and then counted.

6. QUANTIFYING ENDOTHELIAL CELL MIGRATION

Endothelial cell migration is critical to processes like wound healing and angiogenesis. A number of method have been developed to measure cell migration. In this chapter, we will focus on approaches to investigate cell migration in uniform conditions; although there are a number of methods impose chemical gradients for the study of chemotaxis and haptotaxis. Most notably is the Boyden Chamber assay, in which cell migration is measured on the basis of the number of cells that migrate from a chamber containing no chemical factor through a filter into a chamber containing a chemotactic cue (Boyden, 1962). More recently, several methods have been developed to measure cell migration in chemical gradients created with microfluidic platforms (Irimia *et al.*, 2007; Saadi *et al.*, 2006; Schaff *et al.*, 2007; Wu *et al.*, 2006). Microfluidics allows for precise control over the imposed gradient and most often permit the simultaneous observation of cell migration relative to the chemical gradient with a standard inverted microscope.

Here, two approaches are presented: one to measure collective motion and the second to measure individual cell motion. The first is the traditional wound-healing assay, in which the ability of endothelial cells to migrate into an imposed wound is measured. The second is a method to quantitatively measure migration speed and persistence of individual cells.

6.1. Collective cell migration: The wound-healing assay

The wound-healing assay is used to study the ability of cells to initiate migration once a denuded area is created in a confluent culture. The method has been in use for more than 40 years (Todaro *et al.*, 1965) and has been useful in characterizing a number of factors involved in cell migration, including the role of ECM proteins, the role of cell–cell connections, and the role of various intracellular proteins in mediating cell directionality. In this assay, a confluent monolayer of cells is "scratched" away, and cell migration is measured on the basis of the amount of time it takes the bordering cells to repopulate the denuded area. The wound-healing assay is particularly relevant to the healing of the endothelium that occurs *in vivo*. When the endothelium is injured because of a wound or denuded because of balloon angioplasty, for instance, endothelial cells migrate as a sheet into the injured area to re-endothelialize the area. The wound-healing assay is a relatively easy, straightforward method to study endothelial cell migration that can be accomplished with tools readily available in most cell biology laboratories.

1. Culture dishes are coated with the ECM protein of choice and blocked with BSA as described previously.
2. Endothelial cells are plated on the dishes and grown to confluence. Because the wound healing is due to both cell migration and cell proliferation, actinomycin C can be added to the medium after the cells reach confluence at a concentration of 1 ng/ml to inhibit proliferation.

   NOTE: The assay can be performed on native or transfected cells. If transfected cells are to be used, the cells should be transfected with the plasmid of interest and a reporter plasmid such as GFP and grown to confluence before scratching the monolayer.
3. Scratch a "wound" in the monolayer by dragging a (p200 or p1000) pipette tip in a straight line across the monolayer (Fig. 3.3).
4. Mark the location of the wound with a marker on the underside of the dish. This will make it easier to find the wound in the following steps.

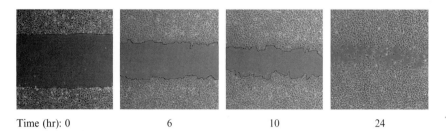

Time (hr):   0   6   10   24

Figure 3.3   Example of images acquired during a would healing experiment. Bovine aortic endothelial cells were scraped from a dish with a pipet tip and images were taken at 0,6,12, and 24 h by use of phase-contrast microscopy as the cells repopulate the wound. To measure the rate of healing, the area between the wound edges is measured and compared relative to the area of the original wound at $t = 0$. Images provided by Joseph Califano.

5. Aspirate the media from the dish and replace with fresh, warmed media.
6. Use the markings created in step 5 as reference points to find an area of the wound that will be imaged throughout the experiment and acquire an image of the wound under phase-contrast microscopy.
7. Return the dish to the incubator for 6 h.
8. Locate the same area imaged in step 6 with the reference markings and take an image with phase-contrast microscopy.
9. Repeat steps 7 and 8 until the wound has completely filled in. This time will be based on the conditions and cell type, but typically takes approximately 24 h.

To quantify cell migration, the area of the initial wound is compared with the area of the healing wound at various time points after the scratch is imposed (See Fig. 3.3), where

$$\%\text{Healed} = [(\text{Area of original wound} - \text{Area of wound during healing})/\text{Area of orginal wound}] \times 100.$$

This can be plotted against time to determine the rate of healing, a measure of cell migration. To calculate the area, automated programs have been written to calculate the denuded area in a given field of view (Bindschadler and McGrath, 2007; Sottile *et al.*, 2007). Alternately, this can be done manually by use of program such as ImageJ. Although some have quantified wound healing on the basis of the distances between the two wounded edges, we have found that measurements of area result in much less error in the sampling.

Although this method is relatively easy, it does injure cells at the border of the wound. To minimize this injury, techniques to create denuded areas without scratching the surface have been developed (Kumar *et al.*, 2005). In these methods, typically a barrier is placed in culture while the cells grow to confluence. It is then removed to allow cells to migrate into the area previously occupied by the barrier. These methods attempt to eliminate injuring the cells at the wound edge that can introduce debris that could affect migration.

6.2. Individual cell motions: Calculation of cell speed

An important aspect of cell migration is not only whether cells are capable of migrating, but also how fast they migrate. However, the process of calculating cell speed by use of time-lapse microscopy is not as straightforward as observing a cell at $t = 0$ and then at $t = 1$ h and on the basis of the distance traveled, calculating the migration speed. During normal chemokinesis, cells migrate in a random walk (Lauffenburger and Linderman, 1993). As depicted in Fig. 3.4, if cell movement were measured as the distance between the starting point A and the finishing point B, all of the information about the cell's path would be lost, and the total distance traveled would be incorrectly reported. Therefore, it is important to choose an interval that is appropriate for the cell and conditions. For endothelial cells, we have found that 5- to 10-min intervals are optimal. Given the ability to observe cells at 37 °C and 5% CO_2 as described previously, cell migration speed and persistence time can be calculated as follows:

1. Treat culture plates with ECM proteins and block with BSA as described previously.
2. Plate the cells sparsely and return the dish to an incubator for 6 h, allowing the cells to adhere and spread.
3. During the last hour of incubation, turn on the microscope incubator to allow the temperature and CO_2 to equilibrate.

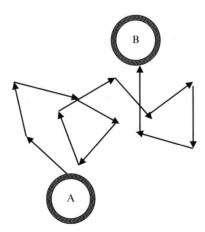

Figure 3.4   Cartoon of a sample trajectory of a cell migrating on a 2D substrate, starting at point A and stopping at B during the course of observation.

4. Place the cells on the microscope stage and image the cells by use of a 10× objective under phase-contrast microscopy. The cells should be spare such that cell–cell collisions during migration are minimal.
5. Take one image every 5 to 10 min for the duration of the experiment, which typically ranges from 6 to 24 h, while shuttering the light to prevent exposing the cells unnecessarily.
6. Track the motion of individual cells from frame to frame by either:
   a. Tracking the position of the nucleus (i.e., the centroid of the cell)
   b. Tracing the cell outline and use a program such as ImageJ to calculate the center of the cell on the basis of the outline.
7. Use these pixel coordinates, converted to micron displacements, to calculate the mean-squared displacement ($<d^2>$) for the range of time intervals.
8. The speed, S, and direction persistence time, P, can be determined by fitting the mean-squared displacement ($<d^2>$) and the time interval, t, to the persistent random walk equation: $<d^2> = 2S^2P(t-P(1-e^{-(t/P)}))$ with nonlinear least squares regression analysis (Lauffenburger and Linderman, 1993).

The Matlab code to perform this analysis is included below. This type of analysis has been used extensively to characterize the motion of cells (Peyton and Putnam, 2005; Stokes et al., 1991) and particles (King, 2006) in 2D. A more detailed description of the derivation of the equation can be found in Lauffenburger and Linderman (1993). The code requires input of x,y coordinates, entered as micron measurements, which are loaded into the code from sheet 1, column 1, and column 2, respectively, of an Excel file. The following code is commented in areas designated by "%."

```
%File name is cellmig.m
%uses a second file "SSEcellmig.m" included below
global ts MSDs
x=Sheet1(:,1); y=Sheet1(:,2);
%x and y are the location of the cell in microns,
  %converted from pixels
dt=10; %10 minutes between images
n=length(x);
Nfit=50;
MSD=zeros(n-2,1);
% calculate the mean squared displacement for range of
  %time intervals
for i=1:n-2
MSD(i)=0;
for j=1:n-i-1
MSD(i)=MSD(i)+(x(j+i)-x(j))^2+(y(j+i)-y(j))^2;
end
MSD(i)=MSD(i)/(n-i);
```

```
end
t=[1:length(MSD)]*dt;
ts=t(1:Nfit);
MSDs=MSD(1:Nfit);
% to determine S and P, perform nonlinear least squares
   % regression of migration model to MSD data.
   Uses separate %m-file "SSEcellmig.m" included below.
c=fminsearch(@SSEcellmig,[.005;400]);
migration_speed=c(1) %in units of microns/min
persistence_time=c(2) %in units of min
% plot of cell path in x-y space
Fig3.1
plot(x,y,'b')
xlabel('x (\mum)')
ylabel('y (\mum)')
% plot of MSD as a function of dt, with best-fit model
Fig.3.2
plot(t,MSD, 'bo',t,2*c(1)^2*(c(2)*t-c(2)^2*(1-exp
   (-t/c(2)))), 'b-')
xlabel('t (min)')
ylabel('MSD (\mum^2)')
% File name is SSEcellmig.m
% SSEcellmig.m is a function to determine the Sum of
   % the Squared Errors, for the nonlinear regression of
   % migration model.
function temp=SSEcellmig(c)
% time and MSD are passed between this function and main
   %program cellmig as global variables
global ts MSDs
S=c(1); P=c(2);
temp=0;
% sum the squared difference between the model and the data
   % for i=1:length(ts)
temp=temp+(2*S^2*(P*ts(i)-P^2*(1-exp(-ts(i)/P)))-
   MSDs(i))^2;
end
```

REFERENCES

Asthagiri, A. R., Nelson, C. M., Horwitz, A. F., and Lauffenburger, D. A. (1999). Quantitative relationship among integrin-ligand binding, adhesion, and signaling via focal adhesion kinase and extracellular signal-regulated kinase 2. *J. Biol. Chem.* **274,** 27119–27127.

Bader, B. L., Rayburn, H., Crowley, D., and Hynes, R. O. (1998). Extensive vasculogenesis, angiogenesis, and organogenesis precede lethality in mice lacking all alpha v integrins. *Cell* **95**, 507–519.

Berrier, A. L., and Yamada, K. M. (2007). Cell-matrix adhesion. *J. Cell. Physiol.* **213**, 565–573.

Bindschadler, M., and McGrath, J. L. (2007). Sheet migration by wounded monolayers as an emergent property of single-cell dynamics. *J. Cell. Sci.* **120**, 876–884.

Boyden, S. (1962). The chemotactic effect of mixtures of antibody and antigen on polymorphonuclear leucocytes. *J. Exp. Med.* **115**, 453–466.

Brown, M. A., Wallace, C. S., Anamelechi, C. C., Clermont, E., Reichert, W. M., and Truskey, G. A. (2007). The use of mild trypsinization conditions in the detachment of endothelial cells to promote subsequent endothelialization on synthetic surfaces. *Biomaterials* **28**, 3928–3935.

Cuvelier, D., Thery, M., Chu, Y. S., Dufour, S., Thiery, J. P., Bornens, M., Nassoy, P., and Mahadevan, L. (2007). The universal dynamics of cell spreading. *Curr. Biol.* **17**, 694–699.

Davis, G. E., and Senger, D. R. (2005). Endothelial extracellular matrix: Biosynthesis, remodeling, and functions during vascular morphogenesis and neovessel stabilization. *Circ. Res.* **97**, 1093–1107.

Dembo, M., and Wang, Y. L. (1999). Stresses at the cell-to-substrate interface during locomotion of fibroblasts. *Biophys. J.* **76**, 2307–2316.

Dobereiner, H. G., Dubin-Thaler, B., Giannone, G., Xenias, H. S., and Sheetz, M. P. (2004). Dynamic phase transitions in cell spreading. *Phys. Rev. Lett.* **93**, 1–4.

Dubin-Thaler, B. J., Giannone, G., Dobereiner, H. G., and Sheetz, M. P. (2004). Nanometer analysis of cell spreading on matrix-coated surfaces reveals two distinct cell states and STEPs. *Biophys. J.* **86**, 1794–1806.

Eliceiri, B. P., and Cheresh, D. A. (2001). Adhesion events in angiogenesis. *Curr. Opin. Cell. Biol.* **13**, 563–568.

Friedlander, M., Brooks, P. C., Shaffer, R. W., Kincaid, C. M., Varner, J. A., and Cheresh, D. A. (1995). Definition of two angiogenic pathways by distinct alpha v integrins. *Science* **270**, 1500–1502.

Galbraith, C. G., and Sheetz, M. P. (1997). A micromachined device provides a new bend on fibroblast traction forces. *Proc. Natl. Acad. Sci. USA* **94**, 9114–9118.

Georges, P. C., Miller, W. J., Meaney, D. F., Sawyer, E. S., and Janmey, P. A. (2006). Matrices with compliance comparable to that of brain tissue select neuronal over glial growth in mixed cortical cultures. *Biophys. J.* **90**, 3012–3018.

Giannone, G., Dubin-Thaler, B. J., Dobereiner, H. G., Kieffer, N., Bresnick, A. R., and Sheetz, M. P. (2004). Periodic lamellipodial contractions correlate with rearward actin waves. *Cell* **116**, 431–443.

Guo, W. H., Frey, M. T., Burnham, N. A., and Wang, Y. L. (2006). Substrate rigidity regulates the formation and maintenance of tissues. *Biophys. J.* **90**, 2213–2220.

Irimia, D., Charras, G., Agrawal, N., Mitchison, T., and Toner, M. (2007). Polar stimulation and constrained cell migration in microfluidic channels. *Lab. Chip.* **7**, 1783–1790.

Johnson, K. R., Leight, J. L., and Weaver, V. M. (2007). Demystifying the effects of a three-dimensional microenvironment in tissue morphogenesis. *Methods Cell. Biol.* **83**, 547–583.

Jun, H. W., and West, J. L. (2005). Endothelialization of microporous YIGSR/PEG-modified polyurethaneurea. *Tissue Eng.* **11**, 1133–1140.

Kandow, C. E., Georges, P. C., Janmey, P. A., and Beningo, K. A. (2007). Polyacrylamide hydrogels for cell mechanics: Steps toward optimization and alternative uses. *Methods Cell. Biol.* **83**, 29–46.

King, M. R. (2006). Anisotropic Brownian diffusion near a nanostructured surface. *J. Colloid Interface Sci.* **296**, 374–376.

Klein, E. A., Yung, Y., Castagnino, P., Kothapalli, D., and Assoian, R. K. (2007). Cell adhesion, cellular tension, and cell cycle control. *Methods Enzymol.* **426,** 155–175.

Kumar, G., Meng, J. J., Ip, W., Co, C. C., and Ho, C. C. (2005). Cell motility assays on tissue culture dishes via non-invasive confinement and release of cells. *Langmuir* **21,** 9267–9273.

Lamalice, L., Le Boeuf, F., and Huot, J. (2007). Endothelial cell migration during angiogenesis. *Circ. Res.* **100,** 782–794.

Lauffenburger, D. A., and Linderman, J. J. (1993). "Receptors: Models for Binding, Tracking and Signalling." Oxford University Press, New York.

Lo, C. M., Wang, H. B., Dembo, M., and Wang, Y. L. (2000). Cell movement is guided by the rigidity of the substrate. *Biophys. J.* **79,** 144–152.

McClay, D. R., Wessel, G. M., and Marchase, R. B. (1981). Intercellular recognition: quantitation of initial binding events. *Proc. Natl. Acad. Sci. USA* **78,** 4975–4979.

Miranti, C. K. (2002). Application of cell adhesion to study signaling networks. *Methods Cell Biol.* **69,** 359–383.

Nelson, C. M., Raghavan, S., Tan, J. L., and Chen, C. S. (2003). Degradation of micropatterned surfaces by cell-dependent and -independent processes. *Langmuir* **19,** 1493–1499.

Paszek, M. J., Zahir, N., Johnson, K. R., Lakins, J. N., Rozenberg, G. I., Gefen, A., Reinhart-King, C. A., Margulies, S. S., Dembo, M., Boettiger, D., Hammer, D. A., and Weaver, V. M. (2005). Tensional homeostasis and the malignant phenotype. *Cancer Cell* **8,** 241–254.

Peyton, S. R., and Putnam, A. J. (2005). Extracellular matrix rigidity governs smooth muscle cell motility in a biphasic fashion. *J. Cell. Physiol.* **204,** 198–209.

Pless, D. D., Lee, Y. C., Roseman, S., and Schnaar, R. L. (1983). Specific cell adhesion to immobilized glycoproteins demonstrated using new reagents for protein and glycoprotein immobilization. *J. Biol. Chem.* **258,** 2340–2349.

Raeber, G. P., Lutolf, M. P., and Hubbell, J. A. (2005). Molecularly engineered PEG hydrogels: a novel model system for proteolytically mediated cell migration. *Biophys. J.* **89,** 1374–1388.

Reinhart-King, C. A., Dembo, M., and Hammer, D. A. (2003). Endothelial cell traction forces on RGD-derivatized polyacrylamide substrata. *Langmuir* **19,** 1573–1579.

Reinhart-King, C. A., Dembo, M., and Hammer, D. A. (2005). The dynamics and mechanics of endothelial cell spreading. *Biophys. J.* **89,** 676–689.

Saadi, W., Wang, S. J., Lin, F., and Jeon, N. L. (2006). A parallel-gradient microfluidic chamber for quantitative analysis of breast cancer cell chemotaxis. *Biomed. Microdevices* **8,** 109–118.

Sayers, R. D., Raptis, S., Berce, M., and Miller, J. H. (1998). Long-term results of femorotibial bypass with vein or polytetrafluoroethylene. *Br. J. Surg.* **85,** 934–938.

Schaff, U. Y., Xing, M. M., Lin, K. K., Pan, N., Jeon, N. L., and Simon, S. I. (2007). Vascular mimetics based on microfluidics for imaging the leukocyteendothelial inflammatory response. *Lab. Chip.* **7,** 448–456.

Sottile, J., Shi, F., Rublyevska, I., Chiang, H. Y., Lust, J., and Chandler, J. (2007). Fibronectin-dependent collagen I deposition modulates the cell response to fibronectin. *Am. J. Physiol. Cell Physiol.* **293,** C1934–C1946.

Stokes, C. L., Lauffenburger, D. A., and Williams, S. K. (1991). Migration of individual microvessel endothelial cells: Stochastic model and parameter measurement. *J. Cell Sci.* **99**(Pt 2), 419–430.

Tan, J. L., Tien, J., Pirone, D. M., Gray, D. S., Bhadriraju, K., and Chen, C. S. (2003). Cells lying on a bed of microneedles: an approach to isolate mechanical force. *Proc. Natl. Acad. Sci. USA* **100,** 1484–1489.

Todaro, G. J., Lazar, G. K., and Green, H. (1965). The initiation of cell division in a contact-inhibited mammalian cell line. *J. Cell. Physiol.* **66,** 325–333.

Wang, Y. L., and Pelham, R. J., Jr. (1998). Preparation of a flexible, porous polyacrylamide substrate for mechanical studies of cultured cells. *Methods Enzymol.* **298,** 489–496.

Wong, J., Velasco, A, Rajagopalan, P, and Pham, Q. (2003). Directed movement of vascular smooth muscle cells on gradient-compliant hydrogels. *Langmuir* **19,** 1908–1913.

Wu, M., Roberts, J. W., Kim, S., Koch, D. L., and DeLisa, M. P. (2006). Collective bacterial dynamics revealed using a three-dimensional population-scale defocused particle tracking technique. *Appl. Environ. Microbiol.* **72,** 4987–4994.

Yang, J. T., Rayburn, H., and Hynes, R. O. (1993). Embryonic mesodermal defects in alpha 5 integrin-deficient mice. *Development* **119,** 1093–1105.

Yeung, T., Georges, P. C., Flanagan, L. A., Marg, B., Ortiz, M., Funaki, M., Zahir, N., Ming, W., Weaver, V., and Janmey, P. A. (2005). Effects of substrate stiffness on cell morphology, cytoskeletal structure, and adhesion. *Cell. Motil. Cytoskeleton* **60,** 24–34.

AN OPTIMIZED THREE-DIMENSIONAL IN VITRO MODEL FOR THE ANALYSIS OF ANGIOGENESIS

Martin N. Nakatsu *and* Christopher C. W. Hughes

Contents

Abstract

Angiogenesis is the formation of new blood vessels from the existing vasculature. It is a multistage process in which activated endothelial cells (EC) degrade basement membrane, sprout from the parent vessel, migrate, proliferate, align, undergo tube formation, and eventually branch and anastomose with adjacent vessels. Here we describe a three-dimensional *in vitro* assay that reproduces each of these steps. Human umbilical vein endothelial cells (HUVEC) are cultured on microcarrier beads, which are then embedded in a fibrin gel. Fibroblasts cultured on top of the gel provide factors that synergize with bFGF and VEGF to promote optimal sprouting and tube formation. Sprouts appear around day 2, lumen formation begins at day 4, and at day 10 an extensive anastomosing network of capillary-like tubes is established. The EC express a similar complement of genes as angiogenic EC *in vivo* and undergo identical morphologic changes during tube formation. This model, therefore, recapitulates

Department of Molecular Biology and Biochemistry, University of California Irvine, Irvine, California

Methods in Enzymology, Volume 443
ISSN 0076-6879, DOI: 10.1016/S0076-6879(08)02004-1

in vivo angiogenesis in several critical aspects and provides a system that is easy to manipulate genetically, can be visualized in real time, and allows for easy purification of angiogenic EC for downstream analysis.

1. OVERVIEW

1.1. Angiogenesis

The establishment of new blood vessels is critical for several physiologic and pathologic processes. Under normal conditions, angiogenesis plays a significant role during embryogenesis and wound healing, whereas in pathologic settings, uncontrolled and prolonged angiogenesis can support inflammatory diseases and the growth of solid tumors (Conway *et al.*, 2001). An intricate balance of signals from growth factors, mural cells, and extracellular matrix (ECM) proteins is critical for the proper development of new blood vessels; however, the interplay between these is only just beginning to be understood. Angiogenesis involves endothelial cells (EC) undergoing several morphologic transitions and changes in gene expression profile, and the process can be thought of as a series of steps, including degradation of the basement membrane by secreted proteases, migration (sprouting) of the EC, alignment, proliferation, lumen formation, branching, and anastomosis (Conway *et al.*, 2001; Folkman, 1975; 1985; Yancopoulos *et al.*, 2000). The final stages include recruitment of mural cells such as pericytes or smooth muscle cells, the synthesis and arrangement of a mature basement membrane, and eventually a return of the EC to a quiescent state. In a tissue where angiogenesis is ongoing, all of these stages will be occurring simultaneously, rendering problematic, for example, the temporal analysis of gene expression. For such studies, good *in vitro* models are essential, because only here can conditions be controlled to the extent that coordination of these events becomes possible.

Numerous extracellular stimuli can promote angiogenesis. Angiogenic factors such as vascular endothelial growth factor (VEGF) and basic fibroblast growth factor (bFGF) are two potent cytokines that are involved during several stages of angiogenesis (Friesel and Maciag, 1995; Gale and Yancopoulos, 1999; Yancopoulos *et al.*, 2000). VEGF in particular is a critical angiogenic factor and on binding of VEGF to its receptor VEGFR-2, EC responses such as proliferation, differentiation, and migration are elicited (Ferrara *et al.*, 2003). Developing sprouts are composed of EC with several unique phenotypes, and these cells respond to VEGF quite differently. The sprout is led by a tip cell that is highly migratory, displays numerous lamellipodia and filopodia, but lacks a lumen and does not usually divide. Tip cells respond to a gradient of VEGF by migration, but not proliferation. Tip cells also express a distinct set of genes and, in particular, are enriched for VEGFR2, PDGFB, Unc5b, dll4, and jagged-1. Trailing the tip cell are trunk cells. These display fewer membrane evaginations, are capable of forming intracellular and intercellular lumens, and have lower migratory capacity than tip cells. In response to VEGF, trunk cells undergo proliferation. Several genes are more highly expressed in trunk cells than tip cells, including tight junction proteins such as claudin 5 and extracellular matrix/basement membrane proteins such as collagen IV. Recently, a number of studies have shed light on the mechanism of EC tubulogenesis, and it seems that lumen formation begins when intracellular pinocytotic vesicles fuse to form a vacuole or intracellular lumen. Vacuoles in adjacent cells then fuse to form an intercellular lumen. This basic mechanism was first worked out *in vitro* by use of EC embedded in collagen gels (Bayless and Davis, 2003) and has recently been observed *in vivo* during development of the intersegmental vessels in zebrafish (Kamei *et al.*, 2006).

In addition to secreted factors such as VEGF, bFGF, and the angiopoietins, extracellular matrix (ECM) proteins are also key contributors to vessel morphology, because ECM ligands bind to specific integrin receptors and promote EC adhesion and migration (Davis and Senger, 2005; Ingber and Folkman, 1989). EC express integrins capable of binding several ECM proteins, including laminin, collagen, fibrin, fibronectin, vitronectin, and thrombospondins (Stupack and Cheresh, 2002). Finally, recent reports have focused on the importance of mural cells (pericytes, smooth muscle cells and fibroblasts) in the proper development and stabilization of vessels (Armulik *et al.*, 2005; Darland and D'Amore, 2001; Kitahara *et al.*, 2005; Montesano *et al.*, 1993; Tille and Pepper, 2002; Velazquez *et al.*, 2002; Watanabe *et al.*, 1997). During angiogenesis, mural cells are often recruited to the developing vessel following a gradient of PDGFB secreted by tip cells (Armulik *et al.*, 2005). The full complement of stromal-derived factors that are necessary for maturation of developing vessels has not yet been determined.

1.2. *In vitro* angiogenesis models

Although *in vitro* angiogenesis models do not model all of the complex interactions found *in vivo*, most notably the remodeling as a result of the onset of blood flow, much can be learned from these studies, because recent work has shown that the same sets of genes and morphologic processes are required during *in vitro* tubulogenesis as are used during angiogenesis *in vivo* (Aitkenhead *et al.*, 2002; Bell *et al.*, 2001). Physiologic, 3D models include aortic rings embedded in gels (Nicosia and Ottinetti, 1990; Zhu and Nicosia, 2002) and isolated EC grown in gels (Bayless and Davis, 2003; Nakatsu *et al.*, 2003a; Nehls and Drenckhahn, 1995; Passaniti *et al.*, 1992). These provide a setting that is closer to the local environment *in vivo*, where EC are surrounded by matrix and can form tubelike structures in response to growth factors and other matrix proteins. In the aortic ring assay, explants of rat or mouse aortas are embedded in fibrin or collagen gels and supplemented with growth medium. Over several days, microvessels form and undergo extensive sprouting and branching. Factors that promote or inhibit angiogenesis can be directly tested to examine their effects on sprouting. Mural cells can also be cocultured with the aortic rings, and these can provide additional signals (Montesano et al., 1993; Nakatsu et al., 2003a; Nicosia and Ottinetti, 1990). Recently, ES cells have been differentiated into embryoid bodies that when cultured in collagen gels in the presence of VEGF generate EC that sprout and form capillary-like tubes (Feraud et al., 2001). The advantage to this model is that gene expression can be modified at the ES cell level, and its effect on downstream morphologic changes can then be followed as EC develop and sprout. Isolation of pure EC from either of these models is difficult, however, because the aortas also contain smooth muscle cells and fibroblasts, and the ES cells can also differentiate into several other cell types besides EC.

Matrigel is an extracellular matrix preparation derived from the mouse Engelbreth-Holm-Swarm sarcoma and is enriched in collagen IV, nidogen, and laminin. When plated onto Matrigel, EC form a network of cords; however, although widely used, the Matrigel assay suffers from a number of serious flaws: the EC do not make intercellular lumens, neither sprouting nor proliferation occurs—only alignment—and none of the morphologic changes require mRNA or protein synthesis. In addition, a number of nonendothelial cells, including fibroblasts and U87-MG glioblastoma cells, also align into similar cords on Matrigel (Donovan et al., 2001).

Collagen gels have been widely used in angiogenesis assays, and cells of either microvascular or macrovascular origin can be induced to form tube-like structures in response to VEGF, bFGF, or both (Davis and Saunders, 2006; Goto et al., 1993; Kamei et al., 2006; Montesano et al., 1983; Pepper et al., 1992). Early work identified fibroblasts as providing critical cofactors, and in many assays PMA is used as a substitute for these unidentified morphogens (Montesano et al., 1993; Velazquez et al., 2002). In the mid 1990s Nehls and Drenckhahn published a method for inducing sprouting of bovine aortic endothelial cells in fibrin gels (Nehls and Drenckhahn, 1995), a substrate that models physiologic wound healing angiogenesis. Collagen is probably a better model for physiologic and developmental angiogenesis, whereas tumor angiogenesis likely occurs on a background of collagen and fibrin. The basic morphologic processes are almost certainly the same in either substrate although the relevant integrins may differ —for example, $\alpha_v\beta_3$ and $\alpha_5\beta_1$ are necessary for sprouting in fibrin, whereas $\alpha_2\beta_1$ is necessary in collagen (Bayless and Davis, 2003; Carnevale et al., 2007). In the fibrin gel model, EC were cultured on Cytodex beads and then induced to sprout into the gel is response to VEGF and bFGF. Although long cords were formed, lumen formation was limited.

An ideal *in vitro* model should provide the following: ease of manipulating gene expression, the ability to isolate cells at high purity for examination of gene expression, the ability to follow development of sprouts and lumens in real time by videomicroscopy. In the field of vascular biology the most widely studied endothelial cells are human umbilical vein EC (HUVEC), because they are readily obtainable, simple to isolate, and easy to culture. As a consequence, there is a huge literature on the biology of these cells, and in numerous assays they have been shown to behave similarly to EC isolated from other vascular beds. For all of these reasons, we have developed a modified procedure that is optimized for HUVEC and based on the procedure described by Nehls and Drenckhahn (Nehls and Drenckhahn, 1995). A critical modification is coculturing myofibroblasts with the EC-coated beads, which allows us to observe the generation of patent, intercellular lumens. By growing the myofibroblasts on top of, rather than in, the gel, they can be easily removed before the isolation of the EC, allowing for collection of protein or RNA solely from the EC. We have also shown that human mesenchymal stem cells can support tubulogenesis (Ghajar *et al.*, 2006) and that the distance of the mural cells from the EC-coated beads is critical, suggesting that the soluble factor(s) is poorly diffusible, either because of large size and/or strong affinity for matrix molecules (Griffith *et al.*, 2005). Here we present a detailed protocol for this *in vitro* angiogenesis assay and provide additional protocols for laser capture microdissection and immunochemical staining of the cultures, as well as retroviral transduction of the EC.

2. FIBRIN BEAD ASSAY

2.1. Introduction

We have optimized a three-dimensional *in vitro* assay that uses human umbilical vein endothelial cells (HUVEC) and recapitulates the major steps of angiogenesis, including sprouting, migration, alignment, proliferation, tube formation, branching, and anastomosis (Nakatsu *et al.*, 2003a). HUVEC at low passage (p2 to 4) are used, because higher passage cells do not sprout well or form lumens. The cells are coated onto Cytodex beads and allowed to attach overnight. Beads are then embedded in fibrin gels. Plating of fibroblasts on top of the fibrin gels is absolutely necessary for sustained sprouting and tube formation. Cultures are maintained in low-serum, defined medium, because we find that high levels of serum can prevent sprouting. Around day 2, we observe budding and sprouting of HUVEC from the beads. Lumen formation usually becomes apparent around day 4, and after several days, numerous vessels with fully developed lumens are visible (Fig. 4.1G). As described previously, the sprouts are led

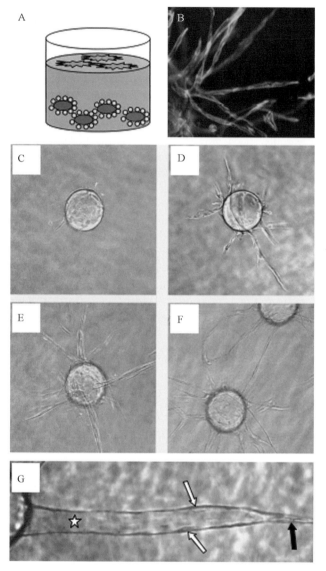

Figure 4.1 EC sprouting and tube formation in fibrin gels. (A) Schematic of HUVEC-coated beads embedded in a fibrin matrix. A fibroblast monolayer is cultured on top of the gel. (B) Immunocytochemical staining for vimentin and DAPI staining for cell nuclei. Cultures were established for 14 days, and fibroblasts were removed before staining. (C) An EC-coated bead at Day 1. Sprouts are beginning to appear. (D) An EC-coated bead at Day 4, displaying sprouting and branching. Intracellular vacuole fusion is also present within the sprouts. (E) An EC-coated bead at Day 6, displaying more EC branching and intercellular lumen formation. (F) EC-coated bead at Day 6, displaying lumen formation and anastomosis. (G) A single vessel at Day 10. The white star indicates the intercellular lumen. White arrows highlight individual EC nuclei within trunk cells, and the black arrow highlights the tip cell. For scale, beads are 100 to 150 μm diameter.

by a single tip cell, and it is the trailing trunk cells that generate lumens and from which new sprouts emerge (Fig. 4.1).

2.2. Protocol

HUVEC are isolated from umbilical cords and routinely grown in M199 supplemented with 10% fetal bovine serum (FBS) (GIBCO, Carlsbad, CA) and endothelial cell growth supplement (ECGS) (BD Biosciences, Bedford, MA) at 37° C and 5% CO_2. Passage 2 to 4 HUVEC are used for all experiments (Passage 0 being the primary culture). Lung fibroblasts (LF) were a kind gift from Dr. Cory Hogaboam (U. Michigan). LF are routinely grown in DMEM (GIBCO, Carlsbad, CA) supplemented with 10% FBS at 37° C and 5% CO_2 and used between P10 and P15. Other fibroblast lines, obtainable from ATCC, can also be used; however, because some lines are considerably better than others, we routinely test more than one line for effectiveness.

2.2.1. Preparing the cells

Expand HUVEC and fibroblasts in M199/10% FBS/Pen-Strep (1:100) 1 to 2 days before beading.

For HUVEC, switch medium to EGM-2 (Clonetics, Walkersville, MD) the day before beading. For fibroblasts, switch to EGM-2 the day before embedding.

Beading requires ~400 HUVEC per bead.

We use 20,000 fibroblasts per well of a 24-well plate. Ninety-six-well plates can also be used, with quantities scaled accordingly.

2.2.2. Cytodex 3 bead preparation

Use Cytodex 3 microcarrier beads (Amersham Pharmacia Biotech, Piscataway, NJ).

1. 0.5 g dry beads are hydrated and swollen in 50 ml PBS (pH = 7.4) for at least 3 h at RT.
2. Use a 50-ml tube and place it on a rocker.
3. Let the beads settle (~15 min). Discard the supernatant, and wash the beads for a few minutes in fresh PBS (50 ml).
4. Discard the PBS and replace with fresh PBS:

$$25 \, \text{ml} \rightarrow 20 \, \text{mg/ml} \Rightarrow 60,000 \, \text{beads/ml} \text{ or } 50 \, \text{ml} \rightarrow 10 \, \text{mg/ml} \Rightarrow 30,000 \, \text{beads/ml}$$

5. Place the bead suspension in a siliconized glass bottle (Windshield Wiper or Sigmacote).

6. Sterilize the beads by autoclaving for 15 min at 115° C.
7. Store at 4° C.

2.2.3. Reagents
2.2.3.1. Fibrinogen solution

Dissolve 2 mg/ml fibrinogen in DPBS in a 37° Cwaterbath. Mix by invert-
ing the tube. Do not vortex.
Note percentage of clottable protein and adjust accordingly.
Pass through a 0.22-μm filter to sterilize.

2.2.3.2. Aprotinin

Reconstitute lyophilized aprotinin at 4 U/ml in DI water.
Sterile filter.
Make aliquots of 1 ml each.
Store at −20° C.

2.2.3.3. Thrombin

Reconstitute in sterile water at 50 U/ml.
Make aliquots of 0.5 ml each.
Store at −20° C.

2.2.4. Coating the beads with HUVEC (Day 1)

1. Trypsinize HUVEC.
2. Allow beads to settle (do not centrifuge), aspirate the supernatant, and
   wash the beads briefly in 1 ml of warm EGM-2 medium.
3. Mix 2500 beads w/1 × 10^6 HUVEC in 1.5 ml of warm EGM-2 medium
   in a FACS tube and place it vertically in the incubator. (This will be
   enough for ∼10 wells. Scale up if needed.)
4. Incubate for 4 h at 37° C, inverting and mixing the tube every 20 min.
   (Good coating is crucial for sprouting. Beads should look like mini golf
   balls, after beading.)
5. After 4 h, transfer the coated beads to a T25 tissue culture flask (Falcon,
   Bedford, MA) and leave overnight in 5 ml of EGM-2 at 37° C and 5%
   CO_2.

2.2.5. Embedding coated beads in fibrin gel (Day 0)

1. Prepare the 2.0 mg/ml fibrinogen solution (see earlier).
Add 0.15 Units/ml of aprotinin to the fibrinogen solution.
2. Transfer coated beads to a 15 mL conical tube and let the beads settle.
   Resuspend beads in 1 ml of EGM-2 and transfer to a 1.5-ml centrifuge
   tube.

3. Wash the beads three times with 1 ml of EGM-2, mixing by pipetting up and down slowly with a P1000 pipette.

4. Count the beads on a coverslip and resuspend in fibrinogen solution at a concentration of 500 beads/ml.

5. Add 0.625 Units/ml of thrombin to each well of a 24-well plate.

6. Add 0.5 ml of the fibrinogen/bead suspension to each well. Change the pipette tip for each well.

7. Mix the thrombin and the fibrinogen/beads by pipetting up and down gently ~four to five times. Avoid creating bubbles in the fibrin gel. Allow the fibrinogen/bead solution to clot for 5 min at room temperature and then at 37° C and 5% CO_2 for 15 min. It is important that the plate *not* be disturbed during the first 5 min of clotting, because sheared fibrin reduces sprouting.

8. Add 1 ml of EGM-2 to each well slowly (drop wise).

9. Seed LF on top of the clot at a concentration of 20,000 cells/well (Fig. 4.1A).

10. Replace with fresh EGM-2 medium every other day until desired growth is achieved (Fig. 4.1B).

NOTES:

Usually, when the fibrin gel is formed, tiny bubbles will be present in the gel. They will disappear in 3 to 4 days.

Sprouting should be apparent between day 2 and 4 (Fig. 4.1C).

Lumen formation begins around day 4 to 5 and sprouts continue to elongate (Fig. 4.1D,E,G).

Newly formed tubes begin to branch around day 4 to 6 (Fig. 4.1D,F).

By day 6 to 7, the microvessel-like structures begin to anastomose with adjoining tubes (Fig. 4.1B,F). Increasing the number of beads per well results in earlier anastomosis.

3. IMMUNOCYTOCHEMISTRY OF ANGIOGENIC SPROUTS *IN VITRO*

For EC nuclei staining, fibrin gels are washed twice with 1× PBS and then fixed overnight in 2% paraformaldehyde. After two more washes with 1× PBS gels are then stained with 4′, 6-diamidino-2-phenylindole (DAPI) (Sigma, St. Louis, MO). For immunostaining, LF are first removed through a brief treatment of the gels with 10× trypsin. Digestion is stopped with serum as soon as all fibroblasts are removed. Gels are then extensively washed with HBSS, 1× (Cellgro, Herndon, VA). Cultures are then fixed for 10 min in 10% formalin and permeabilized with 0.5% Triton X-100 for 5 min. Nonspecific binding is blocked with a solution of 5% BSA in PBS

for 2 h. Primary antibodies are used at a 1/100 dilution in blocking buffer
and incubated overnight at 4° C. After extensive washing, bound antibody
is detected by species-specific Alexa Fluor 488–conjugated or Alexa Fluor
568–conjugated secondary antibodies at a 1/1000 dilution (Molecular
Probes, Carlsbad, CA). Isotype-specific nonbinding antibodies are used
as a control. When high background becomes a problem, we reduce
the concentration of primary or secondary antibody and, if necessary,
increase the incubation and/or washing times. F-actin is stained with
TRITC-phalloidin (Sigma, St. Louis, MO) at a concentration of 0.2 μM.
Phase-contrast and fluorescent images are captured on an IX70 Olympus
microscope coupled with a digital camera. Fluorescent Z-series image
stacks are captured on a two-photon Carl Zeiss MicroImaging LSM 510
Meta microscope and compiled into three-dimensional renderings with
Metamorph software (Universal Imaging Corporation, Downingtown PA).

Examples of immunostaining are shown in Fig. 4.2. EC nuclei were first
stained with DAPI, followed by double staining for the junctional protein
β-catenin (red) and the intermediate filament protein vimentin (green)
(Fig. 4.2A–C). A montage of all three figures reveals the different expression
pattern of these markers within the developing vessel (Fig. 4.2D). Thus,
expression of various markers can be readily detected in this system.

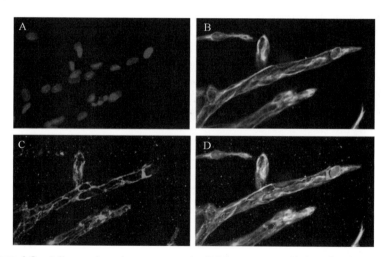

Figure 4.2 Adherens junctions connect the EC in a sprout. Fibrin gel cultures were
established on chamber slides for 10 days. Fibroblasts were then removed and the gels
fixed in paraformaldehyde overnight at 4° C before staining. (A) Cell nuclei are revealed
by DAPI staining. (B) Intermediate filaments containing vimentin are apparent in both
tip and trunk cells. (C) Staining for β-catenin reveals adherens junctions between individual EC, including between the tip cell and the first trailing trunk cell. The presence
of adherens junctions in the tip cell suggests that an epithelial–mesenchymal transition
(EMT) is not occurring. (D) Merged images show cell nuclei (blue), vimentin (green),
and adherens junctions (red). (See color insert.)

We next captured fluorescent optical image stacks along the z-axis of the cultures to create 3D representations of the vessels. The nuclei were stained by DAPI (green), and vessel walls were stained for vimentin (orange). Wide, hollow lumens are clearly visible, surrounded by a single layer of endothelial cells. Again, these images confirm that the lumens present in the *in vitro* assay are intercellular and not intracellular slits as is often seen in Matrigel assays. Furthermore, our images confirm that the HUVEC are polarized, in that they have an apical membrane, facing the lumen, and a basal membrane, apposed to a collagen IV–rich basement membrane and the fibrin gel.

4. RETROVIRAL TRANSDUCTION OF EC

Although transfection of plasmids can be used to alter gene expression in EC, the transient nature of expression can be problematic. In this angiogenesis system, sprouting is often not seen until day 2, and lumen formation may not be occurring until day 4 or 5. Because the cells are not plated on the beads until at least 1 day after transfection, expression of the exogenous gene is often waning by the time its presence might be relevant to EC morphologic changes. Sprouting can be triggered earlier by use of smaller wells (96-well plates) and reducing the volume of the gels so that the fibroblasts are closer to the beads (Griffith *et al.*, 2005), and this has been used successfully to study the role of some genes; however, in many cases, longer term expression of exogenous genes is desirable. For this reason, we have optimized the use of retrovirally transduced EC in the assay.

The Orbigen Phoenix Retrovirus Expression System (Orbigen, San Diego, CA) can be used to generate retroviruses expressing GFP. The pBMN-GFP is an MMLV-based retroviral plasmid that carries an IRES-GFP cassette for bicistronic expression and the Epstein–Barr virus nuclear antigen (EBNA) for the establishment of stable episomes. The Phoenix-ampho cells are used for retroviral packaging and produce helper-free amphotropic retroviruses. Phoenix cells are transfected with pBMN-GFP with Lipofectamine 2000 following the manufacturer's protocol. After 4 h, medium is replaced with fresh growth medium (DMEM, supplemented with 10% FBS) without antibiotics. Virus supernatants are collected 48 h after transfection and filtered. HUVEC (10% confluence) are infected with fresh virus supernatants (supplemented with 8 μg/ml polybrene) for 6 h. Virus supernatants are then removed and replaced with fresh growth medium (M199, supplemented with 10% FBS and endothelial cell growth supplement) without antibiotics for overnight incubation. The same transduction procedure is then repeated twice more on subsequent days. Transduction efficiency, as assessed by GFP expression, is monitored under fluorescent microscopy, and FACS and is usually greater than 95%.

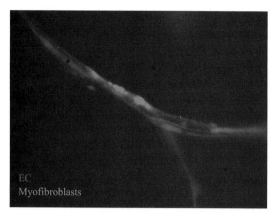

Figure 4.3   Myofibroblast-endothelial interactions can be modeled. EC were transduced with a pBN-mCherry retroviral vector, and myofibroblasts were transduced with a pBN-GFP retroviral vector. Transduced EC were coated on beads and embedded in fibrin gels through which were distributed transduced myofibroblasts. Over several days, myofibroblasts (green) migrate toward and wrap around EC sprouts (red), appearing to behave as pericytes and smooth muscle cells do *in vivo*. Photograph courtesy of Xiaofang Chen and Steve George, UC Irvine. (See color insert.)

Cells continue to express GFP at a similar level for at least 14 days. Of note, we have also tested adenoviral transduction but achieved only limited success. In our hands, when cells were transduced with high enough titers to show widespread GFP expression, they no longer sprouted efficiently or made tubes, although they survived well in monolayer cultures. Conversely, if titer was reduced to the point at which sprouting was robust, GFP expression was too low to observe by microscopy. Unlike retroviruses, adenovirus confers a load on the cell as a result of transcription and translation of adenoviral genes, and this load may hamper the ability of the EC to undergo the necessary morphologic transitions needed for sprouting and tube formation. An example of cells expressing fluorescent tags is shown in Fig. 4.3. EC were transduced with a vector expressing mCherry (red), and these cells were plated onto beads. Myofibroblasts were transduced with a GFP vector (green) and mixed into the fibrin gel along with the beads. Myofibroblasts are seen to wrap around the EC tube in a way reminiscent of smooth muscle cells or pericytes *in vivo*.

5. LASER CAPTURE MICRODISSECTION

As described previously, EC assume different phenotypes at different locations in the developing sprouts: there are distinct tip cells, trunk cells, cells at branch points, and cells undergoing anastomosis. Because gene expression in each of these is likely to differ, it would be useful to have a way of isolating each of these populations. To achieve this goal we have optimized a laser capture microdissection (LCM) method. Bead-containing gels of approximately 500 μl volume are lightly trypsinized to remove fibroblasts (>99% efficient), fixed in 70% ethanol, and then sequentially dehydrated. Once in 100% ethanol, gels are transferred to 2.0-μm PEN-Membrane slides (Leica no. 11505158) and dried down. The dried gels are approximately 50 to 150 μm thick. To date we have only attempted to isolate tip and trunk cells (Fig. 4.4). These are excised by use of a Leica LS-LMD system. Microdissected cells are collected directly into TRIZOL (Invitrogen) to preserve RNA. Samples are frozen at −80° C until roughly 500 cells have been collected. Before RNA purification samples are pooled and 1 μg of tRNA is added to each. RNA is isolated and purified according to standard procedures and cDNA is synthesized for qRT-PCR. Five hundred cells provide enough RNA for 5 to 10 reliable qRT-PCR runs, depending on the abundance of the target gene. Because of limitations on the accuracy of cutting, we obtain enriched populations of tip cells, with some trunk cell contamination, whereas trunk cell preparations are extremely pure.

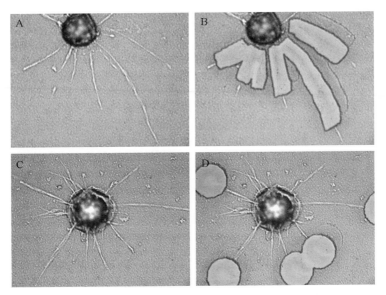

Figure 4.4 Laser capture microdissection of tip and trunk cells. Cultures were established as earlier and grown until day 10 to 14. Fibroblasts were removed, gels were fixed in 70% ethanol, and were then sequentially dehydrated. Tip and trunk cells were excised by use of a Leica LS-LMD system. (A and B) Sprouts before and after capture of trunk cells. (C and D) Sprouts before and after capture of tip cells.

6. Discussion

Sophisticated 3D *in vitro* angiogenesis assays bridge the gap between simple monolayer cultures of EC and complex *in vivo* assays of blood vessel development. 3D assays and *in vivo* approaches each have their place and can provide unique insights into endothelial cell biology. Although the effects of a complex multicellular environment or changes in blood flow can best be studied *in vivo*, gene expression studies and real-time imaging are best performed *in vitro*. The assay described here is a particularly powerful tool, because most of the fundamental processes of angiogenesis are faithfully reproduced: basement membrane is degraded before sprouting of single EC; migration and proliferation of EC occur; lumen formation proceeds by the same process as has been demonstrated *in vivo*; and, in addition, microvessels branch and anastomose. In the procedure described previously, we keep the supporting myofibroblasts separate from the EC-coated beads, because this simplifies isolation of pure populations of angiogenic EC and allows easier visualization of EC morphologic changes. However, we have also modified the assay by including the myofibroblasts in the gel. With the EC and myofibroblasts fluorescently labeled by retroviral transduction, we are able to visualize the myofibroblasts migrating toward the neovessels and being recruited by them, ultimately investing them in a similar way to smooth muscle cells and pericytes *in vivo* (Fig. 4.3).

We have used this method in a number of studies, demonstrating the importance of VEGF concentration in setting vessel diameter (Nakatsu *et al.*, 2003b) and establishing a role for notch signaling in the patterning of tip cells during sprouting (Sainson *et al.*, 2005). This latter study was the first to demonstrate a cell autonomous role for notch signaling in regulating EC proliferation and sprouting during angiogenesis. Importantly, we were the first to show that blocking notch signaling results in an excess of endothelial tip cells and a concomitant increase in sprouting. Ectopic sprouting has subsequently been noted in the retinas of mice lacking the tip cell–enriched notch ligand dll4 (Hellstrom *et al.*, 2007; Lobov *et al.*, 2007; Suchting *et al.*, 2007) and in the intersegmental vessels of dll4 morphant zebrafish embryos (Leslie *et al.*, 2007; Siekmann and Lawson, 2007).

In recent work from our group, we have investigated the gene expression profile of EC during sprouting. We have derived two important insights from this study. First, HUVEC take on a more microvascular phenotype as they undergo angiogenic sprouting in the gels, switching their gene expression profile from one that matches large vessel venular EC more to one that matches microvascular EC (Chi *et al.*, 2003). For example, the large vessel-specific genes integrin α5 and RGS5 are reduced in the cells over time, whereas the microvascular-specific genes collagen IV, α2-macroglobulin, and DSCR1 are upregulated (Holderfield and Hughes, unpublished observations). Second, because the cultures are somewhat synchronized, at least in the first few days, we can correlate gene expression with different stages of angiogenesis. For example, we have identified genes that are expressed during the initial sprouting phase (day 0 to 2) and others that peak in expression at the time that lumen formation begins (day4 to 6). Yet others are only expressed at later times (day 8 to 10) when the vessels are maturing, forming tight junctions and laying down basement membrane.

This culture system also offers the unique ability to collect cells showing different phenotypes within a single developing sprout. Thus, we have used laser capture microdissection to isolate tip and trunk cells, from which we have purified RNA for gene expression analysis. We have found that a number of genes known to be enriched in tip cells *in vivo* are also over-expressed in tip cells *in vitro*, including PDGFB and VEGFR2. We have also found enrichment of the tight junction protein, claudin 5, in trunk cells, consistent with its role in assembly of tight junctions.

In conclusion, this assay is simple to perform and uses readily available cells. Sprouting of EC *in vitro* recapitulates most of the stages seen during *in vivo* angiogenesis, including basement membrane degradation, sprouting, alignment, proliferation, tube formation, branching, and anastomosis. Gene expression also closely mirrors that seen *in vivo*, and we have shown that notch signaling regulates tip cell function the same way *in vitro* as it does *in vivo* settings such as developing in mouse retina and developing zebrafish intersegmental vessels. In addition, the assay allows for isolation of cells at different stages of angiogenesis, and laser capture microdissection can even be used to examine gene expression in unique cells such as those found at the tip of developing sprouts. The use of *in vitro* assays such as the one described here, therefore, provides a unique and powerful complement to *in vivo* studies.

REFERENCES

Aitkenhead, M., Wang, S. J., Nakatsu, M. N., Mestas, J., Heard, C., and Hughes, C. C. (2002). Identification of endothelial cell genes expressed in an *in vitro* model of angiogenesis: Induction of ESM-1, (beta)ig-h3, and NrCAM. *Microvasc. Res.* **63,** 159–171.

Armulik, A., Abramsson, A., and Betsholtz, C. (2005). Endothelial/pericyte interactions. *Circ. Res.* **97,** 512–523.

Bayless, K. J., and Davis, G. E. (2003). Sphingosine-1-phosphate markedly induces matrix metalloproteinase and integrin-dependent human endothelial cell invasion and lumen formation in three-dimensional collagen and fibrin matrices. *Biochem. Biophys. Res. Commun.* **312,** 903–913.

Bell, S. E., Mavila, A., Salazar, R., Bayless, K. J., Kanagala, S., Maxwell, S. A., and Davis, G. E. (2001). Differential gene expression during capillary morphogenesis in 3D collagen matrices: Regulated expression of genes involved in basement membrane matrix assembly, cell cycle progression, cellular differentiation and G-protein signaling. *J. Cell Sci.* **114,** 2755–2773.

Carnevale, E., Fogel, E., Aplin, A. C., Gelati, M., Howson, K. M., Zhu, W. H., and Nicosia, R. F. (2007). Regulation of postangiogenic neovessel survival by beta1 and beta3 integrins in collagen and fibrin matrices. *J. Vasc. Res.* **44,** 40–50.

Chi, J. T., Chang, H. Y., Haraldsen, G., Jahnsen, F. L., Troyanskaya, O. G., Chang, D. S., Wang, Z., Rockson, S. G., van de Rijn, M., Botstein, D., and Brown, P. O. (2003). Endothelial cell diversity revealed by global expression profiling. *Proc. Natl. Acad. Sci. USA* **100,** 10623–10628.

Conway, E. M., Collen, D., and Carmeliet, P. (2001). Molecular mechanisms of blood vessel growth. *Cardiovasc. Res.* **49,** 507–521.

Darland, D. C., and D'Amore, P. A. (2001). Cellcell interactions in vascular development. *Curr. Top. Dev. Biol.* **52,** 107–149.

Davis, G. E., and Saunders, W. B. (2006). Molecular balance of capillary tube formation versus regression in wound repair: Role of matrix metalloproteinases and their inhibitors. *J. Invest. Dermatol. Symp. Proc.* **11,** 44–56.

Davis, G. E., and Senger, D. R. (2005). Endothelial extracellular matrix: Biosynthesis, remodeling, and functions during vascular morphogenesis and neovessel stabilization. *Circ. Res.* **97,** 1093–1107.

Donovan, D., Brown, N. J., Bishop, E. T., and Lewis, C. E. (2001). Comparison of three *in vitro* human 'angiogenesis' assays with capillaries formed *in vivo*. *Angiogenesis* **4,** 113–121.

Feraud, O., Cao, Y., and Vittet, D. (2001). Embryonic stem cellderived embryoid bodies development in collagen gels recapitulates sprouting angiogenesis. *Lab. Invest.* **81,** 1669–1681.

Ferrara, N., Gerber, H. P., and LeCouter, J. (2003). The biology of VEGF and its receptors. *Nat. Med.* **9,** 669–676.

Folkman, J. (1975). Tumor angiogenesis: A possible control point in tumor growth. *Ann. Intern. Med.* **82,** 96–100.

Folkman, J. (1985). Tumor angiogenesis. *Adv. Cancer Res.* **43,** 175–203.

Friesel, R. E., and Maciag, T. (1995). Molecular mechanisms of angiogenesis: Fibroblast growth factor signal transduction. *FASEB J.* **9,** 919–925.

Gale, N. W., and Yancopoulos, G. D. (1999). Growth factors acting via endothelial cell-specific receptor tyrosine kinases: VEGFs, angiopoietins, and ephrins in vascular development. *Genes Dev.* **13,** 1055–1066.

Ghajar, C. M., Blevins, K. S., Hughes, C. C., George, S. C., and Putnam, A. J. (2006). Mesenchymal stem cells enhance angiogenesis in mechanically viable prevascularized tissues via early matrix metalloproteinase upregulation. *Tissue Eng.* **12,** 2875–2888.

Goto, F., Goto, K., Weindel, K., and Folkman, J. (1993). Synergistic effects of vascular endothelial growth factor and basic fibroblast growth factor on the proliferation and cord formation of bovine capillary endothelial cells within collagen gels. *Lab. Invest.* **69,** 508–517.

Griffith, C. K., Miller, C., Sainson, R. C., Calvert, J. W., Jeon, N. L., Hughes, C. C., and George, S. C. (2005). Diffusion limits of an *in vitro* thick prevascularized tissue. *Tissue Eng.* **11,** 257–266.

Hellstrom, M., Phng, L. K., Hofmann, J. J., Wallgard, E., Coultas, L., Lindblom, P., Alva, J., Nilsson, A. K., Karlsson, L., Gaiano, N., Yoon, K., Rossant, J., *et al.* (2007). Dll4 signalling through Notch1 regulates formation of tip cells during angiogenesis. *Nature* **445,** 776–780.

Ingber, D. E., and Folkman, J. (1989). How does extracellular matrix control capillary morphogenesis? *Cell* **58,** 803–805.

Kamei, M., Saunders, W. B., Bayless, K. J., Dye, L., Davis, G. E., and Weinstein, B. M. (2006). Endothelial tubes assemble from intracellular vacuoles *in vivo*. *Nature* **442,** 453–456.

Kitahara, T., Hiromura, K., Ikeuchi, H., Yamashita, S., Kobayashi, S., Kuroiwa, T., Kaneko, Y., Ueki, K., and Nojima, Y. (2005). Mesangial cells stimulate differentiation of endothelial cells to form capillary-like networks in a three-dimensional culture system. *Nephrol. Dial. Transplant.* **20,** 42–49.

Leslie, J. D., Ariza-McNaughton, L., Bermange, A. L., McAdow, R., Johnson, S. L., and Lewis, J. (2007). Endothelial signalling by the Notch ligand Delta-like 4 restricts angiogenesis. *Development* **134,** 839–844.

Lobov, I. B., Renard, R. A., Papadopoulos, N., Gale, N. W., Thurston, G., Yancopoulos, G. D., and Wiegand, S. J. (2007). Delta-like ligand 4 (Dll4) is induced by VEGF as a negative regulator of angiogenic sprouting. *Proc. Natl. Acad. Sci. USA* **104,** 3219–3224.

Montesano, R., Orci, L., and Vassali, P. (1983). *In vitro* rapid organization of endothelial cells into capillary-like networks is promoted by collagen matrices. *J. Cell. Biol.* **97,** 1648–1652.

Montesano, R., Pepper, M. S., and Orci, L. (1993). Paracrine induction of angiogenesis *in vitro* by Swiss 3T3 fibroblasts. *J. Cell. Sci.* **105**(Pt 4), 1013–1024.

Nakatsu, M. N., Sainson, R. C., Aoto, J. N., Taylor, K. L., Aitkenhead, M., Perez-del-Pulgar, S., Carpenter, P. M., and Hughes, C. C. (2003a). Angiogenic sprouting and capillary lumen formation modeled by human umbilical vein endothelial cells (HUVEC) in fibrin gels: The role of fibroblasts and Angiopoietin-1. *Microvasc. Res.* **66,** 102–112.

Nakatsu, M. N., Sainson, R. C., Perez-del-Pulgar, S., Aoto, J. N., Aitkenhead, M., Taylor, K. L., Carpenter, P. M., and Hughes, C. C. (2003b). VEGF(121) and VEGF (165) regulate blood vessel diameter through vascular endothelial growth factor receptor 2 in an *in vitro* angiogenesis model. *Lab. Invest.* **83,** 1873–1885.

Nehls, V., and Drenckhahn, D. (1995). A novel, microcarrier-based *in vitro* assay for rapid and reliable quantification of three-dimensional cell migration and angiogenesis. *Microvasc. Res.* **50,** 311–322.

Nicosia, R. F., and Ottinetti, A. (1990). Growth of microvessels in serum-free matrix culture of rat aorta. A quantitative assay of angiogenesis *in vitro*. *Lab. Invest.* **63,** 115–122.

Passaniti, A., Taylor, R. M., Pili, R., Guo, Y., Long, P. V., Haney, J. A., Pauly, R. R., Grant, D. S., and Martin, G. R. (1992). A simple, quantitative method for assessing angiogenesis and antiangiogenic agents using reconstituted basement membrane, heparin and fibroblast growth factor. *Lab. Invest.* **67,** 519–528.

Pepper, M. S., Ferrara, N., Orci, L., and Montesano, R. (1992). Potent synergism between vascular endothelial growth factor and basic fibroblast growth factor in the induction of angiogenesis *in vitro*. *Biochem. Biophys. Res. Commun.* **189,** 824–831.

Sainson, R. C., Aoto, J., Nakatsu, M. N., Holderfield, M., Conn, E., Koller, E., and Hughes, C. C. (2005). Cell-autonomous notch signaling regulates endothelial cell branching and proliferation during vascular tubulogenesis. *FASEB J.* **19,** 1027–1029.

Siekmann, A. F., and Lawson, N. D. (2007). Notch signalling limits angiogenic cell behaviour in developing zebrafish arteries. *Nature* **445,** 781–784.

Stupack, D. G., and Cheresh, D. A. (2002). Get a ligand, get a life: Integrins, signaling and cell survival. *J. Cell. Sci.* **115,** 3729–3738.

Suchting, S., Freitas, C., le Noble, F., Benedito, R., Breant, C., Duarte, A., and Eichmann, A. (2007). The Notch ligand Delta-like 4 negatively regulates endothelial tip cell formation and vessel branching. *Proc. Natl. Acad. Sci. USA* **104,** 3225–3230.

Tille, J. C., and Pepper, M. S. (2002). Mesenchymal cells potentiate vascular endothelial growth factorinduced angiogenesis *in vitro*. *Exp. Cell. Res.* **280,** 179–191.

Velazquez, O. C., Snyder, R., Liu, Z. J., Fairman, R. M., and Herlyn, M. (2002). Fibroblast-dependent differentiation of human microvascular endothelial cells into capillary-like 3-dimensional networks. *FASEB J.* **16,** 1316–1318.

Watanabe, S., Morisaki, N., Tezuka, M., Fukuda, K., Ueda, S., Koyama, N., Yokote, K., Kanzaki, T., Yoshida, S., and Saito, Y. (1997). Cultured retinal pericytes stimulate *in vitro* angiogenesis of endothelial cells through secretion of a fibroblast growth factor-like molecule. *Atherosclerosis* **130,** 101–107.

Yancopoulos, G. D., Davis, S., Gale, N. W., Rudge, J. S., Wiegand, S. J., and Holash, J. (2000). Vascular-specific growth factors and blood vessel formation. *Nature* **407,** 242–248.

Zhu, W. H., and Nicosia, R. F. (2002). The thin prep rat aortic ring assay: A modified method for the characterization of angiogenesis in whole mounts. *Angiogenesis* **5,** 81–86.

CHAPTER FIVE

In Vitro Three Dimensional Collagen Matrix Models of Endothelial Lumen Formation During Vasculogenesis and Angiogenesis

Wonshill Koh,* Amber N. Stratman,* Anastasia Sacharidou,* *and* George E. Davis*,†

Contents

* Department of Medical Pharmacology and Physiology, School of Medicine and Dalton Cardiovascular Center, University of Missouri–Columbia, Columbia, Missouri
† Department of Pathology and Anatomical Sciences, School of Medicine and Dalton Cardiovascular Center, University of Missouri–Columbia, Columbia, Missouri

Methods in Enzymology, Volume 443
ISSN 0076-6879, DOI: 10.1016/S0076-6879(08)02005-3

Abstract

Discovery and comprehension of detailed molecular signaling pathways underlying endothelial vascular morphogenic events including endothelial lumen
formation are key steps in understanding their roles during embryonic development, as well as during various disease states. Studies that used *in vitro* three-
dimensional (3D) matrix endothelial cell morphogenic assay models, in conjunction with *in vivo* studies, have been essential to identifying molecules and
explaining their related signaling pathways that regulate endothelial cell morphogenesis. We present methods to study molecular mechanisms controlling
EC lumen formation in 3D collagen matrices. *In vitro* models representing
vasculogenesis and angiogenesis, whereby EC lumen formation and tube morphogenesis readily occur, are described. We also detail different methods of
gene manipulation in ECs and their application to analyze critical signaling
events regulating EC lumen formation.

1. INTRODUCTION

Molecular mechanisms controlling vascular morphogenic events have
stimulated major research efforts over the years given their significant role in
normal development, as well as various pathologic conditions (Adams and
Alitalo, 2007; Carmeliet, 2005; Carmeliet and Jain, 2000; Folkman, 1995;
Hanahan, 1997). Numerous studies both *in vivo* and *in vitro* have identified key
molecules regulating new vessel formation, including extracellular matrices
(ECM), integrins, matrix metalloproteinases, cytokines, and growth factors
(Carmeliet and Jain, 2000; Conway *et al.*, 2001; Hanahan, 1997; Yancopoulos
et al., 2000). *In vitro* three-dimensional (3D) matrix endothelial cell morphogenic assay models have emerged as unique and promising tools for explaining
signaling cascades controlling EC vascular morphogenesis, because they allow
us not only to define distinct EC morphogenic steps but also to perform
systemic molecular manipulations during each event (Davis and Bayless,
2003; Davis *et al.*, 2002; Egginton and Gerritsen, 2003; Hoang and Senger,
2005; Ilan *et al.*, 1998; Liu and Senger, 2004; Montesano and Orci, 1985;
Nakatsu *et al.*, 2003; Nicosia and Madri, 1987; Nicosia and Ottinetti,
1990; Nicosia and Villaschi, 1999; Sainson *et al.*, 2005; Zeng *et al.*, 2007).
To assess the role of particular molecules and their related signaling pathways regulating the EC lumen formation process, our laboratory has developed and used *in vitro* models of EC morphogenesis in 3D extracellular matrix

environments composed of either type I collagen or fibrin, because both represent the major matrix environments where vascular events occur (Bayless *et al.*, 2000; Davis and Bayless, 2003; Davis and Camarillo, 1996; Davis *et al.*, 2002; Senger, 1996; Vernon and Sage, 1995). Vascular lumenogenesis, one of the key events during vascular morphogenesis that has received relatively little attention, is increasingly recognized as a critical step during new vessel development to establish a functional vascular system (Adams and Alitalo, 2007; Bell *et al.*, 2001; Davis and Bayless, 2003; Davis *et al.*, 2002; Egginton and Gerritsen, 2003; Kamei *et al.*, 2006; Parker *et al.*, 2004). Our *in vitro* model analysis has shown that EC lumen formation in 3D ECM is mediated by the formation and coalescence of pinocytic intracellular vacuoles, a mechanism that has been implicated in EC lumen and tube development both *in vivo* and *in vitro* (Davis and Bayless, 2003; Davis and Camarillo, 1996; Davis *et al.*, 2002; Folkman and Haudenschild, 1980; Guldner and Wolff, 1973; Meyer *et al.*, 1997; Yang *et al.*, 1999). By incorporating various methods for gene/molecule manipulation, such as the use of receptor blocking antibodies, adenoviral or lentiviral transfection approaches, and siRNA suppression methods in conjunction with our *in vitro* systems, we have found that EC lumen and tube formation in collagen matrices is regulated by $\alpha2\beta1$ integrin-ECM signaling, which, in turn, activates two required Rho GTPases, Cdc42 and Rac1, for these events (Bayless and Davis, 2002; Bayless *et al.*, 2000; Davis and Camarillo, 1996). Another key molecule controlling EC lumen and tube formation in collagen matrices is the cell surface matrix metalloproteinase, MT1-MMP (Bayless and Davis, 2003; Davis and Senger, 2005; Hotary *et al.*, 2000; Saunders *et al.*, 2006). Our *in vitro* observation that EC lumen formation is regulated by vacuole formation and coalescence has been further validated by a recent *in vivo* analysis of vascular development in zebrafish. Visualization of vessel development in zebrafish expressing the GFP-Cdc42 fusion protein has revealed that vascular lumen development in zebrafish is also regulated by intracellular vacuole formation and coalescence (Bayless and Davis, 2002; Kamei *et al.*, 2006).

As master regulators of cytoskeletal organization, Rho GTPases control diverse biologic functions necessary for EC morphogenesis such as migration, proliferation, trafficking events, and cell polarity (Aspenstrom *et al.*, 2004; Hall, 1998; 2005; Ridley, 2001; Schwartz, 2004). However, detailed biochemical analysis of Rho GTPases function during EC morphogenesis has been limited to a few studies that address their critical relevance during these events (Bayless and Davis, 2002; Bryan and D'Amore, 2007; Connolly *et al.*, 2002; Davis and Bayless, 2003; Davis and Senger, 2005; Davis *et al.*, 2002; Hoang *et al.*, 2004; Kiosses *et al.*, 2002). With our *in vitro* EC morphogenesis model, we have recently carried out detailed biochemical analysis of Rho GTPase function during EC lumen and tube formation process in 3D collagen matrices, where we showed that Cdc42 and Rac1 are highly activated over time during EC lumen formation and that they directly interact with downstream targets such as Pak2, Pak4, and Par3 to regulate this process (Koh *et al.*, 2008).

Here, we describe our 3D collagen matrix EC vasculogenesis and angiogenesis models, as well as several gene/molecular manipulation methods that we have used to identify and study molecules and their related signaling pathways that control EC lumen formation and tube morphogenic events. We also detail Rho GTPase analytical approaches to examine Cdc42/Rac1 activation and their interaction with downstream targets during the EC lumen and tube formation process in 3D collagen matrices.

2. *IN VITRO* MODELS OF VASCULOGENESIS AND ANGIOGENESIS IN 3D COLLAGEN MATRICES

Our laboratory has developed microassay systems representing both vasculogenesis and angiogenesis in 3D extracellular matrices that are the two major processes that regulate new vessel development. Vasculogenesis refers to *de novo* development of vessel from a group of endothelial precursor cells (Carmeliet, 2005; Drake, 2003; Hanahan, 1997; Patan, 2000; Risau and Flamme, 1995). Our vasculogenesis model closely mimics this process, where ECs participate in tube development through the formation and coalescence of vacuole and lumen formation, because they are suspended as single cells within 3D collagen matrices (Figs. 5.1 and 5.2). On the other hand, a subset of the ECs invade and undergo tube development in our angiogenesis assay (Davis *et al.*, 2000) in response to S1P (Bayless and Davis, 2003) or SDF-1α (Saunders *et al.*, 2006), two known stimuli for EC invasion and tube morphogenesis in our system (Fig. 5.3). Thus, these models mimic angiogenesis, where new vessels form by sprouting from a preexisting vessel wall (see Figs. 5.1 and 5.3) (Carmeliet, 2005; Hanahan, 1997; Patan, 2000). Our vasculogenesis assay model can be further used to study EC tube maturation and stabilization in addition to analyzing EC lumen and tube formation by suspending pericytes with ECs in 3D collagen matrices, where pericytes are recruited to EC-lined tubes (see Fig. 5.1) (Saunders *et al.*, 2006). These interactions between two different cell types can be visualized more easily by use of stable lines of ECs and pericytes that express GFP or mRFP labeling through lentiviral transfection (Kamei *et al.*, 2006; Saunders *et al.*, 2006). The effect of EC–pericyte interactions on tube stabilization compared with that of ECs alone in our vasculogenesis model can be further assessed by inducing tube regression after allowing ECs to form a network of tubes in 3D collagen matrices (Davis *et al.*, 2001; Saunders *et al.*, 2005; 2006). Our previous work has shown that a number of serine proteases regulate EC regression in our *in vitro* system in a MMP-1– and MMP-10–dependent manner (Davis *et al.*, 2001; 2002;

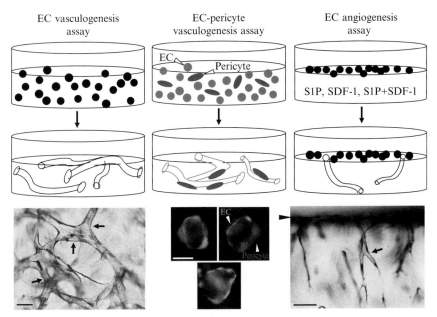

Figure 5.1 *In vitro* models of EC tube morphogenesis mimicking vasculogenesis and angiogenesis in 3D collagen matrices. Left panels, In the vasculogenesis-like assay with ECs alone, ECs undergo lumen and tube formation over a 3-day period. Arrows indicate EC-lined tubes. Bar equals 25 μm. Middle panels, In the EC-pericyte coculture vasculogenesis-like assay containing both ECs and pericytes, pericytes are recruited along EC-lined tubes over 72 h. Arrowheads indicate ECs or Pericytes. Bar equals 25 μm. Right panels, In the angiogenesis-like assay, ECs invade into collagen matrices forming luminal structures with either sphingosine-1-phosphate (S1P) (right lower panel), stromal-derived factor-1-α (SDF-1), or the combination of these two factors. Arrow indicates an EC sprout with a lumen. Arrowhead indicates the position of the EC monolayer. Bar equals 50 μm.

Saunders *et al.*, 2005). Cells that are manipulated before the assay by various methods, including adenoviral transfection or siRNA treatment, are also compatible with our system, which can be used to analyze various steps of vascular events, such as EC lumen/tube formation, EC tube stabilization, and EC tube regression (Bayless and Davis, 2002; 2004; Saunders *et al.*, 2005; 2006).

2.1. Study of EC lumen and tube morphogenesis with an *in vitro* EC vasculogenesis model

1. Trypsinize HUVECs (passages 2 to 6) and resuspend in 1× medium 199 (M199) for a concentration of 10^7 cells/ml. Mix well with a pipette to prevent any clumps of HUVECs. Place it on ice.

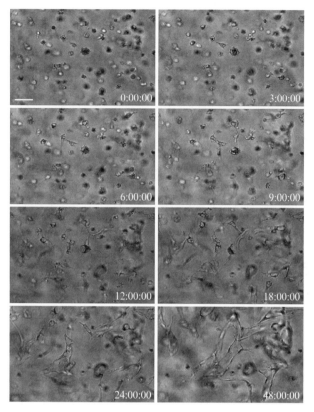

Figure 5.2 Time-lapse analysis of EC lumen and tube formation in 3D collagen matrices in assays that mimic vasculogenesis. Time-lapse images illustrating EC vacuole, lumen, and tube formation are shown at the indicated time points over a 48 h time period of culture. Bar equals 100 μm.

2. It is important to keep all reagents cold to prevent collagen from polymerizing. To prepare 1 ml of 3.75 mg/ml collagen gel solution, combine the following: 525 μl 7.1 mg/ml type I collagen in 0.1% acetic acid, 58.5 μl 10× M199 (Invitrogen), 3.15 μl 5 N NaOH, and 213 μl 1× M199 (Invitrogen). To prepare 1 ml of 2.5 mg/ml collagen gel solution, combine the following: 350 μl 7.1 mg/ml Type I collagen, 39 μl 10× M199, 2.1 μl 5 N NaOH, and 409 μl 1× M199. Mix collagen gel solution thoroughly.

3. Add 200 μl HUVECs to the collagen solution for a final concentration of 2 × 10^6 cells/ml. Gently pipette to mix with collagen solution without making any air bubbles. The cell-collagen mix is then added at 28 μl per well in 96 half-area well clear flat bottom TC-treated microplate (Corning). After every third to fourth well, tap the plate gently in each side to evenly spread out the cell-collagen mix within each well.

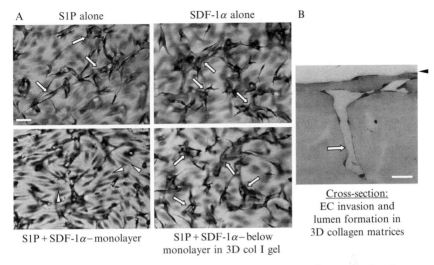

Figure 5.3   EC invasion and lumenal morphogenesis in 3D collagen matrices in assays that mimic angiogenic sprouting. (A) A subset of ECs invades and undergoes tube morphogenesis (arrows) in response to sphingosine-1-phosphate (S1P), stromal-derived factor-1-α (SDF-1α), or both factors over a 24-h time period as indicated in the different panels. In the lower panels, the left photograph is focused on the monolayer surface, whereas the right photograph is focused below the monolayer within the 3D collagen matrix illustrating EC invasion and lumen formation. Arrowheads indicate EC monolayer on top of collagen matrices. Bar equals 67 μm. (B) Cross-section of invading ECs undergoing lumen formation and tube morphogenesis in 3D collagen matrices. A plastic thin section was prepared, stained with toluidine blue, and photographed. Arrow indicates an invading EC sprout with a lumen and arrowhead indicates the position of the EC monolayer. Bar equals 25 μm. These experimental approaches represent a convenient assay for both EC invasion and its coupling with lumen and tube morphogenesis. The addition of both factors synergizes together to induce invasion and lumenogenesis. In particular, SDF-1α significantly strongly enhances lumen formation during these assays.

4. Place the plate in 37 °C with 5% CO_2 incubator for 30 min to allow collagen to polymerize and equilibrate.
5. Add 100 μl feeding media containing 1× M199 with reduced-serum II supplement (see preparation procedure following) at a 1:250 dilution, 40 ng/ml VEGF-165 (Upstate Biotechnology), 40 ng/ml FGF-2 (Upstate), 50 μg/ml ascorbic acid (Sigma Aldrich), and 50 ng/ml 12-O-tetradecanoyl-phorbol-13-acetate (TPA) (Sigma Aldrich) to each well.
6. Incubate the plate in 37 °C with 5% CO_2 and allow ECs to undergo lumen and tube formation.
7. To prepare 20 ml reduced-serum II supplement (RSII):
   a. Add 2.5 g bovine serum albumin (essentially fatty acid free and globulin free, min. 99%) (Sigma Aldrich) and 17 ml sterile water to a small beaker with stir bar. Stir gently until all bovine serum albumin is dissolved.

   b. Add 100 mg sodium selenite (Sigma Aldrich) and 4 ml sterile water to a 15-ml tube to make 25 mg/ml stock.

   c. Add 25 mg recombinant human insulin (sodium, Zn free) (Sigma Aldrich), 1.25 ml DMEM (Invitrogen), and 60 to 65 μl 1 M HCl to a 15-ml tube for a total volume of 1.31 ml. Solution will turn to a bright clear yellow color.

   d. Add 1.31 ml of insulin solution, 1 μl of 25 mg/ml sodium selenite stock, and 25 mg human holo-Transferrin (Sigma Aldrich) to the beaker containing dissolved bovine serum albumin for a total volume of 18.3 ml. Mix well until all is dissolved.

   e. Prepare 5 mg/ml oleic acid stock solution by adding 100 mg sodium oleate (Sigma Aldrich) to 20 ml 100% EtOH. The recipe calls for 21.4 mg oleic acid, which is 4.28 ml of the 5 mg/ml oleic acid stock solution. To a 24-ml glass vial, add 1 ml of the oleic acid solution and dry down under a stream of nitrogen. Repeat until all 4.28 ml oleic acid has dried down.

   f. To the glass vial containing the 24.1 mg dried oleic acid, add 10 ml of the bovine serum mixture solution from step 4. Mix well until oleic acid is dissolved. Add the remaining 8.3 ml the bovine serum mixture solution and mix well. Add 1.7 ml sterile water for a final volume of 20 ml.

   g. Use a syringe and a 0.2-μm syringe filter to sterile filter the RSII solution into a 50-ml tube.

   h. Make 1-ml aliquots and store at $-20\,^{\circ}$C. Thawed RSII is maintained at $4\,^{\circ}$C and can be kept at this temperature for several months for use in our culture systems.

2.2. Study of EC tube stabilization with an *in vitro* EC-pericyte vasculogenesis model

1. Trypsinize HUVECs and bovine retinal pericytes. These pericytes were prepared as described from adult bovine retinas (Nayak *et al.*, 1988). Resuspend each in 1× M199 for a concentration of 10^7 cells/ml. Mix well with a pipette to prevent any clumps of cells. Place them on ice.
2. Prepare 1 ml collagen gel solution as described previously.
3. Add 200 μl HUVECs and 40 to 50 μl pericytes (20 to 25% of HUVECs) to the collagen solution. Mix well and place 28 μl per well in 96-half area well plate. After every third to fourth well, tap the plate gently in each side to evenly spread out the cell-collagen mix within each well.
4. Incubate and add feeding media as described previously.

2.3. Study of EC tube regression with an *in vitro* EC or EC-pericyte vasculogenesis model

1. Prepare 1 ml collagen gel solution with EC alone or EC-pericytes as described previously.

2. The cell-collagen mix is then added at 15 μl per well in 384-well tissue culture plates (VWR). After every third to fourth well, tap the plate gently in each side to evenly spread out the cell-collagen mix within each well. Set up at least eight wells per condition.
3. Incubate the plate in 37 °C with 5% CO_2.
4. Add 60 μl feeding media as described previously with or without 2 μg/ml plasminogen (American Diagnostica) or 0.5 μg/ml plasma kallikrein (Enzyme Research Laboratories). Addition of media denotes time zero.
5. Monitor cultures every 4 h for gel contraction.
6. As gels start contracting, gel area is recorded with an ocular grid. Calculate percent collagen gel contraction as follows: [(original area−current gel area)/original area] × 100. When contraction is complete, cultures are fixed or extracted. Save culture media to examine MMP expression and activity by Western blot analysis. Add 15 μl culture media and 30 μl 3× Laemmli sample buffer with 7.5% beta-mercaptoethanol to a 1.7-ml Eppendorf tube. Boil at 100 °C for 10 min. Samples are now ready for SDS-PAGE gel analysis.

2.4. Study of EC invasion and lumen formation with an *in vitro* angiogenesis model

1. Prepare 1 ml collagen gel solution as described previously. Mix collagen gel solution thoroughly. Add 200 ng/ml SDF-1α (R&D Systems), 1 μM S1P (Avanti Polar Lipids), or both molecules to the collagen gel solution and mix thoroughly.
2. The collagen gel solution is then added at 28 μl per well in 96-half area well plate. After every third to fourth well, tap the plate gently in each side to evenly spread out the cell-collagen mix within each well.
3. Place the plate in 37 °C with 5% CO_2 incubator for 45 to 60 min and allow collagen gel to be equilibrated.
4. HUVECs (50,000 cells/well) are then added on top of collagen gel in 100 μl 1× M199 containing reduced serum supplement II at a 1:250 dilution, 40 ng/ml FGF-2, 50 μg/ml ascorbic acid, and 50 ng/ml TPA.

2.5. Data analysis

Cultures can be fixed and stained at any time points of your interest. For fixation and staining, remove the culture medium and add 140 μl 3% glutaraldehyde in PBS and fix for at least 2 h. To stain the fixed culture, remove glutaraldehyde solution and add 100 μl 1% toluidine blue in 30% methanol. We usually leave the dye for 30 min to 1 h before destaining with water. We invert plates in a plastic container and float them on water and agitate the water on a stir plate with stir bar over a 2- to 3-h period.

Vasculogenesis assays are quantified in two ways either by counting the number of EC lumens or by tracing the area of EC lumens. Eight to ten fields per each well are selected (three to four wells per condition) and photographed with an inverted microscope equipped with a camera and software at 10× magnification. Each picture is taken at a similar z-plane. Pictures are then loaded onto Metamorph® software. Metamorph® software allows you to calibrate pixel measurements and convert these measurements to distance on the basis of the millimeter-to-pixel ratio designated by each manufacturer. As you trace the EC lumen structure, Metamorph® automatically calculates the area of each traced luminal structure (Fig. 5.4), and the number of lumen structures traced can be counted. The number of EC lumens can also be counted manually with an eyepiece equipped with an ocular grid. The latter approach will take less time than tracing each luminal area with Metamorph® software, which provides us with more detailed information on EC lumen and tube formation. We have recently shown that siRNA suppression of Cdc42, Pak2, or Pak4 blocks EC lumen and tube formation in 3D collagen matrices (Koh *et al.*, 2008). The inhibitory effect of these siRNAs can be assessed in more detail by determining EC luminal area over time as shown in Fig. 5.4. This type of analysis leads to a more comprehensive examination on how these genes regulate EC lumen formation.

Angiogenesis assays are quantified by counting the number of ECs invading into the collagen matrix. For each well, eight to ten fields at 10× magnification are selected, and the number of invading cells per high-power field can be counted manually with an eyepiece equipped with an ocular grid.

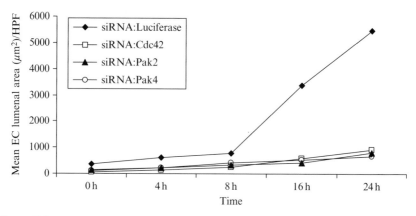

Figure 5.4 SiRNA suppression of Cdc42 and its effectors markedly inhibits EC lumen formation as analyzed with time-lapse photomicroscopy. An inhibitory effect of Cdc42, Pak2, or Pak4 siRNA on EC lumen formation was assessed by tracing EC luminal areas at the indicated time points over a 24-h time period with images obtained by time-lapse analysis. Luciferase siRNA was used as control. Specific knockdown of genes was assessed with Western blot analysis that showed marked suppression of protein expression of each of the targets in a selective manner.

In addition to fixing the culture, cultures can also be lysed at any time point to examine the expression of protein levels regulated during EC morphogenesis. To make lysates, culture media are removed, and collagen gels are picked out with fine forceps. Collagen gels are added to a preheated 1.7-ml Eppendorf tube containing $1.5\times$ Laemmli sample buffer with 7.5% beta-mercaptoethanol (use 75 μl sample buffer per gel). Tubes are boiled for 10 min at $100\,^{\circ}$C. Samples are now ready for SDS-PAGE gel analysis; we usually load 30 to 40 μl of each sample per lane on SDS-PAGE minigels.

2.6. Time-lapse imaging of vascular morphogenesis

Time-lapse imaging of EC morphogenesis allows us to visualize EC lumen formation and progression in great detail. A series of still images from EC tube formation time-lapse analysis is shown in Fig. 5.2. Our time-lapse analysis is performed with a fluorescence-inverted microscope (Eclipse TE2000-U; Nikon) equipped with a digital camera (Photometrics Cool-Snap HQ2), Metamorph software, and a temperature/CO_2–controlled chamber.

1. Set up EC vasculogenesis assay in 96-half area well as described previously.
2. Water is added in surrounding wells and in between wells to keep the plate hydrated.
3. A thin glass plate is placed over the wells that will be imaged. This prevents condensation from accumulating on the plastic lid and altering the light and imaging ability.
4. The temperature-controlled chamber is set at $37\,^{\circ}$C, with a continuous 5% CO_2 infusion and allowed to equilibrate before inserting the 96-half area well plate.
5. By use of the imaging software, bright-field or fluorescence image fields are selected. Fields toward the bottom of the gel, although not on the plastic, are favored to reduce background and increase the clarity of the image. Once selected, the imaging software allows the stage to return to previously selected xyz dimensions at any given time interval to capture pictures.
6. Once images for all time points have been captured, a stack of these files can be generated to create the final movie with Metamorph software.

3. Molecular/Genetic Manipulation of ECs

Our *in vitro* systems are highly compatible with a number of different molecular/genetic approaches, and this has allowed us to identify and study molecules/genes and their related signaling pathways controlling EC tube morphogenic events (Bayless and Davis, 2002; 2003; 2004; Bayless *et al.*, 2000; Bell *et al.*, 2001; Davis and Camarillo, 1996; Saunders *et al.*, 2005; 2006). Blocking antibodies or chemical inhibitors can be used to test general function of a certain group of molecules. The specific roles of genes can be further analyzed in more detail by transfecting cells with adenoviruses carrying recombinant proteins or with siRNAs targeting specific genes.

3.1. Use of blocking antibodies or chemical inhibitors

We have shown that integrins and MT1-MMP are required for EC lumen and tube formation with blocking antibodies or chemical inhibitors such as blocking integrin antibodies, GM6001, TIMP-1, -2, and -3 (Bayless and Davis, 2003; Saunders *et al.*, 2006). To analyze the function of molecules by blocking antibodies or chemical inhibitors, simply add an agent in a desired concentration (determined in dose-response experiments) as you prepare feeding media (step 5 in section 2.1 or step 4 in section 2.3). This is straightforward in our system, because the assays are performed in 96-well plates. Add 100 μl media to each well.

3.2. Adenoviral transfection

Previously, we have used adenoviral approach to evaluate function of Cdc42/Rac1 and MT1-MMP in EC lumen and tube formation, as well as that of MMP-1 and MMP-10 in EC tube regression (Bayless and Davis, 2002; Saunders *et al.*, 2005; 2006). The following procedure describes the adenoviral transfection of ECs grown in a 75-cm^2 surface area. For transfection in cells grown on different flasks, scale the components up or down on the basis of the surface area of the tissue flask used. Detailed protocols for recombination and virus production have been described previously (He *et al.*, 1998).

1. Make sure cells are 90 to 95% confluent on the day of transfection.
2. Wash cells twice with 10 ml M199.
3. Add 7 ml media and virus.
4. Incubate for 6 h.
5. Aspirate media and add your choice of growth media.
6. Incubate for 24 h before setting up the experiment.

3.3. siRNA transfection

Our laboratory has developed and used an efficient and highly effective siRNA suppression approach to identify numerous genes that regulate EC tube and lumen formation (Saunders *et al.*, 2005; 2006). The following procedure describes the siRNA transfection of ECs grown in a 25-cm^2 surface area. For transfection in cells grown on different flasks, scale all the components up or down on the basis of the surface area of the tissue flask. After siRNA transfection, the level of gene knockdown is assessed by Western blot analysis, reverse-transcriptase PCR, or real-time PCR.

1. Make sure that ECs are grown to 90 to 95% confluency on the day of transfection.
2. Add 566 μl Opti-Mem I (Invitrogen) and 17.5 μl siPORT^{TM} *Amine* (Ambion) to a 1.7-ml Eppendorf tube. Vortex for 10 sec and quick spin. Leave at room temperature for 30 min.
3. Add 16.5 μl of siRNA prepared at 40 μM for a final concentration of 50 to 200 nM. For many genes, we have successfully used pools of siRNA (Dharmacon Smartpools) to selectively knockdown genes (Saunders *et al.*, 2005; 2006). Gently mix tubes and incubate at room temperature for 20 min.
4. Wash cells twice with 5 ml of DMEM-LG solution (Invitrogen) containing no antibiotics.
5. After second wash, add 3 ml DMEM (no antibiotics) with 1% fetal bovine serum (Invitrogen) to a flask.
6. Quick spin the siRNA solution and add all 600 μl to a flask.
7. Incubate at 37 °C with 5% CO_2 for 4 to 6 h.
8. Wash cells twice with 5 ml of DMEM solution containing no antibiotics and feed cells with 5 ml of feeding media of your choice containing no antibiotics.
9. Incubate at 37 °C for 48 h.
10. After 48 h, repeat procedures 2–8.
11. After completing second transfection, incubate for 24 h before setting up the experiments.

4. RHO GTPASE ACTIVATION ASSAY DURING EC TUBE MORPHOGENESIS IN 3D COLLAGEN MATRICES

The standard protocol for assaying Rho GTPase activation involves the use of a Cdc42/Rac-binding domain from human p21 activated kinase 1 protein (PAK), termed PAK-binding domain (PBD), which interacts only with GTP-bound form of Rho GTPases (Azim *et al.*, 2000). The PAK-PBD protein can be readily purchased, which is often supplied in a form of GST-fusion protein for a simple pull down assay. Therefore, we adapted the regular pull-down assay approach to show the activation of Cdc42 and Rac1 during EC lumen and tube formation in 3D collagen matrices over an extended time (Koh *et al.*, 2008). We have developed a simple and efficient detergent lysis method that allows us to extract ECs out of 3D collagen matrices while they are undergoing EC lumen and tube formation, which then can be incorporated to the PAK-PBD pull-down assay to examine the activation state of Rho GTPases in 3D environment.

1. Set up vasculogenesis assay as described previously and stop it at any time points of your interest. For examining the EC lumen formation process, we usually stop at 0, 4, 16, and 24 h.

2. At each time point, collagen gels are removed from 10 microwells with a forceps and added to a 1.7-ml Eppendorf tube containing cold 500 μl detergent lysis buffer (1% Triton X-100 in 1× TBS, pH 8.0, containing 1 mM CaCl$_2$, 1 mM MgCl$_2$, complete EDTA-free protease inhibitor cocktail tablets (Roche Diagnostics), and 100 nM GTPrS (Calbiochem) and 75 μl 1 mg/ml high-purity collagenase (Sigma Aldrich). We use 1 mini complete EDTA-free protease inhibitor cocktail tablet for every 10 ml lysis buffer.

3. Cell lysis solution is placed on a rocker at 4 °C until all collagen is dissolved. Clarify cell lysis solution by centrifugation at 14,000 rpm for 20 min at 4 °C. Transfer 50 to 100 μl of the supernatant to a new 1.7-ml Eppendorf tube and add the equal amount of 2× Laemmli sample buffer containing 7.5% beta-mercaptoethanol. Boil for 10 min at 100 °C. This will be used as a total Cdc42/Rac1 loading control.

4. Wash twice 30-μl PAK-GST protein beads slurry (Cytoskeleton) with 200 μl washing buffer (1:10 dilution of lysis buffer in 1× TBS, pH 8.0). Add the rest of supernatant of cell lysates to the beads and incubate for 45 min at 4 °C.

5. Beads-cell lysis mix is washed four times with 200 μl washing buffer.

6. Add 90 μl 1× sample buffer containing 7.5% beta-mercaptoethanol to the beads. Quick vortex and boil for 10 min at 100 °C. Pellet beads at 5000 rpm for 1 min. The supernatant contains the GTP-bound Cdc42/Rac1 fraction.

7. Samples are now ready for SDS-PAGE gel analysis; we usually load 30- to 40-μl samples to each well. Blot for Cdc42 or Rac1 to examine the level of activated Cdc42/Rac1 as well as total Cdc42/Rac1.

5. BIOCHEMICAL ANALYSIS OF INTERACTIONS BETWEEN CDC42 AND ITS DOWNSTREAM EFFECTORS DURING EC LUMEN AND TUBE FORMATION IN 3D COLLAGEN MATRICES

We have found that Cdc42 regulates EC lumen and tube formation through its activation of various downstream effectors (Bayless and Davis, 2002). Expression of GFP-Cdc42 fusion protein is markedly localized around EC vacuole and lumen membranes both *in vivo* and *in vitro*, which allows us to track the progression of EC lumen and tube formation in detail

(Bayless and Davis, 2002; Kamei *et al.*, 2006). For further biochemical analysis of Cdc42 and its downstream effectors, we have developed a new molecular by adding an S-tag to the GFP-Cdc42 construct and creating adenoviral vector carrying this recombinant protein (Koh *et al.*, 2008). The S-tag fusion system has emerged as an attractive protein analysis tool with high binding specificity and rapid protein purification (Raines *et al.*, 2000). Therefore, by allowing ECs expressing S-GFP-Cdc42 adenoviral vector, we have been able to carry out detailed molecular analysis of interactions between Cdc42 and its effector proteins during EC tube and lumen formation in 3D collagen matrices (Koh *et al.*, 2008). The general approach that we describe should be applicable to many other proteins to assist in identifying binding partners and signaling networks during complex events such as EC lumen and tube formation in 3D collagen matrices.

5.1. Generation of S-GFP-Cdc42 vector

Standard restriction digestion cloning was performed to clone GFP-Cdc42 fusion construct into pTriEX-2 neo vector (Novagen) by use of primers 5′ *BamHI* AGGGATCCGATGGTGAGCAAGGGCGAGGAG and 3′ *BamHI* AGG GATCCTTAGAATATACAGCACTTCC. S-GFP-Cdc42 construct was then cloned into pShuttle-CMV by use of primers 5′ *KpnI* AGGGTA CCGCCACCATGAAAGAAACCGCTGCTGCG and 3′ *XbaI* AGTCTAGATTAGAATATACAGCACTTCCTTTTGGG. Recombinant adenoviral production was carried out as previously described (He *et al.*, 1998).

5.2. The S-tag pull-down assay

1. Infect ECs with S-GFP-Cdc42 adenovirus a day before the experiment as described previously. After 24 h of infection, set up vasculogenesis assay.
2. At each desired time point, collagen gels are removed from 20 microwells with a forceps and are added to a 1.7-ml Eppendorf tube containing 1 ml cold lysis buffer (1% Triton X-100 in 1× TBS, pH 8.0, containing 1 mM CaCl$_2$, 1 mM MgCl$_2$, complete EDTA-free protease inhibitor cocktail tablets (Roche Diagnostics), and 100 nM GTPrS (Calbiochem) and 150 μl 1 mg/ml high-purity collagenase (Sigma Aldrich).
3. Cell lysis solution is placed on a rocker at 4 °C until all collagen is dissolved. Clarify cell lysis solution by centrifugation at 14,000 rpm for 20 min at 4 °C.
4. Wash twice 200 μl blank-Sepharose 4B beads slurry (Sigma-Aldrich) with 500 μl washing buffer (1:10 dilution of lysis buffer in 1× TBS, pH 8.0). Preclear the supernatant by incubating with blank-sepharose 4B beads for 2 h at 4 °C.

5. Pellet beads at 5000 rpm for 1 min. Wash twice 250 μl S-protein agarose beads slurry (Novagen) with 500 μl washing buffer. Add the precleared supernatant to S-protein agarose beads and incubate for 45 min at 4 °C.
6. Beads-cell lysis mix is washed four times with 500 μl washing buffer.
7. Add 250 μl 1× sample buffer containing 7.5% beta-mercaptoethanol to the beads. Quick vortex and boil for 10 min at 100 °C. Pellet beads at 5000 rpm for 1 min.
8. Samples are now ready for SDS-PAGE gel analysis; we usually load 30- to 40-μl samples to each well. Blot for Cdc42 downstream proteins of your interest to identify Cdc42-associated target during EC lumen formation in 3D collagen matrices.

6. CONCLUSIONS

It is becoming increasingly clear that balanced approaches to investigate molecular mechanisms controlling vasculogenesis and angiogenesis are necessary to explain the complex control of these events. *In vitro* models of these processes are powerful tools to rapidly identify novel mechanisms and genes that regulate vascular morphogenesis. In the methods presented here on the basis of our work, we have described approaches that allow for an analysis of key steps in vascular morphogenesis, including EC intracellular vacuole and lumen formation, EC invasion and sprouting, EC-pericyte coassembly, and EC tube regression. These strategies have allowed our laboratory to: (1) identify novel mechanisms of EC lumen formation through intracellular vacuole formation and coalescence (Bayless and Davis, 2002; Bayless *et al.*, 2000; Davis and Bayless, 2003; Davis and Camarillo, 1996; Davis *et al.*, 2002; Kamei *et al.*, 2006) and its dependence on integrin signaling (Bayless *et al.*, 2000; Davis and Bayless, 2003; Davis and Camarillo, 1996); (2) identify the critical role of Cdc42 and Rac1 in vacuole and lumen formation (Bayless and Davis, 2002; Davis and Bayless, 2003); (3) identify differentially expressed and novel genes such as capillary morphogenesis gene-2 (CMG-2) during these events (Bell *et al.*, 2001); (4) investigate molecules such as S1P (Bayless and Davis, 2003) and SDF-1α that control MT1-MMP–dependent EC invasion in 3D collagen matrices (Saunders *et al.*, 2006); (5) identify novel mechanisms controlling vascular tube disassembly including those dependent on MMP-1 and MMP-10 (Davis and Saunders, 2006; Saunders *et al.*, 2005; 2006) and microtubule depolymerization (Bayless and Davis, 2004); and (6) demonstrate novel mechanisms controlling pericyte-induced EC tube stabilization through delivery of the MMP and ADAM proteinase inhibitor, TIMP-3, which is selectively expressed by pericytes (Saunders *et al.*, 2006). We have also used and adapted many new molecular and imaging techniques into our vascular morphogenic microassay systems such as DNA microarray, siRNA suppression technology, viral vector gene delivery of fusion proteins, secreted proteins and signaling molecules (wild-type versus mutants), and time-lapse light and fluorescence microscopy, which are discussed previously and/or referred to in our publications on this work. We believe that our work and that of others illustrates the critical importance of quality *in vitro* models of vascular morphogenesis in 3D extracellular matrices to identify new mechanisms regulating these events. In particular, this is because these assay systems allow for precise control of cell types, culture conditions, growth factors, and composition of extracellular matrices. By varying such conditions and, particularly, through the use of microwell formats, it is possible to rapidly identify fundamental regulators of vessel formation, stabilization and regression.

ACKNOWLEDGMENTS

We thank Ms. Anne Mayo for excellent technical assistance. This work was supported by NIH grants HL59373 and HL79460 to G. E. Davis.

REFERENCES

Adams, R. H., and Alitalo, K. (2007). Molecular regulation of angiogenesis and lymph angiogenesis. *Nat. Rev. Mol. Cell. Biol.* **8,** 464–478.

Aspenstrom, P., *et al.* (2004). Rho GTPases have diverse effects on the organization of the actin filament system. *Biochem. J.* **377,** 327–337.

Azim, A. C., *et al.* (2000). Determination of GTP loading on Rac and Cdc42 in platelets and fibroblasts. *Methods Enzymol.* **325,** 257–263.

Bayless, K. J., and Davis, G. E. (2002). The Cdc42 and Rac1 GTPases are required for capillary lumen formation in three-dimensional extracellular matrices. *J. Cell Sci.* **115,** 1123–1136.

Bayless, K. J., and Davis, G. E. (2003). Sphingosine-1-phosphate markedly induces matrix metalloproteinase and integrin-dependent human endothelial cell invasion and lumen formation in three-dimensional collagen and fibrin matrices. *Biochem. Biophys. Res. Commun.* **312,** 903–913.

Bayless, K. J., and Davis, G. E. (2004). Microtubule depolymerization rapidly collapses capillary tube networks in vitro and angiogenic vessels *in vivo* through the small GTPase Rho. *J. Biol. Chem.* **279,** 11686–11695.

Bayless, K. J., *et al.* (2000). RGD-dependent vacuolation and lumen formation observed during endothelial cell morphogenesis in three-dimensional fibrin matrices involves the alpha(v)beta(3) and alpha(5)beta(1) integrins. *Am. J. Pathol.* **156,** 1673–1683.

Bell, S. E., *et al.* (2001). Differential gene expression during capillary morphogenesis in 3D collagen matrices: Regulated expression of genes involved in basement membrane matrix assembly, cell cycle progression, cellular differentiation and G-protein signaling. *J. Cell Sci.* **114,** 2755–2773.

Bryan, B. A., and D'Amore, P. A. (2007). What tangled webs they weave: Rho-GTPase control of angiogenesis. *Cell Mol. Life Sci.* **64,** 2053–2065.

Carmeliet, P. (2005). Angiogenesis in life, disease and medicine. *Nature* **438,** 932–936.

Carmeliet, P., and Jain, R. K. (2000). Angiogenesis in cancer and other diseases. *Nature* **407**, 249–257.

Connolly, J. O., *et al.* (2002). Rac regulates endothelial morphogenesis and capillary assembly. *Mol. Biol. Cell.* **13**, 2474–2485.

Conway, E. M., *et al.* (2001). Molecular mechanisms of blood vessel growth. *Cardiovasc. Res.* **49**, 507–521.

Davis, G. E., and Bayless, K. J. (2003). An integrin and Rho GTPase-dependent pinocytic vacuole mechanism controls capillary lumen formation in collagen and fibrin matrices. *Microcirculation* **10**, 27–44.

Davis, G. E., *et al.* (2002). Molecular basis of endothelial cell morphogenesis in three-dimensional extracellular matrices. *Anat. Rec.* **268**, 252–275.

Davis, G. E., *et al.* (2000). Capillary morphogenesis during human endothelial cell invasion of threedimensional collagen matrices. *In Vitro Cell Dev. Biol. Anim.* **36**, 513–519.

Davis, G. E., and Camarillo, C. W. (1996). An alpha 2 beta 1 integrin-dependent pinocytic mechanism involving intracellular vacuole formation and coalescence regulates capillary lumen and tube formation in three-dimensional collagen matrix. *Exp. Cell Res.* **224**, 39–51.

Davis, G. E., *et al.* (2001). Matrix metalloproteinase-1 and -9 activation by plasmin regulates a novel endothelial cell-mediated mechanism of collagen gel contraction and capillary tube regression in three-dimensional collagen matrices. *J. Cell Sci.* **114**, 917–930.

Davis, G. E., and Saunders, W. B. (2006). Molecular balance of capillary tube formation versus regression in wound repair: Role of matrix metalloproteinases and their inhibitors. *J. Invest. Dermatol. Symp. Proc.* **11**, 44–56.

Davis, G. E., and Senger, D. R. (2005). Endothelial extracellular matrix: biosynthesis, remodeling, and functions during vascular morphogenesis and neovessel stabilization. *Circ. Res.* **97**, 1093–1107.

Drake, C. J. (2003). Embryonic and adult vasculogenesis. *Birth Defects Res. C Embryo Today* **69**, 73–82.

Egginton, S., and Gerritsen, M. (2003). formation: *in vivo* versus *in vitro* observations. *Microcirculation* **10**, 45–61.

Folkman, J. (1995). Angiogenesis in cancer, vascular, rheumatoid and other disease. *Nat. Med.* **1**, 27–31.

Folkman, J., and Haudenschild, C. (1980). Angiogenesis *in vitro*. *Nature* **288**, 551–556.

Guldner, F. H., and Wolff, J. R. (1973). Seamless endothelia as indicators of capillaries developed from sprouts. *Bibl. Anat.* **12**, 120–123.

Hall, A. (1998). Rho GTPases and the actin cytoskeleton. *Science* **279**, 509–514.

Hall, A. (2005). Rho GTPases and the control of cell behaviour. *Biochem. Soc. Trans.* **33**, 891–895.

Hanahan, D. (1997). Signaling vascular morphogenesis and maintenance. *Science* **277**, 48–50.

He, T. C., *et al.* (1998). A simplified system for generating recombinant adenoviruses. *Proc. Natl. Acad. Sci. USA* **95**, 2509–2514.

Hoang, M. V., and Senger, D. R. (2005). *In vivo* and *in vitro* models of Mammalian angiogenesis. *Methods Mol. Biol.* **294**, 269–285.

Hoang, M. V., *et al.* (2004). Rho activity critically and selectively regulates endothelial cell organization during angiogenesis. *Proc. Natl. Acad. Sci. USA* **101**, 1874–1879.

Hotary, K., *et al.* (2000). Regulation of cell invasion and morphogenesis in a three-dimensional type I collagen matrix by membrane-type matrix metalloproteinases 1, 2, and 3. *J. Cell Biol.* **149**, 1309–1323.

Ilan, N., *et al.* (1998). Distinct signal transduction pathways are utilized during the tube formation and survival phases of in vitro angiogenesis. *J. Cell Sci.* **111(Pt 24)**, 3621–3631.

Kamei, M., *et al.* (2006). Endothelial tubes assemble from intracellular vacuoles *in vivo*. *Nature* **442**, 453–456.

Kiosses, W. B., *et al.* (2002). A dominant-negative p65 PAK peptide inhibits angiogenesis. *Circ. Res.* **90**, 697–702.

Koh, W., Mahan, R. D., and Davis, G. E. (2008). Cdc42-and Rac 1-mediated endothelial lumen formation requires Pak2, Pak4 and Par3, and PKC-dependent signaling. *J. Cell Sci.* **121**, 989–1001.

Liu, Y., and Senger, D. R. (2004). Matrix-specific activation of Src and Rho initiates capillary morphogenesis of endothelial cells. *FASEB J.* **18**, 457–468.

Meyer, G. T., *et al.* (1997). Lumen formation during angiogenesis *in vitro* involves phagocytic activity, formation and secretion of vacuoles, cell death, and capillary tube remodelling by different populations of endothelial cells. *Anat. Rec.* **249**, 327–340.

Montesano, R., and Orci, L. (1985). Tumor-promoting phorbol esters induce angiogenesis *in vitro*. *Cell* **42**, 469–477.

Nakatsu, M. N., *et al.* (2003). Angiogenic sprouting and capillary lumen formation modeled by human umbilical vein endothelial cells (HUVEC) in fibrin gels: the role of fibroblasts and Angiopoietin-1. *Microvasc. Res.* **66**, 102–112.

Nayak, R. C., *et al.* (1988). A monoclonal antibody (3G5)-defined ganglioside antigen is expressed on the cell surface of microvascular pericytes. *J. Exp. Med.* **167**, 1003–1015.

Nicosia, R. F., and Madri, J. A. (1987). The microvascular extracellular matrix. Developmental changes during angiogenesis in the aortic ring-plasma clot model. *Am. J. Pathol.* **128**, 78–90.

Nicosia, R. F., and Ottinetti, A. (1990). Growth of microvessels in serum-free matrix culture of rat aorta. A quantitative assay of angiogenesis *in vitro*. *Lab. Invest.* **63**, 115–122.

Nicosia, R. F., and Villaschi, S. (1999). Autoregulation of angiogenesis by cells of the vessel wall. *Int. Rev. Cytol.* **185**, 1–43.

Parker, L. H., *et al.* (2004). The endothelial-cell-derived secreted factor Egfl7 regulates vascular tube formation. *Nature* **428**, 754–758.

Patan, S. (2000). Vasculogenesis and angiogenesis as mechanisms of vascular network formation, growth and remodeling. *J. Neurooncol.* **50**, 1–15.

Raines, R. T., *et al.* (2000). The S.Tag fusion system for protein purification. *Methods Enzymol.* **326**, 362–376.

Ridley, A. J. (2001). Rho proteins: Linking signaling with membrane trafficking. *Traffic* **2**, 303–310.

Risau, W., and Flamme, I. (1995). Vasculogenesis. *Annu. Rev. Cell Dev. Biol.* **11**, 73–91.

Sainson, R. C., *et al.* (2005). Cell-autonomous notch signaling regulates endothelial cell branching and proliferation during vascular tubulogenesis. *FASEB J.* **19**, 1027–1029.

Saunders, W. B., *et al.* (2005). MMP-1 activation by serine proteases and MMP-10 induces human capillary tubular network collapse and regression in 3D collagen matrices. *J. Cell Sci.* **118**, 2325–2340.

Saunders, W. B., *et al.* (2006). Coregulation of vascular tube stabilization by endothelial cell TIMP-2 and pericyte TIMP-3. *J. Cell Biol.* **175**, 179–191.

Schwartz, M. (2004). Rho signalling at a glance. *J. Cell Sci.* **117**, 5457–5548.

Senger, D. R. (1996). Molecular framework for angiogenesis: A complex web of interactions between extravasated plasma proteins and endothelial cell proteins induced by angiogenic cytokines. *Am. J. Pathol.* **149**, 1–7.

Vernon, R. B., and Sage, E. H. (1995). Between molecules and morphology. Extracellular matrix and creation of vascular form. *Am. J. Pathol.* **147**, 873–883.

Yancopoulos, G. D., *et al.* (2000). Vascular-specific growth factors and blood vessel formation. *Nature* **407**, 242–248.

Yang, S., *et al.* (1999). Functional roles for PECAM-1 (CD31) and VE-cadherin (CD144) in tube assembly and lumen formation in three-dimensional collagen gels. *Am. J. Pathol.* **155**, 887–895.

Zeng, G., *et al.* (2007). Orientation of endothelial cell division is regulated by VEGF signaling during blood vessel formation. *Blood* **109**, 1345–1352.

CHAPTER SIX

IN VITRO DIFFERENTIATION OF MOUSE EMBRYONIC STEM CELLS INTO PRIMITIVE BLOOD VESSELS

Svetlana N. Rylova,* Paramjeet K. Randhawa,* *and* Victoria L. Bautch*,†

Contents

Abstract

Mouse embryonic stem (ES) cells, derived from the inner cell mass of blastocyst stage embryos, undergo programmed differentiation *in vitro* to form a primitive vasculature. This programmed differentiation proceeds through similar processes of vasculogenesis and angiogenesis found during early vascular development *in vivo*. Partially differentiated ES cell clumps or embryoid bodies (EBs) first form blood islands that are subsequently transformed into a network of primitive blood vessels that contain lumens. Therefore, vascular differentiation of ES cells is an ideal model to study and manipulate early vascular development. Here we provide protocols for the routine maintenance of mouse ES cells

* Department of Biology, The University of North Carolina at Chapel Hill, Chapel Hill, North Carolina
† Carolina Cardiovascular Biology Center, The University of North Carolina at Chapel Hill, Chapel Hill, North Carolina

Methods in Enzymology, Volume 443
ISSN 0076-6879, DOI: 10.1016/S0076-6879(08)02006-5

and *in vitro* differentiation. We also include protocols for establishing transgenic ES cell lines and visualization of blood vessels by use of endothelial
specific molecular markers.

1. INTRODUCTION

Mouse embryonic stem (ES) cells isolated from the inner cell mass of
the blastocyst-stage embryo can be propagated *in vitro* to preserve their
pluripotency or manipulated to give rise to multiple cell lineages. This
unique property of ES cells has been extensively explored in the past decade
to study early stages of mammalian embryonic development. In particular,
differentiated ES cells have been instrumental in the understanding of early
vascular development and hematopoiesis (Bautch, 2002; Doetschman *et al.*,
1985; Keller, 2005; Risau *et al.*, 1988; Wang *et al.*, 1992).

Differentiation of mouse ES cells *in vitro* can be achieved by removal of
differentiation inhibitory factors from the media. In this scenario, ES cells
undergo programmed differentiation, which results in the formation of
predetermined cell types. These cell types include endoderm and several
mesoderm derivatives, such as vascular endothelial cells and some hematopoietic cells (erythrocytes and macrophages). This process is very similar to
differentiation in the mouse yolk sac, the first site of vascular and hematopoietic development in the embryo (Keller *et al.*, 1993; 1995; Vittet *et al.*,
1996; Wang *et al.*, 1992; Wiles and Keller, 1991). In addition, differentiated
ES cells give rise to cardiac mesoderm that is not present in the yolk sac. We
have established a method of differentiation of aggregated mouse ES cells
into embryoid bodies (EBs), which are partially differentiated in suspension,
then attached to a substratum. This method reproducibly gives rise to
endothelial cells (15 to 20% of total cells) that are organized into a highly
branched primitive vascular network (Fig. 6.1) (Bautch, 2002; Kearney and
Bautch, 2003; Wang *et al.*, 1992).

Vascular development in differentiated ES cell cultures proceeds through
the same processes of vasculogenesis and angiogenesis found *in vivo* (Risau,
1997). In the course of vasculogenesis, mesodermal precursor cells called
angioblasts differentiate into early endothelial cells that then coalesce into a
primitive vascular plexus. Hemangioblasts, common progenitors of both
hematopoietic and endothelial cells, also contribute to vascular development
(Choi *et al.*, 1998). Endothelial cells of the primitive plexus then proliferate
and migrate, giving rise to a highly branched vascular network that can
contain primitive erythrocytes.

We have previously shown that differentiating cultures of attached EBs
initially form blood islands, reminiscent of those found in the yolk sac of
developing embryos (Bautch *et al.*, 1996). Electron microscopy confirmed

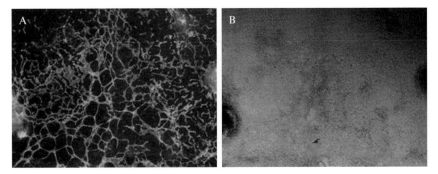

Figure 6.1 Programmed differentiation of ES cells *in vitro* results in the formation of primitive blood vessels. Attached ES cell cultures were differentiated until day 8, then fixed and stained with antibodies against PECAM to visualize blood vessels. (A) Immunofluorescence staining of differentiated ES cultures at day 8 with anti-PECAM antibodies; (B) phase-contrast image of corresponding area shows that other nonendothelial cell types are formed during ES cell differentiation.

that the blood islands are three-dimensional sacs lined by the endothelial cells, with evident cell–cell junctions. These sacs have lumens that are filled with hematopoietic cells. These blood islands later transform into a network of primitive blood vessels. Importantly, programmed differentiation of vascular endothelial cells in our model occurs in the same cellular microenvironment found during vascular development *in vivo*.

This vascular differentiation model has several important characteristics. First, the differentiated ES cell–derived primitive vessels do not experience blood flow. This precludes the study of remodeling, but it does allow us to identify flow-independent events during early vascular development. Second, even though vascular differentiation in our model occurs on reattachment of EBs, detailed analysis of vessel morphology by use of electron and confocal microscopy confirms that forming blood vessels have lumens (Fig. 6.2), allowing for three-dimensional imaging of vascular development *in vitro*. In summary, *in vitro* vascular differentiation of mouse ES cells closely recapitulates early vascular development *in vivo*, thus providing an ideal tool to study this process by use of molecular, cellular, and pharmacologic tools.

Another feature of this vascular differentiation model is the ability to easily analyze genetic manipulations. ES cells lacking the VEGF receptors Flt-1 (VEGFR-1), Flk-1 (VEGFR-2), or vascular endothelial growth factor A (VEGF) have severe defects in vascular development *in vitro*, mimicking the phenotypes *in vivo* (Bautch *et al.*, 2000; Carmeliet *et al.*, 1996; Ferrara *et al.*, 1996; Fong *et al.*, 1995; Kearney *et al.*, 2004; Schuh *et al.*, 1999; Shalaby *et al.*, 1995; 1997). We successfully used this model to analyze the vascular phenotype of ES cell lines with a deletion of *flt-1*, and we have

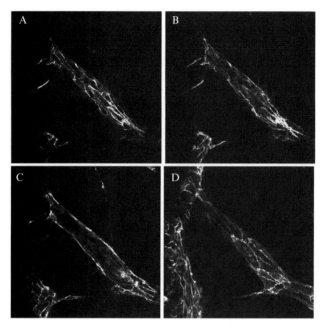

Figure 6.2   Blood vessels formed during programmed differentiation of ES cells contain lumens. Attached ES cell cultures were differentiated until day 8, and blood vessels were visualized with VE-cadherin antibodies. (A to D) Representative images in a z-stack series obtained by use of confocal microscopy. These images indicate that there is a lumen between the upper (A) and the lower (D) layer of endothelial cells in the blood vessel.

recently selectively rescued mutant phenotypes with transgenes (Kappas *et al.*, 2008). These studies provide additional verification of the model.

A number of studies describe blood vessel-specific marker expression in ES cell–derived blood vessels (Bautch *et al.*, 2000; Redick and Bautch, 1999; Vittet *et al.*, 1996). We and others have shown that vascular endothelial cells formed during *in vitro* differentiation of mouse ES cells express many of the same markers found *in vivo*. We routinely use CD31 (PECAM-1) and ICAM-2 antibodies to visualize blood vessels in differentiated ES cultures. Other markers include VE-cadherin, Flk-1, and Flt-1.

In this chapter, we provide basic protocols for maintenance and *in vitro* differentiation of mouse ES cells. In addition, we have included protocols for visualization of blood vessels in differentiated ES cell cultures and for establishing transgenic ES cell lines. We have divided the chapter into sections, each containing a general description followed by protocols and troubleshooting notes.

2. MAINTENANCE OF ES CELLS

ES cells are supplemented with serum and specific factors that prevent differentiation for propagation. The main factor used for this purpose is LIF (leukemia inhibitory factor), which can be obtained from two sources: (1) recombinant LIF, which is commercially available (Sigma, Stem Cell Technologies); and (2) medium conditioned by the 5637 human bladder cancer cell line (ATCC#HTB9), which can be added to ES cell growth medium. In our hands, conditioned medium works better than recombinant LIF and is more economical for routine ES cell maintenance.

Traditionally, ES cells were cultured on top of a feeder layer of mouse embryonic fibroblasts that provide necessary factors to prevent differentiation. However, for *in vitro* differentiation, ES cells must be cultured for several passages off feeder layers. Thus, we routinely culture ES cells without feeder cells and supplement with 5637 medium. ES cells grown on plastic surfaces form colonies that look flat and have some differentiated cells along the periphery compared with ES cell colonies grown on a feeder layer. Nevertheless, they provide consistent and reproducible results for vascular differentiation when used between passages 2 and 50.

2.1. Protocol #1: Passage of ES cells

ES cells should be passed frequently, at least every 2 to 4 days to prevent differentiation (see Note 1). All volumes are given for 6-cm dishes.

1. Aspirate off medium. Wash two times with 5 ml 1× PBS at room temperature.
2. Add 1 ml 0.25× Trypsin-EDTA solution to dish. Place dish in 37 °C incubator until most ES cell clumps dissociate on gentle agitation (3 to 5 min).
3. Stop the reaction by adding 4 ml of trypsin stop solution (20 to 40% FBS in PBS) to the dish. Break up the cell clumps by gentle pipetting up and down.
4. Coat dishes with 0.1 % gelatin solution in PBS for 1 h at 37 °C. Aspirate off gelatin solution, and add 5 ml prewarmed ES medium (contains 65.9% of 5637 conditioned media (Kappas and Bautch, 2007), 17.13% lot selected FBS, 82.5 μM (final concentration) of monothioglycerol (MTG), 15.18% DMEM-H, gentamicin (0.5 μg/ml) (see Note 2).
5. Add 2 to 3 drops of the cell suspension into the dish. Observe the size of the ES cell clumps under a microscope. ES cells should have no more than 4 to 6 cells/clump. If cell clumps are significantly larger, pipette the solution to further break apart cell clumps.

6. Place the dish in a 37 °C incubator with 5% CO_2 and gently move dish in a "back and forth" motion to evenly disperse the ES cells throughout the dish.

NOTES:

1. ES cells maintained without feeder cells normally look somewhat differentiated, especially around the edges. In our experience, passage of ES cultures in this state does not compromise *in vitro* differentiation. However, some ES cell lines, which look particularly differentiated, might benefit from enrichment with ES cells. To achieve this, we plate dissociated ES cells in a dish and let them attach for 30 min at 37 °C. During that time only differentiated cells should attach, whereas ES cells should remain in suspension. We then remove remaining unattached ES cells and plates them into a new dish.

2. We use FBS that has been screened for optimal ES cell maintenance (for protocol, please refer to Kappas and Bautch, 2007). In general, 50 to 75% of FBS lots approved for general tissue culture use are suitable for ES cell propagation.

2.2. *In vitro* differentiation of mouse ES cells

2.2.1. Generation of embryoid bodies (EBs)

We discuss protocols that allow ES cells to undergo a programmed differentiation that leads to formation of a primitive, lumenized vasculature (Bautch, 2002; Kearney and Bautch, 2003). For successful *in vitro* differentiation, ES cells are first aggregated into clumps that differentiate in suspension to form EBs. Subsequently, EBs are attached to tissue culture–treated plastic and spread while continuing a programmed differentiation (Fig. 6.3). Embryoid bodies can be generated by two different methods. One is treatment of ES cells with the enzyme Dispase, which results in detachment of ES cell colonies from the plastic and normally can provide a large amount of material for *in vitro* differentiation. Another alternative is the hanging drop method. It requires dissociation of ES cell colonies into a single cell suspension, followed by setting up small drops containing a specified

| Dispase treatment
day 0 | EBs in suspension
day 0–3 | Attached EBs
day 3 | Primitive vasculature
day 6–8 |

Figure 6.3   Schematic representation of programmed differentiation of ES cells *in vitro*. ES cell aggregates are detached at day 0 by Dispase treatment and allowed to differentiate into EBs in suspension. On day 3 EBs are attached to tissue culture dishes and differentiated until day 8. Formation of primitive blood vessels can be observed between days 6 and 8.

number of ES cells. The hanging drop method generates EBs of homogenous size and provides better synchronization of differentiation program; however, it is less suitable for scale-up. Here we describe the dissociation by Dispase method for EB generation, and for the hanging drop protocol see (Kearney and Bautch, 2003).

2.3. Protocol #2: Generation of EBs for *in vitro* differentiation with dissociation by dispase

All volumes are given for one 6-cm dish of ES cell colonies.

1. Choose the dish that has the best ES cell clumps for differentiation. ES cell clumps should be round and slightly differentiated on the very edge, and shiny and undifferentiated in the middle. ES cell clumps are collected from dishes after incubation at 37 °C for 5 to 6 days without feeding after normal passage.
2. Aspirate off media from ES cell dish. Wash two times with 5 ml of cold 1× PBS.
3. Add 1 ml of cold Dispase (Grade II stock, 2.4 U/ml, Boehringer-Mannheim) diluted just before use, and let dish sit at room temperature for 1 to 2 min.
4. When most of the cell clumps have detached from the dish, use a 5-ml pipette to gently transfer the cells into a 50-ml tube containing 35 ml of room temperature 1× PBS. Rinse the dish with 5 ml 1× PBS and add the rinse to the 50-ml tube. Cap tube, and invert the tube once gently to mix.
5. Let the tube sit until the cell clumps have settled to the bottom of the tube (approximately 10 min).
6. Aspirate all but 4 to 5 ml of PBS and add another 35 ml room temperature 1× PBS. Gently swirl the tube to redistribute the cell clumps. Cap tube, and invert gently to mix.
7. Let the tube sit until the cell clumps have settled to the bottom of the tube (10 min). Aspirate the PBS (leaving 2 to 3 ml PBS/ES cell clump solution at the bottom), and gently add 5 ml prewarmed (37 °C) differentiation medium down the side of the tube. Differentiation media contains DMEM-H supplemented with 20% lot selected FBS, 150 μM MTG (see Note 1).
8. Pipette 10 ml of prewarmed differentiation medium into a labeled Kord–Valmark bacteriologic 10-cm petri dish (see Note 2). By use of a 25-ml pipette (minimizes mechanical disruption of ES cell clumps), transfer the contents of the 50-ml tube (cell clumps/PBS/media) to the 10-cm dish.
9. Transfer the EBs to a new petri dish with fresh differentiation medium on day 2 by use of a Medidropper.

NOTES:

1. We use lot-selected FBS that ensures the best vascular differentiation of ES cells (for detail protocol see Kappas and Bautch [2007]).

2. It is crucial to use bacteriologic dishes at this point. They prevent attachment of ES cell clumps, allowing them to differentiate in suspension into EBs. Despite that, some ES clumps will still stick to the bottom of the dish, and they should be discarded after the medium is changed on day 2.

2.4. Differentiation of embryoid bodies (EBs)

On culturing ES cell clumps in suspension for 2 to 3 days, they round up and start a differentiation program to form EBs containing endoderm, mesoderm, and angioblast precursors. EBs can be reattached anytime between day 0 and day 4 after Dispase treatment, but in our hands, the best differentiation is achieved with EBs attached on day 3. Because dissociation by Dispase generates EBs of various sizes, selecting medium sized EBs that look "shiny" under the phase-contract microscope yields better results and more synchronized vascular differentiation. The density of EBs should be approximately 10 to 20 EBs/per well of 24-well dish. This density will promote optimal spreading of EBs, and they will cover most of the surface of the well by day 8 of differentiation. Vascular differentiation *in vitro* is normally assessed on day 8 after Dispase treatment. However, for certain applications, *in vitro* differentiation can be carried out until day 10 to 12. Care should be taken to provide fresh medium during the course of differentiation. In general, changing medium every 2 days is sufficient if EBs are seeded at the density described earlier.

2.5. Protocol #3: *In vitro* differentiation of EBs

The day of the Dispase treatment is day 0.

1. Set up reattachment cultures on day 3 after the Dispase treatment (see Note 1). Add 1.5 ml of prewarmed differentiation medium to each well of a 24-well tissue culture plate that is to be seeded with EBs.
2. Use a sterile Medidropper to transfer EBs from the dish to the wells of a 24-well plate. Generally, dispense between 10 and 20 EBs per well (see Note 2). Holding the dish up and looking at it from underneath is helpful when determining the number of EBs in a well.
3. Ensure that the EBs are spread evenly in the well by gently shaking/moving the plate, or if necessary, use a pipetter with a sterile tip to gently pipette the medium up and down in the well.
4. Incubate at 37 °C in a humidified incubator with 5% CO_2. Attachment generally occurs within a few hours.
5. Feed the attached cultures every other day. To feed, aspirate off medium, and then slowly add 1.5 to 2 ml fresh prewarmed differentiation medium down the sidewall of the well so as not to disturb the attached cultures (see Note 3, 4).

NOTES:

1. EBs can be attached at any point from right after Dispase treatment to day 4. By day 5, EBs usually start becoming cystic. We routinely set up attachment cultures at day 3, which reproducibly results in good vascular differentiation.
2. EBs can be plated into any size of dishes. We plate EBs into 24-well plates for antibody detection by immunofluorescence and use larger dishes for RNA or protein analysis. Adjust EB numbers for the surface area of the well or dish.
3. Providing fresh medium to differentiating EBs ensures optimal conditions for vascular differentiation. Changing the medium every 2 days is usually sufficient; however, if the medium color turns light orange or yellow, it should be changed more frequently.
4. With this protocol, we typically monitor vascular development in cultures that have been differentiated for 8 days (day of Dispase treatment is day 0). Angioblast formation is generally observed at day 4 to 6, whereas vessel formation occurs at day 6 to 8.

3. GENERATION OF TRANSGENIC ES CELL LINES

The boom in recombinant technology during the past decades resulted in the creation of different methods for generating loss-of-function or gain-of-function mutations for particular genes. ES cell lines with heterozygous or homozygous deletion of certain genes were made by means of homologous recombination (Fung-Leung and Mak, 1992). Other approaches include generation of transgenic ES cell lines that overexpress genes of interest. Moreover, tissue-specific expression of transgenes is achieved by use of tissue-specific promoters. We have successfully created several transgenic ES cell lines expressing reporters that are used for real-time imaging of vascular development in differentiated ES cultures (Kearney et al., 2004; Zeng et al., 2007). At least two approaches exist for production of transgenic ES cell lines, namely, random integration or targeted integration into a specific locus. For random integration, we designed constructs expressing cytoplasmic or nuclear (fused to Histone2B) GFP under the PECAM promoter-enhancer with a drug resistance gene (either neomycin or hygromycin) (Fig. 6.4). The constructs were linearized and electroporated into ES cells followed by selection of drug-resistant colonies.

For targeted integration we chose the ROSA26 locus. The mouse ROSA26 locus has become a preferred site for integration of transgenes, because it can be targeted with a high efficiency, it is expressed in most cell types, and it is not subject to gene silencing (Irion et al., 2007; Soriano, 1999).

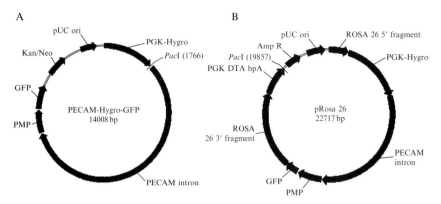

Figure 6.4 Constructs for random (A) and targeted (B) integration of transgenes into DNA. (A) PGK-Hygro-hygromycin resistance gene driven by PGK (phosphoglycerate kinase) promoter, for positive selection; PMP-PECAM minimal promoter; GFP–reporter gene. (B) pROSA26 5′ and 3′ fragments are homologous to the sequences in the ROSA26 locus and serve as sites of homologous recombination for targeted integration of the gene of interest; PGK-Hygro-hygromycin resistance gene driven by PGK promoter, for positive selection; PMP-PECAM minimal promoter; GFP-reporter gene; PGK DTA bpA-diphtheria toxin, driven by PGK promoter, serves for negative selection, because it is only expressed if homologous recombination did not occur.

Our gene of interest was inserted together with the PECAM promoter/enhancer sequence, a drug-resistance cassette, and a diphtheria toxin expression cassette (for negative selection) into a ROSA targeting vector (see Fig. 6.4). On electroporation into ES cells, the gene of interest integrates into the ROSA locus through homologous recombination. Diphtheria toxin driven by the PGK promoter should be expressed only in cells with random integration of the targeted transgene, providing a negative selection.

In our hands, random integration yields ES clones with a range of gene expression, which can be advantageous when different levels of transgene expression are desired (Kappas *et al.*, unpublished results). On the other hand, expression of transgenes from the ROSA locus under the PECAM promoter/enhancer results in overall lower levels of expression, but with equivalent levels of transgene expression among different ES clones. This can be useful if the effects of different genes on vascular development are to be compared (Kappas *et al.*, 2008). However, in our hands, certain approaches have worked for some transgenes but not for others. For example, by use of random integration we created ES cells lines with vascular endothelial–specific expression of a GFP reporter either in the cytoplasm or in the nucleus (Kearney *et al.*, 2004; Zeng *et al.*, 2007). However, the same approach did not work for a dsRed reporter transgene. Subsequently, we achieved endothelial-specific expression of dsRed by adding insulator sequences in front of the PECAM promoter (G. Zeng and V. L. B., personal communication; West *et al.*, 2002).

3.1. Protocol #4: Electroporation of DNA constructs into ES cells

1. Digest DNA with a restriction enzyme to linearize the plasmid. Choose an enzyme that cuts once and outside of the promoter/transgene/drug resistance sequences. Run an aliquot of the digest on the gel to check for complete digestion. Use 20 μg of DNA for a 20-kb construct.
2. Add SDS to 0.5% final concentration, add proteinase K to 0.2 μg/μl final concentration, adjust volume up to 500 μl with 10 mM Tris/ 0.1 mM EDTA solution. Incubate at 37 °C for 30 min. to inactivate the restriction enzyme.
3. Add 1/10 volume of 3 M NaAc and 2 volumes of 100% ethanol. Precipitate on ice for 30 min or store overnight at −20 °C.
4. Centrifuge DNA at 12,000g for 15 min, wash with 70% ethanol, and centrifuge another 10 min. Under the tissue culture hood, remove the supernatant and let dry for approximately 10 min, then resuspend in sterile PBS to 1 μg/μl.
5. Dissociate ES cells with trypsin as described earlier (see Note 1). Stop the reaction by adding trypsin stop solution. Count the cells and then spin them down.
6. Remove the supernatant, and then resuspend the cells in cold PBS to a concentration of 2 × 10^6 cell/0.5 ml and mix with DNA.
7. Transfer cell/DNA mix into prechilled cuvettes and incubate on ice for 20 min. Make a control sample of cells without DNA.
8. Electroporate with following settings: 250 V, 10 msec on ECM 830 electroporator (BTX Harvard Apparatus). Pulse once and incubate on ice for 20 min (see Note 2).
9. Transfer the sample into a 50-ml tube with 10 ml of medium, then plate in two 10-cm dishes (contain 25% or 75% of cells). Bring the volume up to 10 ml with ES medium.
10. After 24 h, change to ES medium containing the drug for selection. We use Geneticin (G418) or hygromycin at 200 μg/ml.
11. Change selection medium every other day. Clones should form during next 12 to 14 days. There should be no clones in the control plate without DNA.
12. Pick the clones when they reach 1 mm in diameter, generally on day 12 to 14. Rinse the dish with the PBS, add fresh PBS so it only covers the surface (5 ml for a 10-cm dish). Pick 20 to 30 colonies with fresh sterile pipet tips and put them into a prepared 48-well plate containing 20 μl of trypsin-EDTA in each well. Incubate for 20 to 30 min at 37 °C.
13. Stop the reaction by adding an equal volume of FBS, and transfer cells to the wells of the 24-well plate coated with gelatin.
14. Expand the clones by passing them into a 6-well plate after 2 to 3 days and to 6-cm dishes after another 2 to 3 days.

15. If the transgene was fused to GFP and expresses in ES cells, identify positive clones by use of fluorescent microscopy. In addition, screen the colonies by *in vitro* differentiation.
16. Freeze the positive clones. Generally, freeze 2 vials of cells per 6-cm dish of ES cells. Prepare 2× freezing medium (20% DMSO/80% FBS), chill on ice. Detach cells from tissue culture plates as described previously. Spin cells down in a clinical centrifuge on speed #3 for 5 min. Resuspend cells in FBS at half the final volume. Cool cells on ice for 2 to 5 min. Drop wise, with gentle swirling of tube, slowly add an equivalent volume of cold freezing medium. Distribute 1 ml volume of the cells to the cryotube. Freeze cells overnight at −70 °C in a Nalgene freezer box for cryofreezing. Move the cells to liquid nitrogen vessels the next day.

NOTES:

1. In general, three 10-cm dishes of ES cells provide enough cells for one electroporation.
2. Electroporation can be performed with other compatible electroporator models. For example, initially we used a BioRad Gene Pulser II, and it worked as efficiently as the BTX ECM 830 model.

3.2. Immunodetection of blood vessels

Primitive vascular networks formed in the course of programmed differentiation of attached EBs can be visualized by immunofluorescence detection with antibodies for endothelial specific markers. A number of specific molecular markers are used to detect endothelial cells. These include Flk-1 (Ly-73, B-D Pharmingen), VE-cadherin (11D4.1, B-D Pharmingen), CD31 or PECAM-1 (Mec13.3, B-D Pharmingen), and ICAM-2 (3C4, B-D Pharmingen). We and others have shown that Flk-1 and PECAM are expressed on mouse angioblasts and endothelial cells. In addition, PECAM is expressed in ES cells and some hematopoietic cells (Redick and Bautch, 1999). Endothelial ICAM-2 and VE-cadherin are only detectable later, when endothelial cells coalesce to form primitive vessels.

3.2.1. Protocol #5: Visualization of ES cell–derived blood vessels with antibodies against PECAM or ICAM-2

For immunostaining, EBs are plated at day 3 after Dispase treatment into 24-well plates. We normally stain 4 wells for each ES cell line. Differentiated ES cell cultures are processed for immunostaining on day 8 of differentiation.

1. Rinse plates with 1× PBS two times.
2. Aspirate PBS. Fix in an ice-cold mix of acetone/methanol (1:1) for 5 min at room temperature (RT) (see Note 1).

3. Wash two times with PBS, 5 min each (see Note 2).
4. Block in staining media (3% FBS in PBS/0.1% sodium azide) for 1 h at 37 °C.
5. Incubate with primary antibodies (Ab) diluted in staining medium for 1 h at 37 °C or overnight at +4 °C (0.25 ml volume is sufficient for one well of 24-well plate). Use rat anti-mouse PECAM Ab (Mec 13.3, BD Pharmingen) at 1:1000 dilution or rat anti-mouse ICAM-2 Ab (3C4, B-D Pharmingene) at 1:500 dilution.
6. Wash two times with PBS, 5 min each.
7. Incubate with secondary Abs diluted in staining medium for 1 h at 37 °C. For secondary Abs, we use goat anti-rat Ab conjugated to Alexa fluor 488 or 555 (Molecular Probes, Invitrogen).
8. Wash with PBS two times and observe under a fluorescent microscope (see Note 3).

Stained ES cultures can be stored at +4 °C for several weeks.

NOTES:

1. Alternately, differentiated ES cultures can be fixed in 4% fresh paraformaldehyde (PFA) for 6 min at RT. This fixation procedure normally yields weaker PECAM staining, and cells are more loosely attached to the surface of the well. However, some antibodies (i.e., VE-cadherin) require PFA for visualization.
2. Care should be taken during all washing steps. To avoid detachment of EBs from the plastic, add PBS slowly along the wall and aspirate it carefully without touching the bottom of the well.
3. For imaging with confocal microscopy, EBs can be attached and differentiated in slide flasks (NUNC) or chamber slides (Lab-Tek chamber slide, NUNC).

 ## 2. CONCLUDING REMARKS

Programmed differentiation of mouse ES cells results in the formation of primitive blood vessels in a process very similar to early vascular development *in vivo*. This approach provides a reliable model for analysis of both pharmacologic and genetic manipulations of early mammalian vascular development.

ACKNOWLEDGMENTS

We thank Dr. Gefei Zeng and Rebecca Rapoport for valuable discussions. This work was supported by grants from the NIH (HL83262, HL86564, and HL43174) to V. L.

REFERENCES

Bautch, V. L. (2002). Embryonic stem cell differentiation and the vascular lineage. *Methods Mol. Biol.* **185**, 117–125.

Bautch, V. L., Redick, S. D., Scalia, A., Harmaty, M., Carmeliet, P., and Rapoport, R. (2000). Characterization of the vasculogenic block in the absence of vascular endothelial growth factor-A. *Blood* **95**, 1979–1987.

Bautch, V. L., Stanford, W. L., Rapoport, R., Russell, S., Byrum, R. S., and Futch, T. A. (1996). Blood island formation in attached cultures of murine embryonic stem cells. *Dev. Dyn.* **205**, 1–12.

Carmeliet, P., Ferreira, V., Breier, G., Pollefeyt, S., Kieckens, L., Gertsenstein, M., Fahrig, M., Vandenhoeck, A., Harpal, K., Eberhardt, K., Declercq, C., Pawling, C., Moons, J., Collen, L., Risau, D., and Nagy, W. (1996). Abnormal blood vessel development and lethality in embryos lacking a single VEGF allele. *Nature* **380**, 435–439.

Choi, K., Kennedy, M., Kazarov, A., Papadimitriou, J. C., and Keller, G. (1998). A common precursor for hematopoietic and endothelial cells. *Development* **125**, 725–732.

Doetschman, T. C., Eistetter, H., Katz, M., Schmidt, W., and Kemler, R. (1985). The *in vitro* development of blastocyst-derived embryonic stem cell lines: formation of visceral yolk sac, blood islands and myocardium. *J. Embryol. Exp. Morphol.* **87**, 27–45.

Ferrara, N., Carver-Moore, K., Chen, H., Dowd, M., Lu, L., O'Shea, K. S., Powel-Braxton, L., Hillan, K. J., and Moore, M. W. (1996). Heterozygous embryonic lethality induced by targeted inactivation of the VEGF gene. *Nature* **380**, 439–442.

Fong, G. H., Rossant, J., Gertsenstein, M., and Breitman, M. L. (1995). Role of the Flt-1 receptor tyrosine kinase in regulating the assembly of vascular endothelium. *Nature* **376**, 66–70.

Fung-Leung, W.-P., and Mak, T. W. (1992). Embryonic stem cells and homologous recombination. *Curr. Opin. Immunol.* **4**, 189–194.

Irion, S., Luche, H., Gadue, P., Fehling, H. J., Kennedy, M., and Keller, G. (2007). Identification and targeting of the ROSA26 locus in human embryonic stem cells. *Nat. Biotechnol.* **25**, 1477–1482.

Kappas, N. C., and Bautch, V. L. (2007). Maintenance and *in vitro* differentiation of mouse embryonic stem cells to form blood vessels. *In* Current Protocols in Cell Biology, (J. S. Bonifacino, *et al.*, Eds.), Vol. 3, pp. 23.3.1–23.3.20. John Wiley & Sons, Inc., New York.

Kappas, N. C., Zeng, G., Chappell, J. C., Kearney, J. B., Hazarika, S., Kallianos, K. G., Patterson, C., Annex, B. H., and Bautch, V. L. (2008). The VEGF receptor Flt-1 spatially modulates Flk-1 signaling and blood vessel branching. *J. Cell Biol.* **181**, 847–858.

Kearney, J. B., and Bautch, V. L. (2003). In vitro differentiation of mouse ES cells: Hematopoietic and vascular development. *Methods Enzymol.* **365**, 83–98.

Kearney, J. B., Kappas, N. C., Ellerstrom, C., DiPaola, F. W., and Bautch, V. L. (2004). The VEGF receptor flt-1 (VEGFR-1) is a positive modulator of vascular sprout formation and branching morphogenesis. *Blood* **103**, 4527–4535.

Keller, G. (2005). Embryonic stem cell differentiation: Emergence of a new era in biology and medicine. *Genes Dev.* **19**, 1129–1155.

Keller, G., Kennedy, M., Papayannopoulou, T., and Wiles, M. V. (1993). Hematopoietic commitment during embryonic stem cell differentiation in culture. *Mol. Cell Biol.* **13**, 473–486.

Keller, G. M. (1995). *In vitro* differentiation of embryonic stem cells. *Curr. Opin. Cell Biol.* **7**, 862–869.

Redick, S. D., and Bautch, V. L. (1999). Developmental platelet endothelial cell adhesion molecule expression suggests multiple roles for a vascular adhesion molecule. *Am. J. Pathol.* **154**, 1137–1147.

Risau, W. (1997). Mechanisms of angiogenesis. *Nature* **386,** 671–674.

Risau, W., Sariola, H., Zerwes, H. G., Sasse, J., Ekblom, P., Kemler, R., and Doetschman, T. (1988). Vasculogenesis and angiogenesis in embryonic-stem-cell-derived embryoid bodies. *Development* **102,** 471–478.

Schuh, A. C., Faloon, P., Hu, Q. L., Bhimani, M., and Choi, K. (1999). *In vitro* hematopoietic and endothelial potential of flk-1(−/−) embryonic stem cells and embryos. *Proc. Natl. Acad. Sci. USA* **96,** 2159–2164.

Shalaby, F., Rossant, J., Yamaguchi, T. P., Gertsenstein, M., Wu, X. F., Breitman, M. L., and Schuh, A. C. (1995). Failure of blood-island formation and vasculogenesis in Flk-1-deficient mice. *Nature* **376,** 62–66.

Shalaby, F., Ho, J., Stanford, W. L., Fischer, K. D., Schuh, A. C., Schwartz, L., Bernstein, A., and Rossant, J. (1997). A requirement for Flk1 in primitive and definitive hematopoiesis and vasculogenesis. *Cell* **89,** 981–990.

Soriano, P. (1999). Generalized lacZ expression with the ROSA26 Cre reporter strain. *Nat. Genet.* **21,** 70–71.

Vittet, D., Prandini, M. H., Berthier, R., Schweitzer, A., Martin-Sisteron, H., Uzan, G., and Dejana, E. (1996). Embryonic stem cells differentiate *in vitro* to endothelial cells through successive maturation steps. *Blood* **88,** 3424–3431.

Wang, R., Clark, R., and Bautch, V. L. (1992). Embryonic stem cell-derived cystic embryoid bodies form vascular channels: an *in vitro* model of blood vessel development. *Development* **114,** 303–316.

West, A. G., Gaszner, M., and Felsenfeld, G. (2002). Insulators: Many functions, many mechanisms. *Genes Dev.* **16,** 271–288.

Wiles, M. V., and Keller, G. (1991). Multiple hematopoietic lineages develop from embryonic stem (ES) cells in culture. *Development* **111,** 259–267.

Zeng, G., Taylor, S. M., McColm, J. R., Kappas, N. C., Kearney, J. B., Williams, L. H., Hartnett, M. E., and Bautch, V. L. (2007). Orientation of endothelial cell division is regulated by VEGF signaling during blood vessel formation. *Blood* **109,** 1345–1352.

CHAPTER SEVEN

THE AORTIC RING MODEL OF ANGIOGENESIS

Alfred C. Aplin,* Eric Fogel,[†] Penelope Zorzi,* *and* Roberto F. Nicosia*,[†]

Contents

Abstract

Angiogenesis is regulated by a complex cascade of cellular and molecular events. The entire process can be reproduced *in vitro* by culturing rat or mouse aortic explants in three-dimensional biomatrices under chemically defined conditions. Angiogenesis in this system is driven by endogenous growth factors released by the aorta and its outgrowth. Sprouting endothelial cells closely interact with pericytes, macrophages, and fibroblasts in an orderly sequence of morphogenetic events that recapitulates all stages of

* Department of Pathology, University of Washington, Seattle, Washington
[†] Division of Pathology and Laboratory Medicine, Veterans Administration Puget Sound Health Care System, Seattle, Washington

Methods in Enzymology, Volume 443
ISSN 0076-6879, DOI: 10.1016/S0076-6879(08)02007-7

angiogenesis. This model can be used to study the basic mechanisms of the angiogenic process and to test the efficacy of proangiogenic or antiangiogenic compounds. Aortic cultures can be evaluated with a range of morphologic and molecular techniques for the study of gene expression. In this chapter we describe basic protocols currently used in our laboratory to prepare, quantify, and analyze this assay.

1. INTRODUCTION

Angiogenesis, the process of formation of new blood vessels, is essential for embryonal development, the female ovarian cycle, and wound healing, but also contributes to the progression of many disease processes including cancer and atherosclerosis (Carmeliet, 2005). Research studies conducted during the past 30 years and clinical applications of this work have demonstrated that angiogenesis-dependent disorders can be tamed or even reversed by pharmacologically interfering with the angiogenic process. Drugs capable of selectively blocking the function of vascular endothelial growth factor (VEGF), a key angiogenic regulator, have improved the eyesight of patients with macular degeneration of the retina and enhanced the survival of patients with disseminated colon or lung cancer (Diaz-Rubio, 2006; Waisbourd *et al.*, 2007). Conversely, treatment with angiogenic factors has been proposed as a biologic bypass strategy for patients with ischemic heart or peripheral vascular disease (Lekas *et al.*, 2004).

Angiogenic regulators are routinely tested in the laboratory for efficacy and target specificity before clinical trials. To that end several *in vitro* models have been developed to quantify the angiogenic process and evaluate the mechanism of action of pro-angiogenic and anti-angiogenic drugs (Auerbach *et al.*, 2000; Guedez *et al.*, 2003; Jain *et al.*, 1997).

Among these, the *ex vivo* aortic ring assay of angiogenesis has proven to be particularly reliable and reproducible. Aortic rings dissected from rats or mice and cultured in a collagen or fibrin matrix spontaneously generate outgrowths of branching microvessels. Aortic angiogenesis occurs in the absence of serum or exogenous growth factors and can be easily monitored by direct microscopic observation of the living cultures. The vascular outgrowths are essentially indistinguishable from the blood vessels that form during angiogenesis *in vivo* and are composed of the same cell types (Kawasaki *et al.*, 1989; Nicosia and Ottinetti, 1990a; Nicosia *et al.*, 1983; 1990b; 1992).

The angiogenic response of the aortic rings is a self-limited process regulated at many levels. Paracrine and juxtacrine signaling between endothelial cells, fibroblasts, macrophages, and pericytes play a critical role in the formation, differentiation, and stabilization of vascular tubes (Nicosia and

Villaschi, 1995; Villaschi and Nicosia, 1994). Endogenous soluble factors stimulating these processes include basic fibroblast growth factor (bFGF) (Villaschi and Nicosia, 1993), VEGF (Nicosia *et al.*, 1994; 1997), platelet-derived growth factor (PDGF) (Nicosia *et al.*, 1994), angiopoietins (Iurlaro *et al.*, 2003; Zhu *et al.*, 2002a), and inflammatory cytokines (Aplin *et al.*, 2006). Integrin-mediated adhesive interactions between endothelial cells and the surrounding matrix critically contribute to the growth, morphogenesis, and survival of neovessels (Bonanno *et al.*, 2000; Nicosia and Bonanno, 1991; Nicosia and Tuszynski, 1994; Nicosia *et al.*, 1990a; 1990b; 1993), whereas proteolytic events driven by matrix metalloproteinases regulate endothelial sprouting, as well as vascular regression (Zhu *et al.*, 2000).

The *ex vivo* aortic ring model combines advantages of both *in vivo* and *in vitro* assays. Neovessel growth occurs in a defined environment, and the culture system can be easily adapted to different experimental conditions. The native endothelium of the explants has not been modified by repeated passages in culture and behaves like endothelial cells do *in vivo*. Microvessels are composed of a properly polarized layer of endothelial cells and surrounding pericytes (Nicosia and Villaschi, 1995). The interstitial space between outgrowing microvessels contains fibroblasts. Because of the lack of blood flow, there are no circulating leukocytes, although resident macrophages and dendritic cells migrate out of the aortic adventitia and actively participate in the angiogenic process (Aplin *et al.*, 2006).

Many molecules with both pro-angiogenic and anti-angiogenic activity have been tested in this assay. Aortic angiogenesis can be inhibited with hydrocortisone (Nicosia *et al.*, 1990a), RGD-containing peptides (Nicosia *et al.*, 1991), collagen synthesis inhibitors (Bonanno *et al.*, 2000), metalloproteinase inhibitors (Zhu *et al.*, 2000), antibodies against bFGF or VEGF (Nicosia *et al.*, 1997; Villaschi *et al.*, 1993), as well as AKT and p42/44 MAPK inhibitors (Zhu *et al.*, 2002a). Conversely, pro-angiogenic effects are obtained by supplementing the culture medium with recombinant growth factors such as bFGF, VEGF, PDGF, angiopoietins, insulin-like growth factor-1 (IGF-1), and inflammatory cytokines (Aplin *et al.*, 2006; Iurlaro *et al.*, 2003; Nicosia *et al.*, 1994; Zhu *et al.*, 2002a).

Because of its many advantages, the *ex vivo* aortic ring assay has become quite popular and is now used in many angiogenesis research laboratories. Although this method is relatively simple and easy to set up, it may pose challenges, especially for investigators who are not familiar with its potential pitfalls. This chapter reviews the methods that have proven to be the most user-friendly and reproducible in our laboratory. We describe here basic protocols needed to successfully prepare and properly interpret rat and mouse aortic ring cultures. This update should be useful to both newcomers to the field and investigators generally interested in our latest experience with this assay.

2. AORTIC RING ANGIOGENESIS MODEL: BASIC PROTOCOL

In this section we describe how to isolate interstitial collagen from rat tails, prepare collagen gel cultures of rat aorta, and quantify angiogenesis. Collagen gels are easy to prepare, provide a robust 3D environment for angiogenic sprouting, and represent a physiologic matrix that is commonly found in many vascularized tissues. For consistency, we prefer to prepare our own collagen in house through a protocol that uses tails harvested from young rats. It is also possible to use commercially available collagen (Masson *et al.*, 2002). Additional biomatrices that have been used for this assay in our laboratory (Nicosia *et al.*, 1990a,b) and by others (Kruger *et al.*, 2000; Malinda *et al.*, 1999) include fibrin and Matrigel.

2.1. Preparing collagen solution from rat tails

The following procedure uses tails harvested from 2- to 3-month-old rats (Elsdale and Bard, 1972). Rat tails may be stored at $-20\,°C$ for several months before collagen preparation. With tissue culture grade reagents and aseptic technique, disinfect the skin surface of 3 to 4 tails with 80% ethanol and air-dry each tail in a separate tissue culture dish. Using a scalpel, circumferentially incise the skin at 1- to 2-cm intervals along the length of the tail. Pull the skin and subcutaneous tissue away from the underlying tendons with a hemostat. As the tendons become exposed, cut them off with sterile scissors and place them into a dry culture dish. Continue until all of the collected tendons are pooled into a single dish. With sterile forceps, transfer the tendons to a 100-mm culture dish containing a solution of 0.9% NaCl. With fine microdissection forceps, gently separate the collagen fibers by teasing the tendons apart. Working under a dissecting microscope, remove any blood vessels, which are pinkish red in color and have an obvious lumen. In a typical isolation there will be several segments of blood vessels that need to be removed. Collagen fibers are recognizable by their velvety and ribbon-like appearance; discard any hemorrhagic collagen fibers. Transfer the fibers to a fresh dish of 0.9% NaCl and gently wash them by agitating the dish. Repeat the transfer and washing procedure through a total of eight separate baths of sterile 0.9% NaCl. Transfer the washed fibers to 100-mm dish containing 80% ethanol and incubate for 20 min; the collagen fibers will stiffen as they become dehydrated. Rehydrate the collagen through six baths of sterile distilled water. Transfer fibers to a 250-ml Erlenmeyer flask containing 100 to 150 ml of 0.5 M glacial acetic acid and stir for 48 h at 4 °C. After this treatment, most of the collagen fibers will be dissolved, and the solution will be viscous. Pass the acetic acid solution through three thin layers of sterile gauze into 50-ml centrifuge tubes. Centrifuge at 12,000g for 1 h at 4 °C. Using a 10-ml glass pipette, carefully transfer the supernatant to a sterile 150-ml glass bottle. Determine the concentration of collagen by dispensing 5 ml of acetic acid solution to an aluminum weigh boat that was tared before the procedure. Evaporate the acetic acid on a hot plate and reweigh the aluminum boat. Calculate the weight of collagen present per milliliter of liquid acetic acid solution. If necessary, 0.5 M acetic acid may be used to dilute the collagen stock. We routinely use a concentration of 1.25 mg/ml, but more concentrated solutions can be prepared if desired (Chun *et al.*, 2004). The resulting acetic acid solution may be stored at 4 °C for periods of several months before dialysis.

To prepare the working stock collagen solution, the acetic acid solution is dialyzed with 1-inch-diameter dialysis tubing (6000 to 8000 MWCO). Prepare an approximately 15-inch length of tubing by wetting it inside and out with sterile distilled water, then boil in 1 L of sterile water for 10 min. Boil the tubing two additional times with fresh water each time. Squeeze water out of tubing and allow it to cool to room temperature without letting it dry. Use clips or tie a double knot at one end of the dialysis tubing and carefully fill with 25 to 30 ml of the 1.25 mg/ml acidic collagen solution. Tie off or clip the top end of tubing, leaving an air bubble between the collagen meniscus and the knot. Dialyze overnight at 4 °C against 3.5 L of 0.1× minimal essential medium (Gibco, Gaithersburg, MD), pH 4.0, dialysis solution. Do not put more than two 25 to 30 ml tubes in 3.5 L of dialysis solution. Repeat for another 24 h against fresh dialysis solution. When dialysis is complete, disinfect the tubing around the air bubble with 80% ethanol, make a small slit in the tubing with sterile fine-tip scissors, and extract the collagen with a long needle. Transfer the dialyzed solution to a chilled sterile bottle and keep refrigerated at 4 °C until use. Collagen solutions prepared in this manner should be used within 6 months of dialysis.

2.2. Preparing aortic rings

Rats are sacrificed by CO_2 asphyxiation. Immediately after the animal has died, shave the abdominal skin and disinfect with 80% ethanol. Make a Y-shaped incision with a scalpel along the thoracic and abdominal region. Carefully dissect the skin from the underlying muscle layer and open the abdominal cavity. After cutting the sternal plate and diaphragm with scissors, expose the thoracic cavity by clamping the xyphoid process of the sternum with a hemostat and folding the entire sternal plate over to the right side of the animal. Displace the intestines, stomach, spleen, and liver to the right side. Carefully section the diaphragm in a ventral/dorsal direction paying attention not to cut the diaphragmatic vessels. The thoracic aorta should now be visible along the vertebral column. Ligate the aorta with 4-0 silk suture proximally, below the aortic arch. While holding the suture with forceps, excise the aorta from the posterior mediastinum with small curved scissors and transfer it into a 100-ml Felsen dish (quadrant dish) containing 4 ml of serum-free EBM (endothelial basal medium, Clonetics, San Diego, CA) per compartment. During excision, careful attention should be paid not to stretch or otherwise mechanically damage the aorta.

Further dissection and cleaning of the aorta is carried out within the compartments of a Felsen dish containing serum-free EBM. By use of a dissecting microscope in a tissue culture clean environment, first dissect away the periaortic fibroadipose tissue with Noyes scissors and curved microdissection forceps (Fine Surgical Instruments, Inc. Hempstead, NY). Care must be taken not to stretch, cut, or crush the aortic wall. Remove intraluminal blood clots with microdissection forceps. Carefully cut any small hemorrhagic fibrous tissue away from the aortic wall. As the dissection progresses, transfer the aorta to the successive compartments of the Felsen dish as needed to remove blood and fragments of dissected fibroadipose tissue. Carefully trim any stumps of collateral vessels. When the aorta appears as a well-defined tube clean of blood, transfer it to a clean Felsen dish containing serum-free EBM.

With a No. 22 scalpel blade, cross-section the aorta into rings that are 1- to 2-mm in length. With careful technique, 20 to 24 individual rings can be obtained from each thoracic aorta. Discard the proximal most and distalmost rings and with microdissection forceps, wash the remaining rings through sequential transfer into eight to twelve consecutive baths of serum-free medium.

2.3. Embedding aortic rings in collagen gels

Our original method to study aortic angiogenesis made use of a thick 300-μl gel per aortic ring (Nicosia and Zhu, 2004; Nicosia *et al.*, 1990a). More recently, we have simplified our approach and developed a thin gel modification of this method (Zhu and Nicosia, 2002b). The thin gel aortic ring assay uses 30 μl collagen per ring and can be set up more quickly. In addition, the thin gels are easily permeated by reagents and can be stained by immunocytochemistry as whole mounts (Zhu *et al.*, 2002b).

To prepare the working collagen solution for the experiment, begin by placing all reagents on ice before setting up individual aortic ring cultures. Mix 1 volume 10× Eagle's MEM (Gibco) with 1 volume 23.4 mg/ml $NaHCO_3$ in a sterile tube on ice until the color of the solution changes to orange. Add 8 volumes of dialyzed collagen and mix thoroughly by pipetting, without generating bubbles. Keep the collagen solution on ice.

Clean and washed aortic rings are placed individually into drops of collagen as follows. First, pipette a 30-μl drop of collagen onto the bottom of each well of a 4-well Nunc dish (NUNC # 179820). Working quickly, carefully grasp a single aortic ring with microdissection forceps and gently drag the ring across a sterile plastic surface to remove residual medium. Place the aortic ring into the collagen drop and, with a sterile plastic pipettor tip, position the ring within the gel so that the luminal axis is parallel to the bottom of the culture dish. Using the pipettor tip, gently spread the gel into a thin uniform disc approximately 8 mm in diameter around the explant. Repeat this procedure for each well of the culture dish.

Incubate the dish at 37 °C for 10 min to induce collagen gelation. Once the gel has set, use a micropipetter to carefully add 500 μl of serum-free EBM drop wise to each culture. Incubate the aortic ring cultures in a humidified CO_2 incubator at 37 °C for the duration of the experiment. The cultures may be incubated for a period of one to three weeks, during which the growth medium should be removed and replaced every 2 to 3 days.

2.4. Quantification of angiogenesis

Aortic ring cultures generate networks of branching microvessels surrounded by spindly and round nonendothelial cells (Fig. 7.1), which migrate out of the aortic wall within 1 to 2 days. The neovessel growth phase begins after 2 to 3 days of culture and continues until day 7 to 10. Endothelial sprouts arise primarily from the cut edges of the explants. As neovessels mature, they become surrounded by pericytes. After the initial growth period, neovessels begin to regress through a process of fragmentation, disintegration, and retraction. During this phase, pericytes become detached from the endothelium. Regression of the vasculature is associated with degradation of the collagen matrix, resulting in the formation of a halo of lysis around each aortic explant.

The angiogenic response of the rat aorta can be quantitated by visual counts (Nicosia *et al.*, 1990a; 1993) or by computer–assisted image analysis (Blacher *et al.*, 2001; Blatt *et al.*, 2004; Nissanov *et al.*, 1995). For visual counts, cultures are examined every 2 to 3 days and scored for angiogenic sprouting by use of an inverted microscope with bright-field optics. Vessels are most easily counted with 4 to 10× objectives and a 10× eyepiece. It is necessary to frequently adjust the focus to accurately score all vessels growing in three dimensions. The same person should be responsible for scoring angiogenic outgrowths from day to day during the course of an experiment. The observer should be properly trained to identify bona fide neovessel sprouts and not include counts of fibroblasts or other groups of nonendothelial cells.

Angiogenesis is scored according to the following criteria (Fig. 7.2) (Nicosia *et al.*, 1990a).

1. Microvessels are distinguished from fibroblasts on the basis of their greater thickness and cohesive pattern of growth.
2. The branching of one microvessel generates two additional microvessels.

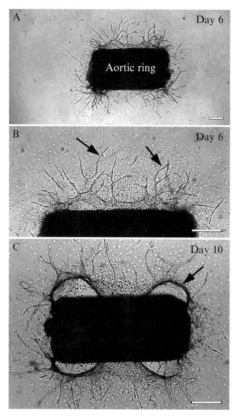

Figure 7.1   Collagen gel cultures of rat aorta at days 6 (A, B) and 10 (C). Arrows show microvessels in B and halos of collagen lysis in C. Scale bar = 400 μm.

3. Each microvascular loop is counted as two vessels, because it usually originates from a pair converging microvessels.

Once all individual cultures have been scored, standard statistical methods are used to analyze data and determine levels of significance between control and treated cultures. A control group consisting of aortic rings dissected from the same aorta must be included in each experiment to minimize interassay variability.

Aortic outgrowths have been successfully quantified also by image analysis. (Blacher *et al.*, 2001; Blatt *et al.*, 2004; Nissanov *et al.*, 1995). Algorithm-based digital methods have the reported advantages of speed, objectivity, and batch processing. Image analysis techniques are most valuable for evaluating large number of cultures for pharmacologic screening and other high-throughput applications. However, in our experience, visual counts remain a very practical, cost-effective, and, ultimately, rapid method to score aortic ring angiogenesis in a basic research laboratory.

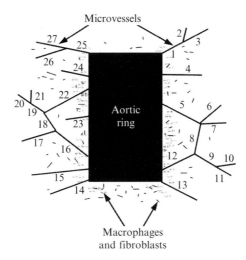

Figure 7.2   Method for visual quantification of angiogenesis in the aortic ring assay. The aortic ring is shown with its luminal axis parallel to the bottom of the dish. Microvessels radiating outwards are counted as shown.

3. PROTOCOLS FOR ANALYSIS OF AORTIC CULTURES

3.1. Whole mount preparations

Immunoperoxidase and fluorescent microscopy of whole mount preparations of aortic rings cultured by the thin gel method allows visualization of specific cell types (Zhu *et al.*, 2002b). Direct immunostaining of thin gels obviates the need for labor-intensive paraffin embedding and tissue sectioning.

Immunostaining combined with confocal microscopy is particularly useful to perform double or triple stains on the intact neovascular outgrowths. Whole mount staining offers several advantages over traditional histologic methods that rely on embedding and sectioning samples. First, whole mounts contain all the neovessels formed in each culture, whereas histologic sections represent a limited sampling of the outgrowths. Second, this method can be used to topographically localize proteins in specific regions of the angiogenic outgrowth, from the root to the tip of the neovessels. Finally, evaluation of intact angiogenic outgrowths allows close correlation between *in situ* expression of molecules of interest and cell behavior. because whole mount preparations are morphologically and structurally identical to the living cultures. This method can be used to identify different cell types by immunohistochemistry (Table 7.1).

Table 7.1 Identification of cell types in aortic cultures

Cell type	Antigen/ marker	Source	Reference
Endothelial	CD31	BD Pharmingen, San Diego, CA	Zhu and Nicosia, 2002
	GS-IB4[a]	Molecular Probes, Eugene, OR	Zhu *et al.*, 2002
	Tie-2	BD Pharmingen	Zhu *et al.*, 2002; Iurlaro *et al.*, 2003
	vWF	Sigma, St. Louis, MO	Nicosia and Ottinetti, 1990
Fibroblast	Vimentin[b]	RDI, Flanders, NJ	Unpublished observations
Macrophage	CD45	AbD Raleigh, NC	Aplin *et al.*, 2006
	CD163 (ED2)	AbD	Aplin *et al.*, 2006
Pericyte	SM α-Actin[c]	Sigma	Zhu *et al.*, 2002
	Desmin	NeoMarkers, Fremont, CA	Unpublished observations
	Calponin	NeoMarkers	Iurlaro *et al.*, 2003
	NG-2	Chemicon, Temecula, CA	Howson *et al.*, 2005

[a] GS-IB4 also labels macrophages.
[b] Vimentin also labels endothelial cells.
[c] SM α-Actin also labels fibroblasts (myofibroblasts).

3.1.1. Horseradish peroxidase immunohistochemistry

Remove growth medium from thin collagen gel cultures and briefly wash each in PBS. Fix with 10% neutral buffered formalin for 10 min and wash the gels three times with distilled water. Before staining with horseradish peroxidase–based methods, quench endogenous peroxidase by incubating the fixed cultures with 0.3% hydrogen peroxide for 30 min. For antibody staining, block cultures with 5% serum (same species as the secondary antibody) in PBS for 1 h at room temperature, rinse briefly in PBS, and react for 1 h with the primary antibody of interest. After antibody incubation, samples should be washed in PBS and incubated for 1 h with appropriate biotin-conjugated secondary antibody. After the secondary antibody incubation, wash samples again in PBS, then incubate with Vectastain ABC reagent (Vector Laboratories, Burlingame, CA) and develop with DAB, according to the manufacturer's recommendations. Lift each gel carefully with a bent spatula, transfer to a glass slide, and mount with an aqueous mounting medium. If desired, counterstain with DAPI or other nuclear stain.

3.1.2. Immunofluorescence and Confocal Microscopy

Perform fixation, washing, as well as blocking and antibody reactions as described in the previous section. Double-stain with a cocktail of antibodies of interest. For studies with the endothelial marker Griffonia Simplicifolia isolectin-B4, add the fluorescently conjugated lectin at the same time as the secondary antibodies. Mount samples as before and examine by fluorescent or confocal microscopy.

3.2. Isolation of RNA and protein

For gene expression studies, we typically pool 4 or more aortic ring cultures per experimental group. Although individual cultures may be used, the amount of protein or nucleic acid recovered from a single culture is in our experience insufficient to consistently obtain enough material for molecular analysis.

Total RNA is isolated by snap freezing aortic cultures in liquid nitrogen followed by manual pulverization. RNA extraction with Trizol® reagent (Invitrogen, Carlsbad, CA) and further purification with the RNAEasy Micro kit (Qiagen, Valencia, CA) typically yields 200 to 500 ng total RNA/aortic culture (Aplin *et al.*, 2006).

To prepare proteins for Western analysis, collagen gel cultures are lysed directly in radioimmunoprecipitation (RIPA) lysis buffer (50 mM Tris, 150 mM NaCl, 0.1 % SDS, 1 % NP-40, 1 mM EDTA; 200 to 500 μl per culture). After centrifugation, supernatants can be analyzed according to standard protocols or frozen at $-80\,^{\circ}$C.

3.3. Quantification of pericytes/mural cells

Supporting mural cells/pericytes are easily visualized in collagen cultures by phase-contrast microscopy (Fig. 7.3).

Pericytes can be readily identified under phase contrast microscopy along the endothelial tubes and counted. The number of pericytes per vessel length represents a useful measure vessel maturity that can be influenced by a number of experimental culture conditions (Zhu *et al.*, 2002a; 2003a).

4. MODIFICATIONS OF BASIC PROTOCOL

4.1. Culture of aortic rings in 96-well plates and thick gels

In most cases, the standard thin gel preparation is all that is required for analyzing angiogenic growth and regression of microvessels. We typically use a 4-well dish with 16-mm wells and 400 to 500 μl of medium per well.

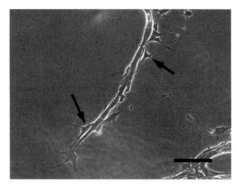

Figure 7.3 Phase-contrast micrograph of neovessel in collagen gel culture of rat aorta. Pericytes are marked by arrows. Scale bar = 100 μm.

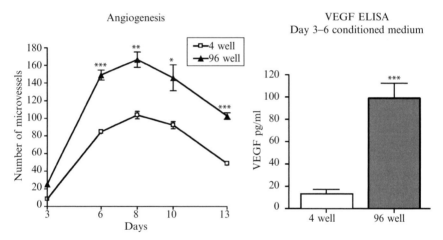

Figure 7.4 Rat aortic rings embedded in collagen generate more microvessels when cultured in 96-well plate than in 4-well dish ($n = 4$). Endogenous VEGF levels are higher in the 96-well plate format ($n = 3$). $\star = p < 0.05$; $\star\star = p < 0.01$; $\star\star\star = p < 0.001$.

However, three-dimensional collagen gel cultures of aortas may be prepared in tissue culture dishes of varying size, depending on the experimental conditions desired. For example, for some applications it may be useful to embed rings in the wells of a 96-well plate. This type of dish has several advantages over the 4-well format: (1) increased concentration of VEGF and other endogenous growth factors as a result of volume reduction (Fig. 7.4); (2) greater angiogenic response of rat aortic rings (Fig. 7.4); and (3) sprouting from mouse rings otherwise unable to generate neovessels in a larger well. In addition, the 96-well plate format is geometrically amenable to high-throughput applications and may be used to measure angiogenic growth as well as neovessel regression (Reed et al., 2007). One disadvantage of this format is that it is more difficult to remove cultures from the wells without damaging them for immunostaining. This can be, however, accomplished by carefully lifting and transferring the culture with fine forceps into an Eppendorf tube before the staining reaction.

When embedding aortic rings in a 96-well dish, 80 μl of collagen gel solution is added per well, and a single ring is placed within the collagen with the luminal axis parallel to the bottom of the dish. After 10 min incubation at 37 °C to induce collagen gelation as in the standard thin prep method, 150 μl of serum-free EBM is added to each culture. The growth medium is replaced every 2 to 3 days. The 96-well dish method results in each ring being cultured in ~50% smaller volume compared with the 4-well dish format.

An additional approach used to prepare aortic ring cultures is the "thick gel method" in which a larger volume (300 μl) of collagen (or other matrices) is used to embed each aortic ring (Nicosia et al., 1990a; 2004). We have used this method for a number of years particularly to study aortic ring cultures long term. However, we have now found that the 96-well plate method provides a comparable thick gel environment, with the added advantages of higher concentration of endogenous growth factors, more intense angiogenic growth, and greater ease of use.

4.2. Cultures of quiescent aortic rings

Aortic rings kept floating in serum-free medium for 14 days become "quiescent" and unable to sprout spontaneously when embedded in collagen. Angiogenic quiescence correlates with a gradual decline in the production of VEGF and other angiogenic factors by the explants (Nicosia et al., 1997; Villaschi et al., 1993). Angiogenesis can be reactivated by treating the rings with VEGF or other recombinant endothelial growth factors. This method can be used to differentiate angiogenic inducers from enhancers. For example, exogenous VEGF and IGF-1 are both capable of promoting angiogenesis in cultures of freshly cut aortic rings, but only VEGF has the capacity to trigger new angiogenic sprouting from quiescent rings (Nicosia et al., 2004; Zhu et al., 2002b). This method also lends itself to studying early molecular events that occur in the vessel wall after angiogenic stimulation and before vessel sprouting (Aplin et al., 2006).

To prepare quiescent aortic rings, coat each well of a 4-well dish with a thin layer of 1.5% agarose (Type VI-A). Rinse wells twice with EBM. Place one aortic ring per well in 500 μl of serum-free EBM and incubate at 37 °C, changing the medium three times a week. Aortic rings do not adhere to the agarose-coated dish and float in the culture medium. On day 13 to 14 aortic rings are embedded in collagen gel and cultured with or without recombinant angiogenic factors as described previously (Nicosia *et al.*, 1997).

4.3. Gene transduction of aortic rings

Aortic rings can be transduced with viral vectors to modulate expression of genes of interest. Vectors used in this model include adenovirus, retrovirus, vaccinia virus (Alian *et al.*, 2002; Hajitou *et al.*, 2002; Zhu *et al.*, 2003a), and adenoassociated virus (Nicosia, unpublished observations). Because the collagen gel may act as a barrier to the viral vector, gene transduction should be carried out before embedding the aortic rings in collagen. For example, aortic rings may be incubated for 24 h in serum-free medium containing approximately 1×10^7 plaque-forming units of adenoviral vector per ring. After gene transduction aortic rings are embedded in collagen and cultured as described previously. Appropriate controls for this experiment include untransduced rings, rings transduced with vector expressing a marker gene (for example green fluorescent protein), and rings transduced with empty viral vector.

5. TROUBLESHOOTING

The average count of healthy aortic cultures ranges from 60 to 130 microvessels at the peak of growth (days 6 to 10). If uniformly poor neovessel growth is observed (0 to 10 microvessels by the 6th day of culture) the assay should be rejected. Possible causes of poor growth include unsatisfactory reagents, temperature, or pH. The collagen solution should not be less than 1.25 mg/ml dry weight, and all reagents should be endotoxin free and tissue culture grade. Vascular outgrowths may be compromised if the incubation temperature drops below 35 °C. It is also important to maintain neutral pH throughout the culture period and minimize the exposure time to alkaline pH during dissection of the aortic explants (Burbridge *et al.*, 1999). Another possible source of poor growth is inadequate preparation and/or mixing of the collagen solution. When preparing the working collagen solution, always begin by mixing the MEM and $NaHCO_3$ before adding the stock collagen solution. Failure to thoroughly mix the reagents completely may result in uneven polymerization and defective gels.

There is an inherent variability in the angiogenic growth of aortic rings, and this is to be expected within limits. For example, a group of four identically treated rings may have the following number of neovessels at the peak of growth: 125, 110, 115, and 90. This is an acceptable result, because it is within the normal range of biologic variability. However, if the counts are instead 110, 10, 85, and 5, there is probably a technical problem with the experiment. In addition to intra-assay variability with rings from the same aorta, there may also be inter-assay variability because of the age and genetic background of the animals (Zhu *et al.*, 2003b). In general, rings from older animals produce fewer vessels than those from younger animals. Similar considerations apply to aortic rings from FVB mice that produce fewer vessels in collagen than rings from 129/SVJ or C57 black mice (Zhu *et al.*, 2003b). For these reasons, it is important to always include an untreated control group for each experiment.

Additional technical problems that may lead to uneven vessel growth include: (1) mechanical damage to the aorta: it is important not to stretch or crush the aorta or the aortic rings at any stage of the procedure; (2) drying of the endothelial lining: the aortic explants must be kept wet with growth medium throughout the dissection procedure; and (3) disruption of the collagen gel: when embedding rings in collagen, work quickly and position rings within 30 sec of placing the drop of collagen solution into the culture well. Once the collagen has set, care should be taken to avoid any physical damage to the gel. Cultures should be handled gently during transfer from bench to incubator or microscope stage. Extra care should be taken not to touch the collagen when replacing medium during the course of an experiment. If the preceding precautions are taken, neovessel growth from aortic rings will be robust, and the angiogenic response will be highly reproducible.

 ## 6. CONCLUDING REMARKS

More than 25 years have passed since we first reported that rat aortic rings were capable of producing an angiogenic response *ex vivo* (Nicosia *et al.*, 1982). Since then, angiogenesis research has enjoyed remarkable growth, and this field has moved from the confines of the experimental laboratory to the patient's bedside. Several new anti-angiogenic drugs have been approved to treat cancer and other angiogenesis-dependent disorders, and many more are being tested in clinical trials. As our knowledge has expanded, it has become clear that angiogenesis is an extremely complex process regulated by a cascade of molecular events, of which we have only a limited understanding. It is reasonable to predict that the molecular pathways of angiogenesis will become clearer with the discovery of new regulatory molecules and the definition of their role in this cascade. Assays that accurately reproduce the angiogenic process will be essential to study the activity and mechanism of action of angiogenic regulators and their pharmacologic inhibitors. To that end, the *ex vivo* aortic ring assay represents a most valuable tool in the armamentarium of the angiogenesis researcher. Technical advances introduced over the years have allowed us to make this assay more practical and user-friendly. The protocol guidelines provided in this chapter should be of help to current and future researchers in their efforts to study angiogenesis, unravel its molecular mechanisms, and develop target specific drugs for the treatment of angiogenesis-dependent disorders.

REFERENCES

Alian, A., Eldor, A., Falk, H., and Panet, A. (2002). Viral mediated gene transfer to sprouting blood vessels during angiogenesis. *J. Virol. Methods* **105,** 1–11.

Aplin, A. C., Gelati, M., Fogel, E., Carnevale, E., and Nicosia, R. F. (2006). Angiopoietin-1 and vascular endothelial growth factor induced expression of inflammatory cytokines before angiogenesis. *Physiol. Genomics* **27,** 20–28.

Auerbach, R., Akhtar, N., Lewis, R. L., and Shinners, B. L. (2000). Angiogenesis assays: Problems and pitfalls. *Cancer Metastasis Rev.* **19,** 167–172.

Blacher, S., Devy, L., Burbridge, M. F., Roland, G., Tucker, G., Noel, A., and Foidart, J. M. (2001). Improved quantification of angiogenesis in the rat aortic ring assay. *Angiogenesis* **4,** 133–142.

Blatt, R. J., Clark, A. N., Courtney, J., Tully, C., and Tucker, A. L. (2004). Automated quantitative analysis of angiogenesis in the rat aorta model using Image-Pro Plus 4.1. *Comput. Methods Programs Biomed.* **75,** 75–79.

Bonanno, E., Iurlaro, M., Madri, J. A., and Nicosia, R. F. (2000). Type IV collagen modulates angiogenesis and neovessel survival in the rat aorta model. *In Vitro Cell Dev. Biol. Anim.* **36,** 336–340.

Burbridge, M. F., West, D. C., Atassi, G., and Tucker, G. C. (1999). The effect of extracellular pH on angiogenesis *in vitro. Angiogenesis.* **3,** 281–288.

Carmeliet, P. (2005). Angiogenesis in life, disease and medicine. *Nature* **438,** 932–936.

Chun, T. H., Sabeh, F., Ota, I., Murphy, H., McDonagh, K. T., Holmbeck, K., Birkedal-Hansen, H., Allen, E. D., and Weiss, S. J. (2004). MT1-MMP-dependent neovessel formation within the confines of the three-dimensional extracellular matrix. *J. Cell. Biol.* **167,** 757–767.

Diaz-Rubio, E. (2006). Vascular endothelial growth factor inhibitors in colon cancer. *Adv. Exp. Med. Biol.* **587,** 251–275.

Elsdale, T., and Bard, J. (1972). Collagen substrata for studies on cell behavior. *J. Cell. Biol.* **54,** 626–637.

Guedez, L., Rivera, A. M., Salloum, R., Miller, M. L., Diegmueller, J. J., Bungay, P. M., and Stetler-Stevenson, W. G. (2003). Quantitative assessment of angiogenic responses by the directed *in vivo* angiogenesis assay. *Am. J. Pathol.* **162,** 1431–1439.

Hajitou, A., Grignet, C., Devy, L., Berndt, S., Blacher, S., Deroanne, C. F., Bajou, K., Fong, T., Chiang, Y., Foidart, J. M., and Noel, A. (2002). The antitumoral effect of endostatin and angiostatin is associated with a down-regulation of vascular endothelial growth factor expression in tumor cells. *FASEB J.* **16,** 1802–1804.

Iurlaro, M., Scatena, M., Zhu, W. H., Fogel, E., Wieting, S. L., and Nicosia, R. F. (2003). Rat aorta–derived mural precursor cells express the Tie2 receptor and respond directly to stimulation by angiopoietins. *J. Cell Sci.* **116,** 3635–3643.

Jain, R. K., Schlenger, K., Hockel, M., and Yuan, F. (1997). Quantitative angiogenesis assays: progress and problems. *Nat. Med.* **3,** 1203–1208.

Kawasaki, S., Mori, M., and Awai, M. (1989). Capillary growth of rat aortic segments cultured in collagen gel without serum. *Acta Pathol. Jpn* **39,** 712–718.

Kruger, E. A., Duray, P. H., Tsokos, M. G., Venzon, D. J., Libutti, S. K., Dixon, S. C., Rudek, M. A., Pluda, J., Allegra, C., and Figg, W. D. (2000). Endostatin inhibits microvessel formation in the *ex vivo* rat aortic ring angiogenesis assay. *Biochem. Biophys. Res. Commun* **268**, 183–191.

Lekas, M., Kutryk, M. J., Latter, D. A., and Stewart, D. J. (2004). Therapeutic neovascularization for ischemic heart disease. *Can. J. Cardiol.* **20** Suppl B, 49B–57B.

Malinda, K. M., Nomizu, M., Chung, M., Delgado, M., Kuratomi, Y., Yamada, Y., Kleinman, H. K., and Ponce, M. L. (1999). Identification of laminin alpha1 and beta1 chain peptides active for endothelial cell adhesion, tube formation, and aortic sprouting. *FASEB J.* **13**, 53–62.

Masson, V., V., Devy, L., Grignet-Debrus, C., Bernt, S., Bajou, K., Blacher, S., Roland, G., Chang, Y., Fong, T., Carmeliet, P., Foidart, J. M., and Noel, A. (2002). Mouse aortic ring assay: A new approach of the molecular genetics of angiogenesis. *Biol. Proced. Online.* **4**, 24–31.

Nicosia, R., and Zhu, W. H. (2004). Rat aortic ring assay of angiogenesis. *In* "Methods in Endothelial Cell Biology." (H. Augustin, Ed.), pp. 125–144. Springer-Verlag, Berlin Heidelberg.

Nicosia, R. F., and Bonanno, E. (1991). Inhibition of angiogenesis *in vitro* by Arg-Gly-Asp-containing synthetic peptide. *Am. J. Pathol.* **138**, 829–833.

Nicosia, R. F., Bonanno, E., and Smith, M. (1993). Fibronectin promotes the elongation of microvessels during angiogenesis *in vitro. J. Cell Physiol.* **154**, 654–661.

Nicosia, R. F., Bonanno, E., and Villaschi, S. (1992). Large-vessel endothelium switches to a microvascular phenotype during angiogenesis in collagen gel culture of rat aorta. *Atherosclerosis.* **95**, 191–199.

Nicosia, R. F., Lin, Y. J., Hazelton, D., and Qian, X. (1997). Endogenous regulation of angiogenesis in the rat aorta model. Role of vascular endothelial growth factor. *Am. J. Pathol.* **151**, 1379–1386.

Nicosia, R. F., Nicosia, S. V., and Smith, M. (1994). Vascular endothelial growth factor, platelet-derived growth factor, and insulin-like growth factor-1 promote rat aortic angiogenesis *in vitro. Am. J. Pathol.* **145**, 1023–1029.

Nicosia, R. F., and Ottinetti, A. (1990a). Growth of microvessels in serum-free matrix culture of rat aorta. A quantitative assay of angiogenesis *in vitro. Lab. Invest.* **63**, 115–122.

Nicosia, R. F., and Ottinetti, A. (1990b). Modulation of microvascular growth and morphogenesis by reconstituted basement membrane gel in three-dimensional cultures of rat aorta: a comparative study of angiogenesis in matrigel, collagen, fibrin, and plasma clot. *In Vitro Cell Dev. Biol.* **26**, 119–128.

Nicosia, R. F., Tchao, R., and Leighton, J. (1982). Histotypic angiogenesis *in vitro*: Light microscopic, ultrastructural, and radioautographic studies. *In Vitro* **18**, 538–549.

Nicosia, R. F., Tchao, R., and Leighton, J. (1983). Angiogenesis-dependent tumor spread in reinforced fibrin clot culture. *Cancer Res.* **43**, 2159–2166.

Nicosia, R. F., and Tuszynski, G. P. (1994). Matrix-bound thrombospondin promotes angiogenesis *in vitro. J. Cell Biol.* **124**, 183–193.

Nicosia, R. F., and Villaschi, S. (1995). Rat aortic smooth muscle cells become pericytes during angiogenesis *in vitro. Lab. Invest.* **73**, 658–666.

Nissanov, J., Tuman, R. W., Gruver, L. M., and Fortunato, J. M. (1995). Automatic vessel segmentation and quantification of the rat aortic ring assay of angiogenesis. *Lab. Invest.* **73**, 734–739.

Reed, M. J., Karres, N., Eyman, D., and Vernon, R. B. (2007). Culture of murine aortic explants in 3-dimensional extracellular matrix: A novel, miniaturized assay of angiogenesis *in vitro. Microvasc. Res.* **73**, 248–252.

Villaschi, S., and Nicosia, R. F. (1993). Angiogenic role of endogenous basic fibroblast growth factor released by rat aorta after injury. *Am. J. Pathol.* **143**, 181–190.

Villaschi, S., and Nicosia, R. F. (1994). Paracrine interactions between fibroblasts and endothelial cells in a serum-free coculture model. Modulation of angiogenesis and collagen gel contraction. *Lab. Invest.* **71,** 291–299.

Waisbourd, M., Loewenstein, A., Goldstein, M., and Leibovitch, I. (2007). Targeting vascular endothelial growth factor: A promising strategy for treating age-related macular degeneration. *Drugs Aging* **24,** 643–662.

Zhu, W. H., Guo, X., Villaschi, S., and Francesco, N. R. (2000). Regulation of vascular growth and regression by matrix metalloproteinases in the rat aorta model of angiogenesis. *Lab. Invest.* **80,** 545–555.

Zhu, W. H., Han, J., and Nicosia, R. F. (2003a). Requisite role of p38 MAPK in mural cell recruitment during angiogenesis in the rat aorta model. *J. Vasc. Res.* **40,** 140–148.

Zhu, W. H., Iurlaro, M., MacIntyre, A., Fogel, E., and Nicosia, R. F. (2003b). The mouse aorta model: Influence of genetic background and aging on bFGF- and VEGF-induced angiogenic sprouting. *Angiogenesis.* **6,** 193–199.

Zhu, W. H., MacIntyre, A., and Nicosia, R. F. (2002a). Regulation of angiogenesis by vascular endothelial growth factor and angiopoietin-1 in the rat aorta model: Distinct temporal patterns of intracellular signaling correlate with induction of angiogenic sprouting. *Am. J. Pathol.* **161,** 823–830.

Zhu, W. H., and Nicosia, R. F. (2002b). The thin prep rat aortic ring assay: A modified method for the characterization of angiogenesis in whole mounts. *Angiogenesis* **5,** 81–86.

CHAPTER EIGHT

AN ASSAY SYSTEM FOR *IN VITRO* DETECTION OF PERMEABILITY IN HUMAN "ENDOTHELIUM"

Manuela Martins-Green,* Melissa Petreaca,[†] *and* Min Yao[‡]

Contents

Abstract

The molecular mechanisms by which endothelial permeability occurs are often studied more readily *in vitro*, underscoring the importance of the use of systems that mimic human endothelium *in vivo*. We present an assay that accurately models human endothelium by use of primary human microvascular endothelial cells (hMVEC), because permeability primarily occurs at the microvascular level, and transwell filter units coated with Matrigel, extracellular matrix that mimics basal lamina, the matrix that is tightly associated with endothelium and is critical for its proper function. As a tracer molecule, we used 3-kDa dextran-FITC to

* Department of Cell Biology and Neuroscience, University of California, Riverside, Riverside, California
† Graduate Program in Cell, Molecular, and Developmental Biology, University of California, Riverside, Riverside, California
‡ Wellman Center for Photomedicine, Massachusetts General Hospital, Harvard Medical School, Boston, Massachusetts

Methods in Enzymology, Volume 443
ISSN 0076-6879, DOI: 10.1016/S0076-6879(08)02008-9

detect leakage through the small gaps present in the early stages of permeability induction. The permeability-inducing agents IL-8 and VEGF were added to the lower chamber of the transwell units to mimic inflammatory conditions *in vivo*. After optimization, we were able to minimize basal permeability and to detect rapid changes in permeability stimulated by IL-8 and VEGF, similar to that observed *in vivo*. Furthermore, we have used this system to delineate the importance of the transactivation of VEGFR2 in IL-8–induced permeability and have confirmed the relevance of this signaling *in vivo*, suggesting that our permeability assay system adequately mimics the *in vivo* situation. Therefore, this system can be used to better understand the molecular mechanisms of human vascular permeability in a more *in vivo*-like setting and, thus, may be used to test effective therapeutics to prevent and treat diseases involving persistent permeability.

1. INTRODUCTION

The microvascular endothelium plays a vital role in regulating the entry of fluids and cells from the bloodstream into surrounding tissue. Nonpathologic increases in endothelial permeability are transient and occur after tissue injury during the early stages of healing, whereas abnormal endothelial permeability is persistent and has been implicated in a number of pathologic conditions, including chronic inflammation, atherosclerosis, diabetic retinopathy, and tumor growth (Carmeliet, 2003; Folkman, 2002). In both normal and pathologic situations, permeability of endothelium is accompanied by inflammation and angiogenesis, suggesting that endothelial permeability may facilitate the movement of inflammatory cells from the blood vessel into surrounding tissue (Lee, 2005). The role of permeability in angiogenesis is less clear; although VEGF–induced angiogenesis is invariably accompanied by permeability (Dvorak *et al.*, 1995; Weis and Cheresh, 2005), inhibition of VEGF–induced permeability does not prevent VEGF-induced angiogenesis (Eliceiri *et al.*, 1999). It has been suggested that increased permeability may result in the formation of a provisional matrix composed of cross-linked fibrin, which facilitates endothelial cell migration and thus angiogenesis (Dvorak *et al.*, 1995). This may, in part, explain the relationship between angiogenesis and permeability. A better understanding of the processes involved in endothelial permeability may provide mechanistic insights into this relationship, as well as lead to more effective therapeutic approaches for the prevention and treatment of diseases involving persistent vascular permeability. Therefore, *in vitro* systems that both mimic *in vivo* microvascular endothelium and can be used to measure changes in endothelial permeability are important in explaining the mechanisms involved in this process.

In the past, several systems have been used to study increase in permeability by use of a variety of endothelial cells (EC) in transwell systems (Behzadian *et al.*, 2003; Breslin *et al.*, 2003; Cullen *et al.*, 2000; Lal *et al.*, 2001; Varma *et al.*, 2002). However, in these systems, the EC used are primarily cells isolated from umbilical veins, or, in some cases, cell lines that have different characteristics than those of the microvessels (Chi *et al.*, 2003), where the greatest increases in permeability occur during inflammatory responses, angiogenesis, and tumor-igenesis (Baldwin and Thurston, 2001; Hashizume *et al.*, 2000; Pettersson *et al.*, 2000; Thurston *et al.*, 1999). Furthermore, the cells are plated directly onto polycarbonate membranes (Chandra *et al.*, 2006; Feistritzer *et al.*, 2005) or on such membranes coated with collagen type I (Chen *et al.*, 2005; Nobe *et al.*, 2005), collagen type IV (Lal *et al.*, 2001), fibronectin (Breslin *et al.*, 2003; Lal *et al.*, 2001; Varma *et al.*, 2002), gelatin (Sun *et al.*, 2006), or a mixture of collagen and fibronectin (Behzadian *et al.*, 2003). In contrast, the *in vivo* endothelium is tightly associated with a basal lamina that consists of multiple extracellular matrix molecules arranged in a sheetlike structure that supports the endothe-lium (Conway *et al.*, 2001). Therefore, systems that do not ensure the formation of such a mature EC monolayer structure associated with an intact basal lamina–like ECM do not fully represent *in vivo* endothelium. The presence of ECM molecules that do not resemble basal lamina is particularly problematic, because cellular signaling occurs also through ECM contact and is not solely limited to cell–cell contact and cytokine-mediated signaling (Bissell and Radisky, 2001).

In addition, earlier studies often apply permeability inducers to the apical, or "lumenal," surface of the EC. However, many permeability-inducing agents originate in the tissue surrounding blood vessels and thus interact initially with the basal surface of the endothelium. Studies have indicated that receptors and other cell surface proteins are differentially localized on the endothelial cell surfaces, with some present on exclusively lumenal or ablumenal surfaces (Miller *et al.*, 1994; Stolz *et al.*, 1992). Thus, the location of the treatment application to the endothelial surface may affect signaling mechanisms that result in permeability and should be considered when measuring permeability *in vitro*.

Therefore, to perform studies that allow us to understand the mechan-isms by which permeability-inducing factors stimulate increase in perme-ability of human microvascular endothelium, we modified previous permeability assays by including new features that more accurately mimic human microvessel endothelium *in vivo*. In this system, the EC are primary human microvascular EC (hMVEC) that are plated on Matrigel (a complex ECM similar to basal lamina *in vivo*) and form a tight monolayer. This system responds well to permeability-inducing agents that impinge on the basal side of the endothelium and can be used effectively to detect rapid changes in endothelial permeability, to study the signaling pathways involved, and to test/develop agents that modulate this process. Further-more, we have used this system to identify a novel signaling pathway important in IL-8–induced endothelial permeability, that of VEGFR2 trans-activation, and have confirmed the importance of VEGFR2 in IL-8–induced permeability *in vivo*, suggesting that signaling mechanisms identified by use of this system are comparable to signaling *in vivo*.

2. GENERATION OF A HUMAN ENDOTHELIUM MODEL FOR PERMEABILITY STUDIES

2.1. Growth of primary human microvascular endothelial cells

2.1.1. Procedure

1. Obtain primary human lung microvascular endothelial cells (hMVEC-L) from Cambrex Clonetics (now Lonza), catalog number CC-2527, and expand as necessary in EGM-2MV (Lonza, catalog number CC-3202) to obtain sufficient cells for experimentation.
2. These primary cells are used between passages 4 and 9 for permeability assays. After passaging more times, the cells exhibit changes in morphology that results in leaky monolayers without any treatment.

2.2. Coating transwell filters with matrigel

2.2.1. Procedure

1. Use Matrigel from BD Biosciences, catalog number 354234. Thaw Matrigel at $4°$ C; aliquot into small tubes and keep aliquots at $-20°$ C until needed.
2. Before use, thaw an aliquot of Matrigel on ice. Note that it is critical that the Matrigel be kept on ice at all times to prevent it from solidifying and likewise critical for it to be kept sterile to prevent contamination.
3. Place a bucket of ice into the tissue culture hood and expose to UV light for at least 30 min. Place thawed Matrigel on ice in the hood.
4. Keeping the Matrigel on ice in the hood, dilute it at a ratio of 2:1 to 3:1 with *cold* sterile, serum-free media (preferably DMEM from Mediatech). The Matrigel should be diluted enough to facilitate pipetting.
5. Use uncoated 3-μm pore size transwell inserts in 24-well plates from BD Discovery Labware (catalog number 354575). In the hood, examine the transwell filters carefully for tears or wrinkles. Transwell filters with a wrinkled appearance are prone to leak and are thus unsuitable for permeability assays.
6. Coat the transwells with Matrigel; working quickly, withdraw 100 μl of diluted Matrigel from the aliquot on ice in the hood and deposit on the upper surface of the filter.

7. After ensuring that the entire filter surface is covered with dilute Matrigel, rapidly remove as much of the Matrigel as possible and either discard or place back in the tube on ice. This coating method causes the formation of a very thin layer of Matrigel that ultimately yields a better endothelial barrier function than that of thicker layers. Matrigel removed from the insert can be reused if rapidly placed on ice.

8. Repeat steps 6 and 7 until all filters needed for the experiment are coated with Matrigel.

9. Place coated transwell filter units into a sterile 24-well culture dish. This divides the transwell unit into two compartments: the upper chamber (lumenal) and the lower chamber (ablumenal). Allow Matrigel on coated filters to solidify at 37° C for 30 min before seeding with endothelial cells.

2.3. Seeding matrigel-coated filters with microvascular endothelial cells

2.3.1. Procedure

1. After coating transwell filters with Matrigel, prepare hMVEC-L for plating. Dissociate with prewarmed trypsin, centrifuge, and resuspend cell pellet in EGM-2MV according to standard tissue culture technique.

2. Count the hMVEC-L with a hemocytometer. Dilute cells to a concentration of 1×10^6 cells/ml in EGM-2MV. Deposit 100 μl of diluted cells (1×10^5 cells total) into the upper chamber of the transwell insert on top of the Matrigel-coated filter. After cells adhere (30 to 60 min after plating), add 200 μl EGM-2MV to the upper chamber and 1 ml to the lower chamber (in the 24-well plate), as diagrammatically represented in Fig. 8.1A.

3. Incubate at 37° C for 24 h.

4. Prepare hMVEC-L for seeding into the transwell filter units a second time, as described in steps 1 and 2. Carefully remove the media from the unit, and deposit 1×10^5 hMVEC-L in 100 μl EGM-2MV as in #2. After cells adhere, add 200 μl to the upper chamber and incubate an additional 24 h at 37° C before performing the permeability assay as described in the following.

2.4. *In vitro* permeability assay

2.4.1. Procedure

1. Twenty-four hours after the second plating, add dextran sulfate-FITC (3 kD, Invitrogen Molecular Probes catalog number D-3306) to the lower chamber to a final concentration of 10 μg/ml. Simultaneously add the permeability-inducing agent of interest also to the lower chamber. Include untreated transwells as negative controls and at least three wells/treatment.

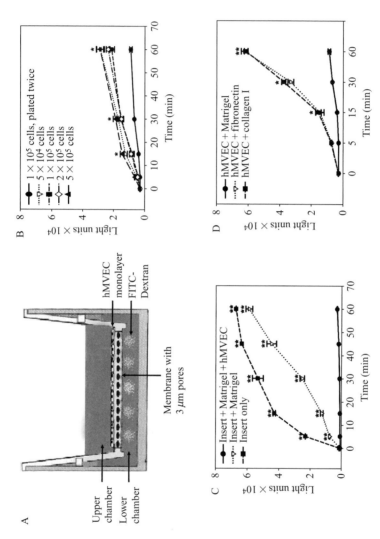

Figure 8.1 Overview and optimization of the permeability assay system. (A) Schematic representation of the transwell assay system. To mimic the endothelium *in vivo*, EC were plated on Matrigel-coated 3-μm pore size filters in the transwell inserts. In the final optimized system, the cells were plated twice, as described in the Material and Methods section, to ensure the formation of a tight hMVEC monolayer. As a permeability indicator, 3-kDa FITC-dextran (5 μg/ml) was used and introduced into the lower chamber along with the permeability-inducing

2. At various time intervals (e.g., 0, 5, 15, 30, 45, 60, 120, and 180 min after treatment), remove 10-μl aliquots of media from the upper chamber and dilute in 90 μl water/well in a 96-well plates.

3. After removing aliquots at the last time point, measure the fluorescence intensity by use of a 96-well plate fluorimeter (e.g., a Victor^{TM} 1420, Perkin-Elmer Life Sciences, Boston, MA) with excitation at 485 nm and emission at 535 nm. Increases in the amount of FITC-dextran diffusing through the endothelial monolayer from the lower chamber to the upper chamber are indicative of increased permeability.

Double-plating of the cells on top of Matrigel-coated filters promotes good survival and stability of the hMVEC monolayer, as shown by the barrier function of the cell monolayer. Single plating of different concentrations of hMVEC on Matrigel-coated filters yields inferior barrier function compared with the double-plating technique (Fig. 8.1B). The observed barrier function results from the endothelial cells rather than the Matrigel or the filter alone (Fig. 8.1C). In addition, endothelial cells plated on filters coated with different ECM molecules, such as collagen I (Sigma; 100 μl of 50 ng/ml collagen type I for 4 to 5 h, followed by air drying overnight) or fibronectin (Upstate; 100 μl of 4 μg/ml fibronectin for 2 h, followed by air drying overnight) results in decreased barrier function compared with cells plated on Matrigel-coated filters (Fig. 8.1D).

2.5. Confirmation of monolayer formation and barrier function by PECAM-1 immunostaining or f-actin staining

The ability of this *in vitro* permeability assay to detect changes in permeability induced by-VEGF (100 ng/ml, Peprotech catalog number 100-20) and IL-8 (50 ng/ml, R & D Systems catalog number 618-IL-050) can be seen in

---

agent. The permeability was quantified as fluorescence units of the FITC that passed from the lower chamber into the upper chamber and was plotted over time after addition of the reagents. (B) Effect of initial cell plating number on the endothelial barrier function. The number of cells plated/well is indicated in the graph. The filters were precoated with Matrigel and the hMVEC cultured for 48 h before initiating the permeability assay. Cells plated twice were plated at 24-h intervals. Data represent mean $\pm$ the standard error. Statistics are shown as comparisons of the group plated twice with 1×10^5 cells each time with all other groups, $\star P < 0.05$. (C) The system was tested with uncoated transwell inserts, Matrigel-coated inserts, or a tight monolayer of hMVEC on top of Matrigel-coated inserts. Data represent mean $\pm$ the standard error. Statistics are shown as comparisons of the insert only and insert coated with Matrigel groups with the hMVEC monolayer on top of Matrigel-coated inserts, $\star\star P < 0.01$. (D) The transwell filters were coated with collagen type I, fibronectin, or Matrigel before cell plating as described in Materials and Methods. Data represent the mean $\pm$ the standard error. Statistics are shown as comparisons of the cells plated on Matrigel with those plated on fibronectin or collagen I, $\star P < 0.05$, $\star\star P < 0.01$. From Petreaca *et al.*, 2007.

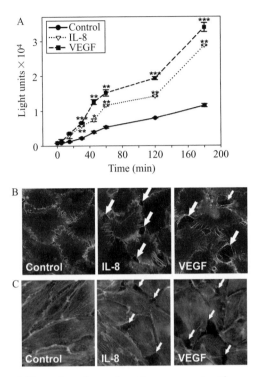

Figure 8.2   Permeability and gap formation after treatment with permeability-inducing agents. (A) The transwell cultures were treated with 50 ng/ml IL-8 or 100 ng/ml VEGF for multiple time points, as indicated. IL-8 stimulates endothelial permeability over time, similarly to VEGF. Each treatment group was performed in duplicate; data are shown as the mean value ± the standard error. Statistics are shown as comparisons between the treatment and control, $\star P < 0.05$, $\star\star P < 0.01$, $\star\star\star P < 0.001$. (B and C) Cultures either left untreated or treated with 50 ng/ml IL-8 or 100 ng/ml VEGF, as indicated, were immunostained with PECAM-1 (B) or were stained with rhodamine-phalloidin (C) to visualize paracellular gap formation and were visualized by fluorescence microscopy. Arrows indicate the position of some of the paracellular gaps. Like VEGF, IL-8 increased gap formation in hMVEC. A and B are from Petreaca *et al.*, 2007.

Fig. 8.2A. The morphologic changes occurring during the increases in permeability result in gap formation between the hMVEC after stimulation with these cytokines. PECAM-1 immunostaining delineates the cell boundaries and helps to visualize paracellular gap formation (Fig. 8.2B).

2.5.1. Procedure for PECAM immunolabeling

1. Plate the hMVEC-L onto Matrigel-coated transwell filters as described previously.
2. Treat hMVEC-L as for the permeability assay, without addition of FITC-dextran, for 30 min. Leave additional wells untreated as controls.

3. Remove media and rinse upper and lower chambers of each transwell unit twice with prewarmed 1× PBS.
4. Fix for 30 min with 4% paraformaldehyde in 1× PBS.
5. Wash, with shaking, 2 × 5 min in 1× PBS.
6. Incubate transwells in 0.1 M glycine in 1× PBS for 10 min to quench free paraformaldehyde radicals.
7. Wash 2 × 5 min with 1× PBS as in #5.
8. Incubate 30 min in 10% goat serum (100 μl/well) to block nonspecific binding of the secondary antibody.
9. Incubate 1 h with 1:200 anti-PECAM mouse antibody (R&D Systems, catalog number BBA7) in 1× PBS containing 1% BSA for 2 h.
10. Wash three times with 1× PBS containing 0.1% BSA.
11. Incubate 1 h with 1:100 goat anti–mouse FITC (Zymed Laboratories, Inc, catalog number 62-6311) in 1% BSA/PBS for 1 h.
12. Wash three times with 1× PBS containing 0.1% BSA.
13. Carefully remove filter from transwell unit with a clean scalpel and place the filter with cells facing up onto a clean glass slide.
14. Mount in Vectashield (Vector Laboratories, catalog number H-1200).
15. Visualize by fluorescence microscopy (we use a Nikon Microphot-FXA).

2.5.2. Procedure for the f-actin staining with rhodamine-phalloidin

1. Follow steps 1 to 5 in the protocol for PECAM immunolabeling.
2. Incubate cells with 1× PBS containing 0.1% Triton X-100 for 10 min to permeabilize the membranes and allow the rhodamine-phalloidin to penetrate inside the cells.
3. Wash two ties for 5 min with 1× PBS.
4. Incubate with 0.165 μM rhodamine-phalloidin (Invitrogen Molecular Probes, catalog number R415) for 20 min.
5. Wash two times for 5 min with 1× PBS.
6. Follow steps 13 to 15 in the procedure for PECAM immunolabeling.

2.6. Use of this *in vitro* permeability assay to answer biologic questions

This assay system has been used to investigate the mechanisms involved in the endothelial permeability induced by IL-8. We observed that IL-8 and VEGF stimulated permeability in a similar manner (Petreaca *et al.*, 2007), and previous studies implicated VEGFR2 transactivation in endothelial cell signaling induced by other GPCR ligands (Miura *et al.*, 2003; Seye *et al.*, 2004; Tanimoto *et al.*, 2002; Thuringer *et al.*, 2002). Therefore, we used a VEGFR inhibitor in conjunction with this *in vitro* permeability assay to determine whether IL-8–induced permeability requires VEGFR2 activity.

The VEGFR inhibitor blocked both IL-8–induced permeability (Fig. 8.3A) and gap formation (Fig. 8.3B), strongly suggesting that this process requires the activation of the VEGF receptor. Further verification that IL-8 transactivates VEGFR2 is shown (Petreaca *et al.*, in press).

2.6.1. Procedure for *in vitro* assay of permeability

1. When performing permeability assays in the presence of an inhibitor, plate endothelial cells on the Matrigel-coated transwell filters as described previously and conduct the *in vitro* permeability assay as described but with additional wells that are preincubated with the inhibitor before initiation of the permeability assay.

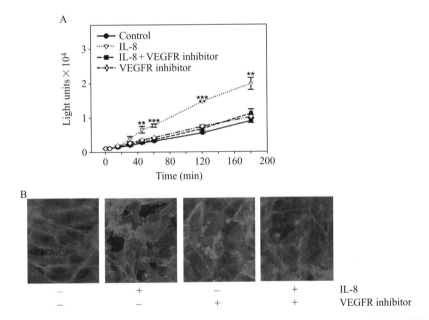

Figure 8.3 Importance of VEGFR in IL-8–induced permeability *in vitro*. (A) hMVEC were plated for the permeability assay as described. Before treatment with 50 ng/ml IL-8, the transwell cultures were incubated with 400 n*M* VEGFR inhibitor for 1 h. The VEGFR inhibitor prevented the permeability induced by both IL-8 (A) and VEGF (data not shown). Each treatment group was performed in triplicate; data are shown as the mean value ± the standard error. Statistics are shown as comparisons of the IL-8–treated group with the control and inhibitor groups, ★★$P < 0.01$, ★★★$P < 0.001$. (B) Cells were preincubated with 600 n*M* VEGFR inhibitor for 15 min, followed by treatment with 100 ng/ml IL-8 for 30 min and staining with rhodamine-phalloidin. F-actin staining and gap formation were visualized by fluorescence microscopy. IL-8–induced actin reorganization and gap formation were prevented with the VEGFR inhibitor. From Petreaca *et al.*, 2007.

2. Add the inhibitor to both upper and lower chambers of the transwell unit for at least 30 min before addition of the FITC-dextran and permeability inducer. It is critical to include inhibitor-only controls to ensure that the inhibitor by itself does not alter barrier function.

3. In Fig. 8.3A, the endothelial transwell system was preincubated with a VEGFR tyrosine kinase inhibitor (400 nM, Calbiochem catalog number 676475) for 1 h before addition of FITC-dextran and IL-8. The VEGFR inhibitor was shown to prevent IL-8–induced permeability.

4. To confirm the role of the molecule of interest in stimulation of permeability, perform rhodamine-phalloidin staining of transwell inserts as described previously to observe endothelial gap formation. This should be performed with some wells preincubated with an inhibitor of the molecule of interest before treatment with the permeability inducer (similar to #1, #2 previously); be sure to include inhibitor-only controls to ensure that the inhibitor does not alter the endothelial monolayer integrity. In Fig. 8.3B, endothelial cells were treated with IL-8 after preincubation with the same VEGFR inhibitor used in 3A. This VEGFR inhibitor prevented IL-8–induced gap formation, confirming the importance of VEGFR in IL-8–induced permeability *in vitro*.

To determine whether the signaling pathways identified with this *in vitro* permeability assay also occur *in vivo*, the Evan's Blue extravasation assay was used. In this assay, the animal is injected with Evan's Blue dye in the tail vein. After the dye moves throughout the circulation, the permeability inducer is injected under the skin at an appropriate location (such as the dorsal skin), and extravasation of the dye is monitored (Eliceiri *et al.*, 1999). By use of this assay, we confirmed that the VEGFR inhibitor blocked IL-8–induced permeability *in vivo* (Fig. 8.4A-B). This not only confirms the importance of the VEGFR in IL-8–induced permeability but also confirms that signaling mechanisms uncovered with our *in vitro* permeability assay are relevant *in vivo* (Petreaca *et al.*, 2007).

2.6.2. Procedure for *in vivo* assay

1. The day before performing the permeability assay, shave hair on the back of the mice that will be used for the assay, with anesthesia, if necessary. Prepare three mice for each treatment group.

2. The next day, with an insulin syringe, inject unanesthetized mice intravenously through the tail vein with 100 μl of 2% Evan's Blue dye in sterile PBS.

3. After dye injection, mark two contralateral regions on the shaved mouse dorsum for treatment. For each mouse, include one control and one treated region so that the levels of extravasation after treatment can be directly compared with the control region of the same mouse. This is critical because of the inherent variation in genetic background and in

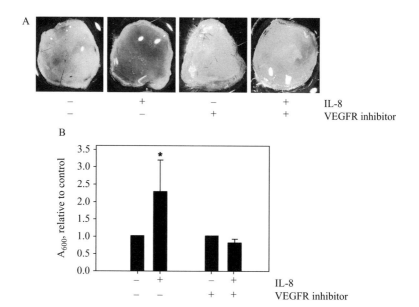

Figure 8.4 Importance of VEGFR in IL-8–induced permeability *in vivo*. (A) Mice were injected with Evans Blue through the tail vein and were then subcutaneously injected with either vehicle (PBS) or with 1 μg IL-8, in the presence and absence of prior treatment with 8 μM VEGFR inhibitor; 30 min after IL-8 treatment, mice were euthanized and perfused with PBS, followed by removal of treated regions of skin with a 7-mm punch biopsy. Skin punches were photographed to visualize Evans Blue dye extravasation into treated or untreated tissues, as indicated. IL-8 treatment increased endothelial permeability *in vivo*, as shown by increased dye extravasation into the tissue; this was blocked in tissues pretreated with the VEGFR inhibitor. (B) Mice were treated and skin punches were isolated as described in (A). Evan's Blue dye was then extracted from the skin punches with formamide, quantified with a spectrophotometer, and calculated relative to the appropriate control (PBS treatment for the IL-8–treated regions; VEGFR inhibitor only for the IL-8 + VEGFR inhibitor–treated regions). Data are shown as the mean $\pm$ the standard deviation, $^\star P < 0.05$, $n = 2$. From Petreaca *et al.*, 2007.

the difficulties encountered in tail vein injection. Both cause variation in the results obtained.

4. Inject the marked contralateral regions of the mouse dorsum subcutaneously with 100 μl of the vehicle (PBS) or with the treatment. For experiments involving an inhibitor, inject the inhibitor diluted in 100 μl 1× PBS into both contralateral regions 30 min before injection of the permeability inducer or PBS.

5. Fifteen to 30 min after injection with the permeability inducer, euthanize mice with CO_2 and perfuse with sterile 1× PBS through the left ventricle of the heart.

6. With a 7-mm biopsy punch (Accupunch), remove the injected region. Photograph skin punches to visualize differences in blue color, and thus extravasation.

7. Extract the Evan's Blue dye from the tissue biopsy samples for quantitative analysis by incubating tissue with 400 μl formamide at 56° C for 24 h.

8. Quantify extracted Evan's Blue with a spectrophotometer with the absorbance at 600 nm and normalize absorbance to tissue area. Calculate the normalized absorbance of the treated skin relative to untreated contralateral PBS control to determine fold change within each mouse. For experiments involving an inhibitor, calculate the normalized absorbance from the inhibitor + inducer relative to the contralateral inhibitor + PBS control to determine fold change.

Photographs of biopsy skin punches show increased Evan's Blue dye extravasation in the IL-8–treated skin compared with the control, inhibitor alone, and inhibitor + IL-8 skin biopsy specimens (Fig. 8.4A). Quantification of Evan's Blue as in steps 7 and 8 demonstrated that IL-8 increased permeability *in vivo*, and that this was prevented with the VEGFR inhibitor (Fig. 8.4B).

3. CONCLUSION

Although a number of permeability studies have been performed with *in vitro* assays that are similar to the one described here, we have made significant modifications to the existing system which provide a number of advantages: (1) unlike other systems, we use primary human microvascular EC, the cells that line microvessels, where most permeability events take place during inflammatory responses (Baldwin and Thurston, 2001; Hashizume *et al.*, 2000; Pettersson *et al.*, 2000; Thurston *et al.*, 1999); (2) the cells were plated on Matrigel, a matrix structure that contains the most abundant molecules present in basal lamina (e.g., laminin, Col IV, entactin), the ECM structure that underlies endothelium, whereas other systems have used collagen I and/or fibronectin, two types of ECM molecules that are present primarily in connective tissue (Culav *et al.*, 1999); (3) a very small form of dextran linked to FITC was used to allow for detection of permeability at very early stages when the gaps between EC are still very small, allowing our system to detect permeability events in minutes; (4) the FITC-dextran and the permeability-inducing agent were placed together in the lower chamber facing the ablumenal side of the EC; during inflammation the factors inducing permeability come from the tissue surrounding the vessels, and hence, impinge on the ablumenal side, that is, the side of EC that contacts the basal lamina. These aspects of our permeability assay ensure that the "endothelium" in this system mimics the endothelium *in vivo*, allowing more detailed and physiologic investigation of permeability mechanisms relevant *in vivo*, particularly those associated with pathologic inflammation and angiogenesis.

On the technical aspects of the work, one unusual step in our procedure is that we plate the cells twice at 24-h intervals to obtain a tight monolayer very quickly. We have developed this procedure because primary hMVEC do not survive well in culture for a long time; our procedure of plating twice accelerates the process of forming a tight monolayer. Furthermore, we applied the permeability detecting dye in the bottom chamber, because sample collection from the bottom chamber slightly displaced the upper chamber, disturbing the cells and leading to less consistent results. We do not see this strategy as problematic, because the sole purpose of this method is to monitor and quantify the passage of a detection agent from one side of the endothelium to the other to measure the permeability of the EC monolayer. This procedure has the additional advantage of requiring the FITC-dextran to diffuse upward, against the force of gravity, thereby reducing the possibility of false-positive results or increased background.

Our results show that plating the cells on collagen I or fibronectin does not render the monolayer of EC as tight as the layer we obtain when plating the cells on Matrigel (Fig. 8.1D). We believe that this is because Matrigel mimics the natural basal lamina, a matrix complex that is tightly associated with endothelial and epithelial sheets and confers on them the ability to develop an apical and a basal side, each with different properties, including the presence of specific growth factor receptors and adhesion molecules. Furthermore, this polarity is also critical for the establishment of proper adhesion junctions, in particular those junctions associated with the apical side of the cells, that are lost during increases in permeability. In short, by plating the cells on Matrigel, we obtain a system that more closely mimics the situation *in vivo*.

We have also confirmed the results of the *in vitro* permeability assay by examining changes in the endothelial monolayer morphology after treatment with PECAM-1 and f-actin staining (Fig. 8.2). After treatment with a permeability inducer, gaps appear between adjacent endothelial cells, correlating increased permeability with endothelial gap formation; the FITC-dextran tracer likely moves through these enlarged paracellular gaps to increase the amount of FITC-dextran passing through the endothelium, resulting in the observed increases in permeability. We have also used our *in vitro* permeability assay to explain novel signaling mechanisms involved in IL-8–induced permeability and have confirmed the importance of one such signaling molecule, the VEGF receptor, in permeability *in vivo* (Figs. 8.3 and 8.4). This strongly suggests that our *in vitro* permeability assay truly mimics the *in vivo* microvascular endothelium, and, as such, can be used to identify novel mechanisms involved in vascular permeability.

In summary, the *in vitro* system described here is capable of detecting very early changes in the permeability of a tight microvascular endothelial layer when the gaps between EC are still very small. This system is simple, sensitive, quantitative, reproducible, and, most importantly, it closely mimics *in vivo* endothelium. For example, the observed permeability increase occurs in minutes, much like *in vivo*. Taken together, our data show that this system can be successfully used to study the effects of various factors on endothelial permeability and the mechanisms involved in this process. In addition, this assay system may also be adapted to study epithelial barrier dysfunction in disorders of the lung, kidney, or intestinal epithelium. Furthermore, this system can potentially be used to test drugs for treatment of pathologic conditions involving increases in endothelial/epithelial permeability.

ACKNOWLEDGMENTS

We thank Qijing Li and Lina Wong, for their work in establishing the *in vitro* permeability assay system, and the colleagues in our laboratory for helpful discussions. This work was supported in part by the American Heart Association.

REFERENCES

Baldwin, A. L., and Thurston, G. (2001). Mechanics of endothelial cell architecture and vascular permeability. *Crit. Rev. Biomed. Eng.* **29,** 247–278.

Behzadian, M. A., Windsor, L. J., Ghaly, N., Liou, G., Tsai, N. T., and Caldwell, R. B. (2003). VEGF-induced paracellular permeability in cultured endothelial cells involves urokinase and its receptor. *FASEB J.* **17,** 752–754.

Bissell, M. J., and Radisky, D. (2001). Putting tumours in context. *Nat. Rev. Cancer* **1,** 46–54.

Breslin, J. W., Pappas, P. J., Cerveira, J. J., Hobson, R. W., 2nd, and Duran, W. N. (2003). VEGF increases endothelial permeability by separate signaling pathways involving ERK-1/2 and nitric oxide. *Am. J. Physiol. Heart Circ. Physiol.* **284,** H92–H100.

Carmeliet, P. (2003). Angiogenesis in health and disease. *Nat. Med.* **9,** 653–660.

Chandra, A., Barillas, S., Suliman, A., and Angle, N. (2006). A novel fluorescence-based cellular permeability assay. *J. Biochem. Biophys. Methods* **70,** 329–333.

Chen, J., Somanath, P. R., Razorenova, O., Chen, W. S., Hay, N., Bornstein, P., and Byzova, T. V. (2005). Akt1 regulates pathological angiogenesis, vascular maturation and permeability *in vivo*. *Nat. Med.* **11,** 1188–1196.

Chi, J. T., Chang, H. Y., Haraldsen, G., Jahnsen, F. L., Troyanskaya, O. G., Chang, D. S., Wang, Z., Rockson, S. G., van de Rijn, M., Botstein, D., and Brown, P. O. (2003). Endothelial cell diversity revealed by global expression profiling. *Proc. Natl. Acad. Sci. USA* **100,** 10623–10628.

Conway, E. M., Collen, D., and Carmeliet, P. (2001). Molecular mechanisms of blood vessel growth. *Cardiovasc. Res.* **49,** 507–521.

Culav, E. M., Clark, C. H., and Merrilees, M. J. (1999). Connective tissues: Matrix composition and its relevance to physical therapy. *Phys. Ther.* **79,** 308–319.

Cullen, V. C., Mackarel, A. J., Hislip, S. J., O'Connor, C. M., and Keenan, A. K. (2000). Investigation of vascular endothelial growth factor effects on pulmonary endothelial monolayer permeability and neutrophil transmigration. *Gen. Pharmacol.* **35,** 149–157.

Dvorak, H. F., Brown, L. F., Detmar, M., and Dvorak, A. M. (1995). Vascular permeability factor/vascular endothelial growth factor, microvascular hyperpermeability, and angiogenesis. *Am. J. Pathol.* **146,** 1029–1039.

Eliceiri, B. P., Paul, R., Schwartzberg, P. L., Hood, J. D., Leng, J., and Cheresh, D. A. (1999). Selective requirement for Src kinases during VEGF-induced angiogenesis and vascular permeability. *Mol. Cell* **4,** 915–924.

Feistritzer, C., Lenta, R., and Riewald, M. (2005). Protease-activated receptors-1 and -2 can mediate endothelial barrier protection: Role in factor Xa signaling. *J. Thromb. Haemost.* **3,** 2798–2805.

Folkman, J. (2002). Role of angiogenesis in tumor growth and metastasis. *Semin. Oncol.* **29,** 15–18.

Hashizume, H., Baluk, P., Morikawa, S., McLean, J. W., Thurston, G., Roberge, S., Jain, R. K., and McDonald, D. M. (2000). Openings between defective endothelial cells explain tumor vessel leakiness. *Am. J. Pathol.* **156,** 1363–1380.

Lal, B. K., Varma, S., Pappas, P. J., Hobson, R. W., 2nd, and Duran, W. N. (2001). VEGF increases permeability of the endothelial cell monolayer by activation of PKB/akt, endothelial nitric-oxide synthase, and MAP kinase pathways. *Microvasc. Res.* **62,** 252–262.

Lee, Y. C. (2005). The involvement of VEGF in endothelial permeability: A target for anti-inflammatory therapy. *Curr. Opin. Invest. Drugs.* **6,** 1124–1130.

Miller, D. W., Keller, B. T., and Borchardt, R. T. (1994). Identification and distribution of insulin receptors on cultured bovine brain microvessel endothelial cells: Possible function in insulin processing in the blood brain barrier. *J. Cell Physiol.* **161,** 333–341.

Miura, S., Matsuo, Y., and Saku, K. (2003). Transactivation of KDR/Flk-1 by the B2 receptor induces tube formation in human coronary endothelial cells. *Hypertension* **41,** 1118–1123.

Nobe, K., Sone, T., Paul, R. J., and Honda, K. (2005). Thrombin-induced force development in vascular endothelial cells: Contribution to alteration of permeability mediated by calcium-dependent and -independent pathways. *J. Pharmacol. Sci.* **99,** 252–263.

Petreaca, M.L., Yao, M., Liu, Y., Defea, K., and Martins-Green, M. (2007). Transactivation of vascular endothelial growth factor receptor-2 by interleukin-8 (IL-8/CXCL8) is required for IL-8/CXCL8-induced endothelial permeability. *Mol. Biol. Cell* **18,** 5014–5023. Epub 2007 Oct 10.

Pettersson, A., Nagy, J. A., Brown, L. F., Sundberg, C., Morgan, E., Jungles, S., Carter, R., Krieger, J. E., Manseau, E. J., Harvey, V. S., Eckelhoefer, I. A., Feng, D., Dvorak, A. M., Mulligan, R. C., and Dvorak, H. F. (2000). Heterogeneity of the angiogenic response induced in different normal adult tissues by vascular permeability factor/vascular endothelial growth factor. *Lab. Invest.* **80,** 99–115.

Seye, C. I., Yu, N., Gonzalez, F. A., Erb, L., and Weisman, G. A. (2004). The P2Y2 nucleotide receptor mediates vascular cell adhesion molecule-1 expression through interaction with VEGF receptor-2 (KDR/Flk-1). *J. Biol. Chem.* **279,** 35679–35686.

Stolz, D. B., Bannish, G., and Jacobson, B. S. (1992). The role of the cytoskeleton and intercellular junctions in the transcellular membrane protein polarity of bovine aortic endothelial cells *in vitro*. *J. Cell Sci.* **103,** 53–68.

Sun, H., Breslin, J. W., Zhu, J., Yuan, S. Y., and Wu, M. H. (2006). Rho and ROCK signaling in VEGF-induced microvascular endothelial hyperpermeability. *Microcirculation* **13,** 237–247.

Tanimoto, T., Jin, Z. G., and Berk, B. C. (2002). Transactivation of vascular endothelial growth factor (VEGF) receptor Flk-1/KDR is involved in sphingosine

1-phosphatestimulated phosphorylation of Akt and endothelial nitric-oxide synthase (eNOS). *J. Biol. Chem.* **277,** 42997–43001.

Thuringer, D., Maulon, L., and Frelin, C. (2002). Rapid transactivation of the vascular endothelial growth factor receptor KDR/Flk-1 by the bradykinin B2 receptor contributes to endothelial nitric-oxide synthase activation in cardiac capillary endothelial cells. *J. Biol. Chem.* **277,** 2028–2032.

Thurston, G., Suri, C., Smith, K., McClain, J., Sato, T. N., Yancopoulos, G. D., and McDonald, D. M. (1999). Leakage-resistant blood vessels in mice transgenically overexpressing angiopoietin-1. *Science* **286,** 2511–2514.

Varma, S., Breslin, J. W., Lal, B. K., Pappas, P. J., Hobson, R. W., 2nd, and Duran, W. N. (2002). p42/44MAPK regulates baseline permeability and cGMP-induced hyperpermeability in endothelial cells. *Microvasc. Res.* **63,** 172–178.

Weis, S. M., and Cheresh, D. A. (2005). Pathophysiological consequences of VEGF-induced vascular permeability. *Nature* **437,** 497–504.

CHAPTER NINE

ASSAYS OF TRANSENDOTHELIAL MIGRATION *IN VITRO*

William A. Muller* *and* F. William Luscinskas[†]

Contents

Abstract

The inflammatory response is critical for our ability to heal wounds and fight off foreign microorganisms. Uncontrolled inflammation is also at the root of most pathologic conditions. Recruitment of leukocytes to the site of inflammation plays a defining role in the inflammatory response, and migration of leukocytes across endothelium is arguably the point of no return of the inflammatory response. Assays to study the transmigration of leukocytes have and will continue to shed light on the regulation of this vital response. Assays of transendothelial migration *in vitro* allow the controlled observation of this phenomenon, as well as experiments to study its regulation. In this chapter, we describe *in vitro* assays of transendothelial migration that have been used

* Department of Pathology, Northwestern University Feinberg School of Medicine, Chicago, Illinois
† Department of Pathology, Brigham and Women's Hospital and Harvard Medical School, Center for Excellence in Vascular Biology, Boston, Massachusetts

Methods in Enzymology, Volume 443
ISSN 0076-6879, DOI: 10.1016/S0076-6879(08)02009-0

successfully in the authors' laboratories for decades and have proven to be reproducible, reliable, and predictive of the behavior of leukocytes and endo-thelial cells in models of inflammation *in vivo*.

1. INTRODUCTION

Leukocyte emigration, a critical step in the inflammatory response, involves a sequential series of leukocyte–endothelial interactions of increas-ing strength that serve to arrest the leukocyte on the endothelium at the site of inflammation (Butcher, 1991) that it may subsequently enter the tissues. Transendothelial migration, or diapedesis, is the step in leukocyte extrava-sation in which the leukocyte squeezes in ameboid fashion across the endothelial cell, usually passing between tightly apposing endothelial cells at their borders (Muller, 2003) or under certain conditions, leukocytes passing through the endothelial cell (reviewed in Rao *et al.* (2007). Many ways are available to model interactions of leukocytes with endothelial cells *in vitro*. Although none of these perfectly recapitulates conditions of the site of inflammation *in vivo*, they do permit a reductionist approach that allows direct observation, quantitative analysis, and manipulation in ways that are currently impossible to attain in the living organism. Furthermore, because human leukocytes and endothelial cells are easily prepared, these methods allow direct experimentation with the actual molecules that would be relevant *in vivo*.

This chapter outlines the assays we have used successfully over the past decades for studying transendothelial migration under static and shear flow conditions. These are by no means the only available methods. However, they have proven reliable, reproducible, and predictive of results later obtained *in vivo*. We also provide methods for the isolation of leukocytes and preparation of the endothelial cells and substrata for these assays.

2. METHODS FOR INVESTIGATING LEUKOCYTE TRANSENDOTHELIAL MIGRATION UNDER STATIC OR FLOW CONDITIONS *IN VITRO*

Transendothelial migration assays carried out under static conditions have become very popular because of the ease with which they can be performed. Although the static model described lacks the physiologic variable of fluid shear, it has the advantage that the endothelial cells are attached to a physiologic basement membrane—a factor that significantly affects endothelial phenotype and may be of even greater importance for the diapedesis event itself. It is suitable for testing multiple variables (e.g., blocking antibodies or inhibitors) at a single time and can even be adapted for high-throughput screening.

The method described here uses endothelial cells plated on hydrated three-dimensional collagen gels generally in 96-well tray format (Muller and Weigl, 1992) and grown to confluence in normal adult human serum in the absence of endogenous growth factors (Fig. 9.1.) Under these

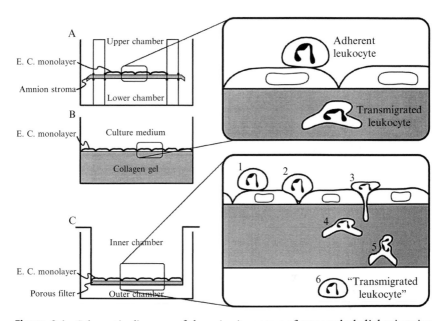

Figure 9.1 Schematic diagram of three *in vitro* assays of transendothelial migration (not drawn to scale.) (A) Endothelial cells (E.C.) cultured on the upper surface of a sheet of amniotic stroma (Furie *et al.*, 1987) that separates two fluid-filled chambers. (B) Endothelial cells cultured on hydrated type I collagen gels overlaid with fibronectin (Muller and Weigl, 1992). Components of the culture medium penetrate into the porous gel. (C) Endothelial cells grown on the upper surface of a porous polycarbonate filter suspended in a larger culture vessel. Culture medium is placed in the inner and outer chambers to reach the apical and basal surfaces of the monolayer, respectively. In methods (A) and (B), adherent leukocytes remaining on the apical surface can be distinguished visually from those that have transmigrated. Procedures have also been devised to strip off any leukocytes remaining bound to the apical surface of the monolayer (Muller and Weigl, 1992). In the filter-chamber method (C), the percentage of the leukocytes added to the upper chamber that appear in the lower chamber is calculated by direct counting. This is technically much easier and faster. However, to be counted as "transmigrated," a leukocyte must (1) attach to the apical surface of the endothelium, (2) migrate to the intercellular junction, (3) diapedese between the endothelial cells, (4) detach from the endothelial cells and penetrate their basal lamina, (5) crawl through the filter itself, and (6) detach from the filter and fall into the chamber below. Reagents that block any of these steps will, therefore, block the readout of transmigration in this system. (This figure is reproduced from Muller [1996] by permission of R.G. Landes Company.)

conditions, endothelial cells grow slowly and show better growth arrest at confluence, better representing the quiescent endothelium lining postcapillary venules at a site of inflammation than a rapidly proliferating monolayer driven by endothelial growth factors.

A variation of this method has been used to study leukocyte adhesion with fluorescently or radioactively labeled leukocytes (Muller and Weigl, 1992). Although this allows rapid assessment of how many leukocytes firmly adhered to the endothelium during the assay, it does not distinguish between those leukocytes that are firmly bound on the apical surface and those that have transmigrated. This is a problem, for example, when assessing the effects of blocking PECAM or CD99, where the arrested leukocytes are firmly adherent but do not migrate across the endothelial monolayer. Thus, we rely on microscopic analysis of the monolayers to distinguish the step in transmigration at which cells are arrested. This is a more tedious analysis but has allowed us to distinguish that interfering with PECAM homophilic interactions arrests leukocytes at the apical surface of the endothelial cells over the cell borders, whereas blocking heterophilic interaction of leukocyte PECAM domain 6 arrests the leukocyte between the basal surface of the endothelial cell and the basement membrane (Liao et al., 1995), as well as to realize that blocking CD99 interactions arrests leukocytes partway through the endothelial junctions (Lou et al., 2007; Schenkel et al., 2002).

Assays that use Transwells® are popular because of their ease of use. These are Boyden chamber–like devices that sit in a tissue culture well. Endothelial cells are grown on top of a porous filter that separates two fluid compartments. Leukocytes are added to the upper chamber and a chemoattractant to the lower chamber. Leukocytes migrate across the endothelial cells, though the filter, and are recovered from the lower chamber and counted. However, because of their cost, analyzing multiple samples simultaneously can be expensive. We have not used this assay extensively, because we find that with prolonged culture endothelial cells grow into the pores, complicating analysis of the assay. In addition, the monolayers are difficult to visualize on these filters before fixation and staining. Most relevant, if one is assessing the regulation of transmigration, recovering fewer leukocytes from the lower well does not distinguish among the many potential steps at which the block might occur as described in Muller (2001) (see Fig. 9.1.).

Cultured endothelial cells produce chemokines and other leukocyte chemoattractants at low levels, even in the absence of activation by inflammatory cytokines (Muller and Weigl, 1992). These accumulate in the culture medium and in the underlying collagen matrix. When the overlying culture medium is washed away, the chemoattractants in the gel diffuse out, creating a chemotactic gradient for the leukocytes to follow into the collagen gel. Activation of endothelial cells with appropriate proinflammatory cytokines (e.g., IL-1β, TNF-α) or bacterial endotoxin (lipopolysaccharide, LPS) induces the expression of additional adhesion molecules and chemokines that enhance the activation of leukocyte adhesion and decrease the time it will take for a population of leukocytes to migrate across the endothelial monolayer.

By forming collagen gels within cloning cylinders on top of coverslips, the static transmigration assay has been adapted to allow visualization of cells in the acts of adhesion, locomotion, and transmigration (Lou *et al.*, 2007; Schenkel *et al.*, 2004). However, these assays still lack fluid shear, which is particularly important for the initial adhesion step.

Transmigration assays carried out under flow conditions mimic *in vivo* fluid dynamics within blood vessels of the microcirculation. In these assays transendothelial migration is examined in the context of the multistep adhesion cascade where leukocytes undergo initial attachment and rolling (step 1) and arrest and spreading (step 2) before transendothelial migration (step 3) (reviewed in Butcher (1991). Hence, the biochemical and biophysical processes that govern these steps can be fruitfully investigated with *in vitro* flow chamber assays. The parallel plate flow chamber is the design most frequently used for this purpose and is the design used in the authors' laboratories (Goetz *et al.*, 1999; Lawrence *et al.*, 1987; Luscinskas *et al.*, 1994). A commercially produced flow chamber (GlycoTech Inc, Rockville MD; http://www.glycotech.com/apparatus/parallel.html) is manufactured in two configurations. Although we have not used this system, it has been used fruitfully by several laboratories to study leukocyte–endothelial cell interactions (see Brown and Larson [(2001] and references therein). We have designed and manufactured a parallel plate chamber and recently reported the details of this chamber (Goetz *et al.*, 1999). In our laboratory's flow chamber leukocyte–endothelial cell adhesive events are monitored in live time and recorded to videotape or by digital videomicroscopy to a PC for later analysis (Lim *et al.*, 2003). A recent advancement is the use of time-lapse digital microscopy in combination with fluorescence microscopy. This has allowed us to investigate of EGFP-tagged molecules transfected into leukocytes or endothelium during the process of leukocyte adhesion and transendothelial migration (Shaw *et al.*, 2001; 2004).

The most important advantage of the flow assay system is that it mimics *in vivo* flow and hence allows one to study the distinct steps of leukocyte initial attachment, rolling, arrest, and transmigration. Second, certain cell types such as unfractionated T cells and eosinophils (Cinamon *et al.*, 2001; Cuvelier *et al.*, 2005) transmigrate significantly better in the presence of shear flow and preincubation of the monolayers with apical chemokines. (However, *in vitro* activated memory T lymphoblasts migrate efficiently under static conditions (Carman *et al.*, 2007).).

Those investigators unable to commit the time and resources to the setup and maintenance of a flow chamber system for studies of transendothelial migration need not be too upset. Everything that we have published that used the static transmigration system described here has been borne out *in vivo*. This includes the role of Mac-1 in locomotion (Phillipson *et al.*, 2006; Schenkel *et al.*, 2004), roles for PECAM (Liao *et al.*, 1997, 1999; Muller *et al.*, 1993), and CD99 (Bixel *et al.*, 2004; 2007; Schenkel *et al.*, 2002) in transmigration, but not adhesion, and the effects of prolonged (days) culture of transmigrated monocytes in our system on their differentiation into immature dendritic cells and monocytes (Randolph *et al.*, 1998; 1999). Furthermore, studies on transmigration of myeloid cells show that flow may increase the rate of transmigration quantitatively, but does not change any aspects *qualitatively* (Cuvelier and Patel, 2001; Kitayama *et al.*, 2000).

Despite the broad applicability for study of molecular mechanism and biophysical features of transmigration, the flow chamber setup described has some disadvantages. First, most flow studies use endothelial cells conditioned in low physiologic shear ($\sim$2 dynes/cm^2) for a very short time (minutes) so that the endothelium has not been preconditioned to physiologic levels of shear stress. Although studies have shown that junctional complexes are reorganized in the early hours after the application of 10 to 12 dynes/cm^2 sheer stress and do not reestablish equilibrium for 2 days (Noria *et al.*, 1999), no information is available on the status of junctional complexes after short application of low shear stress. Second, there is much evidence to suggest that interactions with the basement membrane are at least as important to the physiology of the endothelium as flow along the apical surface. In the static collagen gels system, endothelial cells are grown on a three-dimensional fibrillar matrix, with an endothelial-derived basement membrane. Cells grown on glass coverslips clearly lack this. Third, a limited number of samples can be simultaneously tested, and complete analysis of rolling speed, adhesion, and TEM is time consuming, which reduces the usefulness for large-scale drug screening.

2.1. Preparation of hydrated collagen gels for transmigration assays (Muller *et al.*, 1989)

2.1.1. Reagents

Vitrogen 100 (95 to 98% bovine type I collagen) from Cohesion Corp., Palo Alto (877) 264-3746
10× Medium 199
0.1 N NaOH (sterile filtered)

2.1.2. Procedure

1. Keep all ingredients cold (4 °C). Work in laminar flow hood. Keep the tube in which the reagents are mixed on ice in the hood until you are able to perform all manipulations sufficiently quickly. However,

polymerization of collagen gels is not nearly as rapid as solidification of Matrigel®, for example.

2. Mix the ingredients in the following proportions:
   8 ml Vitrogen
   1 ml 10× M199 or 10× Dulbecco's phosphate-buffered saline
   5 ml 0.1 N NaOH

3. Mix well; avoid making air bubbles. The mixture should have a magenta color from the phenol red. The pH is 9 to 10 (see note 1).

4. Pipette the desired volume of collagen solution into the culture vessel:

6-mm well (96-well microtiter)	50 μl/well
16-mm well (24-well cluster)	300 μl/well
30-mm dish	1 ml/dish
100-mm dish	8 ml/dish

   For smaller volumes, automatic repeating pipetters work well. Pipette slowly, because the solution is viscous, to make sure that the proper volume is dispensed (see notes 2 and 3).

5. Place collagen coated dishes in 37 °C incubator for approximately an hour.

6. The gelled collagen should still be pink from the phenol red and have a whitish tone.

7. Cover with ≥2 volumes of M199 (1×) or culture medium and let stand overnight in the tissue culture incubator to equilibrate with the gel. In this way, the gel and the absorbed medium is at pH 7.4 when the cells are added (see note 4).

8. We routinely coat the gels with human fibronectin (the same as would be used to coat a tissue culture dish) to get somewhat better plating efficiency. Alternately, let the gels stand in culture medium with human serum for an hour or so before plating the cells. In our experience, commercial sources of gelatin contain levels of endotoxin (LPS) too high to use with endothelial cells and leukocytes to avoid unintentional activation.

9. We grow HUVEC on these gels in M199 + 20% normal adult human serum without added growth factors. HUVEC grow slowly under these conditions. (This is what you want: A model of a stable postcapillary endothelium with a low turnover rate.) For best results do not expand cultures when transferring to the gels (i.e., split 1:1). We have split 1:2 and 1:3 with good results at times, but inconsistently. For example, resuspend the HUVEC to 3 × 10^5 cells/ml and plate 100 μl in each well on a 96-well plate (see note 5).

NOTES:

1. The optical properties of the gel are better if it polymerizes at an alkaline pH. The phenol red will indicate an alkaline color; this is OK. There is

plenty of time to neutralize the gel later. We have tested it extensively, and there are no physiologic or biochemical differences between HUVEC grown on collagen that gels at alkaline pH vs neutral pH, as long as the gels are allowed to equilibrate with neutral pH solution later, as described later. However, if you wish to incorporate a pH-sensitive bioactive molecule into the collagen before polymerization, make up the collagen at pH 7.4 with 1 part (1 ml in the previous recipe) NaOH.

2. If you are plating in larger dishes (e.g., 30-mm dishes or larger), wash the culture vessels with sterile fluid (PBS, Hanks', etc.) to wet the bottoms and allow for even spreading of the collagen. Aspirate this just before adding collagen.

3. Most of our experience is with gels made in microtiter wells. We fill the center 60 wells leaving the outer perimeter blank.

4. The company that sells Vitrogen has turned over many times in the past two decades. Although the product seems to be consistent, we have noticed some important differences. In the 1980s and 1990s, HUVEC plated on collagen gels after overnight equilibration in M199 grew well. Starting in 2000, we noticed that the gels had to equilibrate for several days to get the best monolayers. (Between 4 days and 3 weeks did not seem to make a difference.) Starting with lots received in 2005, we have again been able to get nice monolayers on gels equilibrated only overnight. We recommend that you test this timing when you first try plating your cells. If you allow them to equilibrate for several days or more, we have found that placing water or medium in the bare wells surrounding the ones containing the gels will keep the atmosphere in the tray humid enough so that the outer wells will not dry out.

5. When growing cells on collagen gels in microtiter trays, visualizing the cells by routine inverted phase-contrast microscopy is difficult, because the light bouncing off the sides of the well distorts the optics. Only a small circle of cells in the center of the field will be in good phase relief. Once the monolayer is confluent, however, occasional cells that do not cease dividing will be forced out of the monolayer where they will undergo apoptosis. When we see a few round, granular, apoptotic cells on top of our otherwise smooth monolayer, we can tell that the HUVEC have been confluent for several days and are ready to use in transmigration assays. The apoptotic cells wash off easily and do not interfere with the assays.

2.2. Isolation of peripheral blood mononuclear cells (PBMC) (Muller and Weigl, 1992)

This assay is an adaptation of the standard Ficoll-Hypaque procedure optimized to remove platelets that tend to stick to and co-purify with monocytes. It was initially developed for experiments to study the interaction of monocytes with unactivated endothelial cells. Under the conditions of the transmigration assay (see later) lymphocytes do not adhere efficiently to unactivated endothelial monolayers and the few that do adhere do not transmigrate efficiently within the assay time period (20 to 60 min). Therefore, the endothelial cells do a better job of purifying monocytes than any man-made purification scheme (Muller and Weigl, 1992). Monocytes can also be purified by negative selection with antibody cocktails and magnetic bead columns; however, one runs the risk of activating them. Unactivated monocytes will not adhere to the sides of the centrifuge tubes (see step 17). For more quantitative assessment, we examine CD62L and CD11b expression by flow cytometry. The former should remain high, and the latter should not increase during the isolation. A positive control for activation can be obtained by deliberately activating the monocytes with fMLP, for example, for comparison.

2.2.1. Reagents

Ficoll-Paque Plus (endotoxin free) (Pharmacia) 10 to 15 ml/20 ml blood, room temperature.
Hanks' balanced salt solution without divalent cations (HBSS).
EDTA (2.7% = 100 mM), pH 7.4, 1 ml/10 ml of blood.
Bent, siliconized Pasteur pipettes (see note 1).
Hemacytometer
Human serum albumin (HSA), endotoxin-free and suitable for infusion into humans. This can be obtained from a local blood bank or hospital pharmacy.
HBSS + 0.1% human serum albumin (HBSS/HSA) at 4 °C.
Medium 199 + 0.1% human serum albumin (M199/HSA) at 4 °C.

2.2.2. Procedure

1. Place 12 to 15 ml of Ficoll-Paque at the bottom of a 50-ml conical tube. Leave enough room for 40 ml of blood/HBSS. Make one gradient for every 20 ml of blood to be drawn.
2. Load a syringe with 1 ml EDTA for every 10 ml blood to be drawn (see note 2). Draw blood into this syringe.
3. Mix blood/EDTA gently with an equal volume of warm HBSS.
4. Carefully layer the mixture over the Ficoll retaining a crisp interface. Running the mixture slowly down the side of the tube is a good way to avoid disturbing the interface.
5. Centrifuge at 2200 rpm (~1000g), 20 min at room temperature with the brake OFF. At the end of the spin there should be an upper clear orange layer (plasma + HBSS/HSA), a cloudy interface containing PBMC, a clear band of Ficoll-Paque below that, and a pellet of RBC and PMN at the bottom.

6. Remove $\geq$12 ml of plasma from the top of the tube. Transfer it to a 15-ml conical tube on ice. (The plasma will be used in a later step for washing the monocytes.)

7. Rinse the Pasteur pipette several times with HBSS/HSA by pipetting the liquid in and out. Discard the liquid.

8. Rinse the Pasteur pipette several times with plasma from what is now at the top of the 50-ml conical tube. Discard this liquid.

9. Blow the air out of the Pasteur pipette and hold the latex bulb shut as you carefully lower the Pasteur pipette to the interface between the plasma and the Ficoll. Do not blow bubbles into the gradient. There will probably be a crust of white cells loosely adhering to the side of the tube along the interface. This is very rich in PBMC. Harvest this first by gradually releasing pressure on the latex bulb and taking in the cells and surrounding liquid. Transfer the cells to a 50-ml tube containing 40 ml of cold HBSS.

10. Repeat step 9. This time, lower the open end of the pipette to the interface and sweep it along at the level of the interface as you gradually release pressure on the bulb. Aspirate the cells at the interface into the pipette. Transfer them to the tube containing 40 ml of cold HBSS (see note 3). Repeat until the interface is mostly free of leukocytes (cloudiness).

11. Centrifuge the 50-ml tube containing the cells and the 15-ml tube containing the plasma at 1500 rpm for 10 min at 4 °C (see note 4).

12. Aspirate the supernate from the PBMC pellet. Draw 10 ml of platelet-poor plasma from the 15-ml tube taking care not to disturb the platelet pellet.

13. Resuspend the PBMC in the 10 ml plasma. Pipette gently, avoiding air bubbles or directly spraying the stream of plasma onto the PBMC.

14. Transfer 10 to 20 μl to a hemacytometer. Count the cells and assess platelet contamination.

15. Centrifuge at 1000 rpm for 5 min at 4 °C.

16. Resuspend in 10 ml of cold HBSS/HSA.

17. Repeat steps 14 to 16 until only rare platelets are seen in the hemacytometer and monocytes appear free of platelets. This usually takes two to three total washes. Typically, some of the leukocytes are lost with each wash. However, this is usually less than 5 to 10% of the total. Larger losses than this on successive washes could be a sign that the monocytes are being activated during the procedure and sticking to tubes or pipettes.

18. Resuspend to the desired concentration in the desired medium. For the transendothelial migration assay, we resuspend to 2×10^6 cells/ml in M199/HSA.

19. Assess percentage of monocytes (see note 5).

NOTES:

1. Bend the thin end of a short siliconized Pasteur pipette to a right angle in a low Bunsen burner flame. Fire-polish the tip briefly. This will allow you to harvest PBMC from the interface without collecting fluid from the upper or lower layers. Alternately, we have used disposable one-piece plastic pipetting bulbs. However, it is harder to keep the tip of these at the interface. In either case, they must be precoated with HBSS/ HSA and plasma.
2. Depending on the donor, you will isolate 2 to 4 × 10^6 PBMC per ml of blood. We find that 20 ml (one gradient) is more than enough to perform the transmigration assay with 60 HUVEC monolayers on microtiter plates. Often we can get enough for two such plates (120 assays) from one gradient.
3. If you run more than one gradient, they can all be combined into the same tube of HBSS at this point.
4. The PBMC fraction will be considerably contaminated by platelets. This includes platelets adherent to monocytes. To remove them, you will wash in autologous platelet-free plasma containing EDTA in step 15.
5. There are many ways to assess the percentage of monocytes. We have found that allowing an aliquot of the final suspension to sit on the hemacytometer for approximately 5 min will promote spreading of the monocytes to phase dense cells, whereas most lymphocytes will remain round and refractile. Then one simply takes the total cell count and large spread cells = monocytes; small round cells = lymphocytes. Obviously, this is not perfect, but when tested head to head with flow cytometry (LeukoGate®; CD14 and CD45), the results were within experimental error. For most donors the percentage of monocytes will range from 20 to 30% of the total PBMC.

2.3. Isolation of peripheral blood neutrophils (Lou *et al.*, 2007)

For our adhesion and transmigration assays, it is important that the neutrophils (PMN) are activated by contact with the endothelium and not before. Unless care is taken minimize physical disruption or temperature changes to the neutrophil preparation during the isolation procedure, PMN may become activated, and background adhesion and transmigration will be high. In this isolation all steps are carried out at room temperature until the transmigration step, when the cocultures are placed in a 37 °C tissue culture incubator. Low numbers of remaining red blood cells, which do not interfere with the transmigration assay, are not lysed to avoid osmotic shock to the PMN.

As a test of the status of quiescence of isolated PMN, we sometimes plate endothelial monolayers derived from the same cultures as those used for the transmigration assay directly on plastic 96-well plates and assess PMN adhesion with or without prior activation by TNFα or IL-1β. PMN should not adhere to nonactivated endothelial cells, but adhere tightly to activated endothelium.

2.3.1. Reagents

Ficoll-Hypaque ($\rho = 1.077$) Amersham Pharmacia, as for PBMC isolation.
Histopaque 1119 ($\rho = 1.119$) Sigma Aldrich.
Pharmaceutical grade heparin solution.
Hanks' balanced salt solution without divalent cations (HBSS).
Medium 199 (M199).
Precoat all surfaces that will come in contact with the PMN with a solution
  of HBSS + 0.1% HSA (see PBMC isolation protocol). Avoid the use of
  glass. All media are at room temperature.

2.3.2. Procedure

1. Pipette 12 ml Histopaque into a 50-ml conical centrifuge tube.
2. Carefully overlay 10 ml of Ficoll-Paque.
3. Draw heparin into a syringe so that there will be 2 units of heparin/ml of blood.
4. We routinely collect 10 ml of blood for each gradient. At the same time, draw 10 ml of blood into a glass tube to clot for the production of autologous serum.
5. Dilute heparinized blood 1:1 in HBSS.
6. Layer it carefully atop the discontinuous gradient.
7. Centrifuge a 2200 rpm at room temperature for 20 min with the brake off.
8. Harvest neutrophils from the interface between the Ficoll and Histopaque ($\rho = 1.077/1.119$).
9. Dilute into 10 ml of HBSS + 0.1% HSA.
10. Centrifuge at 1000 rpm, 5 min at room temperature.
11. Gently resuspend the PMN pellet in 10 ml HBSS + 0.1% HSA.
12. Repeat step 9.
13. Resuspend in HBSS + 0.1% HSA + 1% heat-inactivated autologous serum to a concentration of 3 to 5×10^6 PMN/ml.
14. Allow PMN to "rest" for 20 to 30 min at room temperature.
15. Dilute PMN in M199 + 20% autologous serum to a concentration of 5×10^5 cells/ml for transmigration. Blocking antibodies or other reagents are added at this step.

16. The transmigration assay is performed as outlined in the following. The PMN are warmed to 37 °C gradually when the culture plate is put into the incubator.

2.4. Transendothelial migration assay (Muller and Weigl, 1992; Muller *et al.*, 1993)

Resting (not cytokine-activated) monolayers will support adhesion and transmigration of monocytes, but almost no lymphocytes. Therefore, for nonactivated monolayers, the entire PBMC fraction can be applied, yet only monocytes will adhere or transmigrate to any significant extent. Cytokine-activated (IL-1β or TNF-α) or LPS HUVEC monolayers will support adhesion of lymphocytes from the PBMC mixture, but only monocytes will transmigrate to any significant extent for up to several hours. Cytokine-activated monolayers are required to support optimal PMN and T-cell adhesion and transmigration.

2.4.1. Reagents

Human umbilical vein endothelial cells (HUVEC) grown to confluence on hydrated collagen gels. (See separate procedure for making gels.) (See note 1.)

Peripheral blood mononuclear cells (PBMC) or neutrophils (PMN). See separate procedures for isolation.

Medium 199 (GIBCO) containing 0.1% human serum albumin (M199/HSA).

Hanks' balanced salt solution (without divalent cations, HBSS).

Dulbecco's phosphate-buffered saline (DPBS) with calcium and magnesium.

10% neutral buffered formalin OR 2.5% glutaraldehyde in 0.1 M sodium cacodylate buffer pH 7.4.

Wright–Giemsa stain (several kits are available commercially under names like "Diff-Quik" or "Hema").

27-guage needle.

Microscope slides and coverslips.

2.4.2. Procedure

1. Resuspend PBMC to a final concentration of 2×10^6/ml in cold M199/HSA (see note 2). Resuspend PMN to a final concentration of 1×10^6/ml in M199/HSA at room temperature.

2. Take HUVEC cultures from the incubator and wash three times with warm M199/HSA. This is best done either by aspirating with a Pasteur pipette connected to a vacuum flask or by inverting the plate and shaking gently. In the latter case, blot gently to remove excess liquid

from the top of the plate. Refill the wells quickly, ideally with a repeating micropipetter. Do not let the monolayers dry out.

Aspirate and refill the wells by placing the pipette tip at an obtuse angle against the wall of the culture vessel to deflect the force of the stream. Never pipette directly onto the endothelial monolayer surface, because the force may dislodge or damage the endothelial cells. We have also found that when a repeating pipetter is used for these washes, placing a small (10- to 200-μl size) disposable pipette tip on the end of the cartridge results in a narrow stream of fluid that has sufficient force to dislodge cells even when directed against the wall of the culture vessel. Therefore, we recommend against capping the cartridge with a disposable pipette tip.

3. Aspirate the last wash and add 100 μl of leukocyte suspension to each well (see note 3).

4. Return the culture plate to the tissue culture incubator for the desired time. Use a separate culture plate for each time point.

5. At the end of the assay, aspirate the remaining supernate. Wash three times with warm DPBS (200 μl per wash) (see note 4).

6. Add fixative (200 μl) to the gels *in situ*. Fix in formalin for routine microscopy; glutaraldehyde for EM, or to obtain firmer (but more brittle) gels. Allow to fix for at least an hour (preferably overnight) before removing monolayers for counting.

   If depth of penetration into the collagen is something you will want to assess, fix in glutaraldehyde, because those gels are resistant to collapse under their own weight or that of the coverslip (see steps 11 and 12), preserving better the original dimensions.

7. Stain the gels with modified Wright–Giemsa stain for several minutes by adding stain and washes directly to the monolayers in the wells. If leukocytes have migrated deeply into the collagen gel, it may require more than 10 min for the stain to penetrate to them.

8. Rinse gels with distilled water.

9. Loosen the gels from the well by inserting a 27-guage needle along the sidewall of the well and "rimming" the gel with the needle. Press outward along the wall while doing this to avoid crushing the gel (see note 5).

10. Remove the gels with fine (No. 5) forceps. Place monolayer side up on a glass slide. Gels can be manipulated by carefully tugging on the edges with fine forceps or the 27-guage needle. Try to smooth out any visible wrinkles in the gel and restore its original shape.

11. Cover with a coverslip. Avoid trapping air bubbles.

12. Examine by light microscopy. Differential interference contrast microscopy (DIC/Nomarsky optics) is best to throw the monolayer into relief. By focusing on the endothelial monolayer, one can determine the number of leukocytes attached to the surface of the HUVEC (in focus

when the surface of the monolayer is in focus) or transmigrated (out of focus when the surface of the monolayer is in focus, but in focus as one brings the focal plane down below the monolayer). You will be able to see collagen strands around the transmigrated leukocytes.

13. We routinely count at least 100 cells in the central field of each monolayer for each of six replicate monolayers per condition. If there are not more than 100 cells in a single field, cells in multiple fields are counted.

14. We express the percent transmigration as the number of leukocytes in each field that have transmigrated (i.e., below the HUVEC monolayer) divided by the total number of leukocytes in that field that had adhered (i.e., above and below the HUVEC monolayer) (see note 6).

NOTES:

1. It is convenient to grow HUVEC on collagen gels in microtiter wells in a 96-well plate format. We routinely culture cells in the center 60 wells on such a plate, allowing six replicates (columns) of ten experimental conditions (rows) at the same time.

2. If you plan to add antibodies or other reagents to block or modify transmigration, we have found it best to resuspend the leukocytes to twice this concentration and mix them with an equal volume of antibody or reagent at twice its final concentration. This avoids exposing leukocytes locally (albeit transiently) to high concentrations of reagent that could be activating or have other untoward effects. Stock antibody concentrations are often 100 to 1000 times the final concentration in the assay.

3. If testing several variables, we find it convenient to test six replicates of each condition, so that all of the wells in a single column receive the same leukocyte sample. Change pipette tips between samples.

4. When specifically trying to distinguish adhesion to the apical surface by integrins from blockade of transmigration by anti-PECAM reagents, for example, we wash two times with warm 1 mM EDTA in HBSS to remove cells bound by divalent cation-dependent interactions. Then we wash two times with warm DPBS.

5. It is good to only stain as many gels as you will count in one sitting, because the stain fades with time. The gels can be restained. Only remove as many gels as you will count on one slide, because they tend to flatten out and dry out once removed from the wells. If they become too dried out or flat, it will become more difficult to distinguish how far into the gel they have migrated.

6. In the absence of inhibitors, >90% of monocytes or neutrophils that adhere to these monolayers generally transmigrate within 20 to 30 min. The time for transmigration per se is rapid (minutes). It takes at least 20 min for the population of leukocytes to settle down on the monolayer.

2.5. Flow chamber studies

The parallel plate flow chamber used in our laboratory consists of an upper and a lower plate made of stainless steel (see Goetz *et al.* [1999] for detailed description and Fig. 9.2). The upper plate has an inlet and outlet for inflow and outflow of buffer, a bubble trap (see note 1), an opening for a temperature probe, and a circular cutout containing a piece of quartz glass. The quartz glass gives efficient passage of a wide spectrum of light, making it possible to perform quantitative fluorescence and bright-field microscopy, and it obviously allows monitoring of adhesive events within the flow chamber (Goetz *et al.*, 1997; Shaw *et al.*, 2001; 2004). The lower plate has a circular cutout, where a 25-mm diameter glass coverslip, which is coated with an endothelial cell monolayer or immobilized adhesion proteins (e.g., P- or E-selectins), is placed. A plastic O-ring is placed between the lower plate of the flow chamber and the coverslip to prevent buffer leaking from the flow chamber. A heating plate of the same dimensions as the flow chamber is used to maintain the appropriate temperature. The upper and lower plates of the flow chamber are separated by a silastic gasket (Allied Biomedical; Paso Robles, CA), which defines the flow area (5.0 × 80 mm, and height, 0.25 mm; see note 2), within the flow chamber. The position and orientation of the silastic gasket are maintained by locator pins, which are embedded in the upper plate of the flow chamber. Buffer is drawn through the flow chamber at defined volumetric flow rates with a syringe pump (Harvard Apparatus, model 44; Natick, MA).

2.5.1. Video recording equipment

The flow chamber is mounted on an inverted microscope equipped with 20× and 40× phase or DIC objectives and camera/videoports. Leukocyte attachment, rolling, spreading, and migration are best observed at higher magnification (40×). Differential interference contrast (DIC or Nomarsky) microscopy is ideal for detecting transmigration of T cells and monocytes. The video port allows live-time viewing by means of a black-and-white video camera connected to a PC equipped with a digital recorder. Alternately, the video camera is connected to a black-and-white monitor and professional grade VCR.

Live cell fluorescence imaging is performed with MetaMorph software (Molecular Devices, Downingtown, PA) to control a digital imaging system coupled to a Nikon model TE2000 inverted microscope (Melville, NY). The details of this system are found in Shaw *et al.* (2004): separate excitation and emission filter wheels (Sutter Instrument Co., Novato, CA) combined with a polychroic beamsplitter (Chroma Technology, Brattleboro, VT) to allow for rapid acquisition of different fluorescence channels. Chroma filters for DsRed and YFP are used for Alexa Fluor-568 (red) and GFP/Alexa

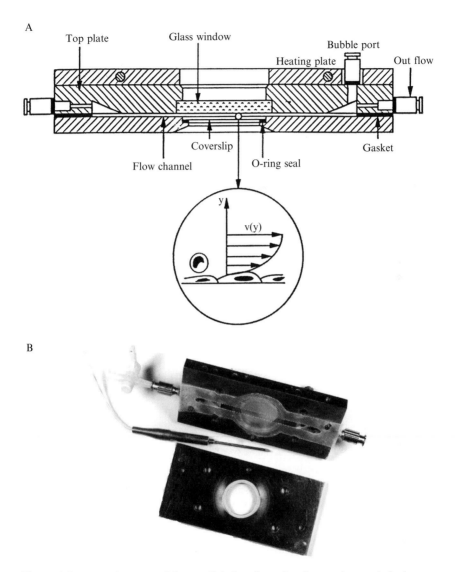

Figure 9.2 (A) Schematic of the parallel plate flow chamber used to study leukocyte transmigration of cytokine-activated HUVEC monolayers under fluid shear stress. (B) Photograph of the top (with gasket in place) and bottom plates that are assembled to create the flow chamber. The temperature probe is placed in the top plate. Reprinted from Goetz *et al.* (1999).

Fluor-488 dyes (green), respectively. Transmitted and fluorescent light illumination are controlled by either a Uniblitz electronic shutter (Vincent Associates, Rochester, NY) or a shutter built in the excitation filter wheel. We have had good success with a high-sensitivity cooled CCD camera

(ORCA-ER, Hamamatsu Co., Bridgewater, NJ) to acquire images with the MetaMorph software.

2.5.2. Flow assay media

Most physiologic cell buffers can be used in the flow assay; however, it is important that the selected buffer contains Ca^{2+}, because some of the adhesion processes (i.e., selectin mediated interactions) are Ca^{2+} dependent. Our current perfusion media consists of DPBS containing 0.75 mM Ca^{2+}, 0.75 mM Mg^{2+}, and 0.5% human serum albumin (Yang et al., 2006).

2.5.2. Methods
2.5.2.1. Preparation for the flow chamber assay

1. Place a finger from a latex glove around microscope objectives for protection from excess flow buffer that causes deposit of salts on microscope components and corrosion of the objectives.
2. Warm the flow buffer to desired temperature by incubating in water bath for several hours (see note 3). Preheat the flow chamber to working temperature. Insert the thermal probe into the upper plate and heat upper plate to the desired temperature.
3. Place the silastic gasket on the underside of the upper plate. The locator pins are used to position the gasket and align the gasket over the quartz glass. Make sure that the gasket is smooth and its width is constant.
4. Place the O-ring into the bottom plate and insert the coverslip containing either an endothelial cell monolayer or immobilized adhesion molecules on top of the O-ring in the bottom plate. The top plate is then placed over the coverslip, and four screws are used to seal the top and bottom plates together. Apply gentle and uniform tightening of these four screws to avoid cell damage or cracking of the glass coverslip. Mount the heating plate on top of assembled chamber.
5. Assemble the inflow and outflow lines to the chamber. Secure outlet to a 50-ml glass syringe pump and set syringe pump to desired flow rate.
6. Place a videotape in the VCR and record the dimensions of the screen with a microscope stage micrometer. Record under objective magnifications to be used in the experiments.

2.5.3. Analyses of leukocyte-endothelial adhesion events

1. Adhesion: Leukocyte adhesion data are reported as a function of the shear stress at the bottom surface of the flow chamber (see note 4). Routinely, we perfuse a small volume of leukocytes (10^6 isolated leukocytes in 100 μl of flow buffer) at 0.5 dynes/cm^2 and then once leukocytes attach, the flow rate is increased to 1 dyne/cm^2 and buffer alone is perfused for an additional 10 min. The number of leukocytes bound per unit area is

determined by counting the adherent cells/mm^2 in five randomly selected microscope fields. Quantification of transendothelial migration with phase contrast or DIC microscopy is as previously described (Rao *et al.*, 2004).

2. Rolling velocities: A general feature of leukocyte–endothelial cell interactions under flow is that the adherent leukocytes routinely exhibit rolling behavior in response to the fluid shear flow that is characterized by a low-velocity, high-variance movement (Goetz *et al.*, 1994) (see note 5). Rolling velocities can be determined either manually (see note 6) or by automated software (see note 7). In either case, the goal is to determine the displacement, d, of the leukocyte over a period of time, t (see note 8). The velocity is then easily calculated by v = d/t.

3. Approaches to block leukocyte rolling, adhesion, or transmigration: A general approach consists of the use of function blocking mAb to known adhesion molecules or the use of siRNA silencing to knockdown specific molecules in either endothelium or leukocyte. Both or either cell types can be pretreated with blocking mAb for at least 30 min before performing the flow study. siRNA knockdown should be optimized for each molecule and cell type.

NOTES:

1. Bubble trap: A third port on top of the plate was added to facilitate removal of air bubbles that arise during assembly of the chamber.

2. The gasket we use is Duralistic I Sheeting of a thickness of 0.25 mm. This thickness defines the approximate height of the flow chamber. From the Duralistic I Sheeting, cut a gasket with an inner dimension of 5 mm (width) by 80 mm (length). The width of the flow field should be large relative to the height (e.g., $\geq$20:1 ratio), so that the flow may be approximated as unidirectional and the fluid velocity is spatially dependent only on the distance from the bottom surface of the flow chamber. Observations should be acquired in the middle of the flow chamber, away from the edges. With the assumption of unidirectional, laminar flow, the equation for the shear rate at the bottom surface of the flow chamber is given by $\gamma = 6Q/wh^2$, where $Q =$ the volumetric flow rate, $w =$ the width of the flow field, and $h =$ the height of the flow field (Lawrence *et al.*, 1987). The shear stress, τ, is related to the shear rate, γ, by $\tau = \gamma\mu$ where μ is the viscosity of the fluid used in the experiments, typically 0.007 poise for a dilute saline solution at 37 °C. Thus, $\tau = 6Q\mu/wh^2$.

3. Our experiments are performed with the flow buffer and flow chamber at 37 °C. When the buffer is preincubated in the 37 °C heating bath for an hour or less, bubbles in the flow chamber are more common than when the preincubation is done for several hours.

4. Adhesion under flow involves a balance between the disruptive force and torque exerted on the leukocyte by the flow of the fluid and a specific

adhesive force and torque mediated by complementary receptor pairs on the surface of the adherent leukocyte and the endothelium (Hammer and Apte, 1992). The force and torque exerted by the flow of the fluid, which can be estimated from various theoretical models (Hammer and Apte, 1992), are related to the shear rate, or shear stress, at the leukocyte-endothelium interface. Therefore, *in vitro* adhesion data are usually reported as a function of the shear stress at the leukocyte-endothelium interface, which is the shear stress at the bottom surface of the flow chamber.

5. The term low velocity refers to the fact that the leukocytes roll at a velocity routinely <10% of the predicted velocity of a noninteracting leukocyte (Goetz *et al.*, 1994). The theoretical velocity of a noninteracting leukocyte can be estimated from the work of Goldman and colleagues (Goldman *et al.*, 1967). This value depends on the radius of the leukocyte, the shear rate at the fluid coverslip interface, and the distance the leukocyte is from the surface of the coverslip. High variance velocity refers to the fact that the leukocytes do not roll with a constant velocity but rather accelerate and decelerate as they roll across the monolayer surface (Goetz *et al.*, 1994).

6. In manual mode, the displacement of the leukocyte can be determined either by recording the experiment with a calibrated reticle in place or by placing a template over the screen during the analysis. One approach is to determine the time required for leukocytes to roll a given distance and calculate the rolling velocity.

7. Automated image analysis can be done with NIH Image or commercially available software (Ed Marcus Laboratories, Newton, MA).

8. As leukocytes roll over the endothelium, they may detach from the endothelium and then reattach several cell diameters downstream from the point of detachment. These are usually referred to as skipping leukocytes and are routinely not included in the data set used to determine rolling velocities.

ACKNOWLEDGMENTS

We thank Jian Shen, C. Forbes Dewey, and Peter Morley (Massachusetts Institute of Technology; Cambridge, MA) for the design, fabrication, and maintenance of the flow chambers; Professor Douglas Goetz (Ohio University) for redesign for the flow chamber during his postdoctoral studies in the Luscinskas laboratory. This work was made possible by grant support from the National Institutes of Health, Bethesda, MD (HL 36028 and HL 53993 to F. W. L. and HL46849, 64774, and 72942 to W. A. M.).

REFERENCES

Bixel, G., Kloep, S., Butz, S., Petri, B., Engelhardt, B., and Vestweber, D. (2004). Mouse CD99 participates in T cell recruitment into inflamed skin. *Blood* **104,** 3205–3213.

Bixel, M. G., Petri, B., Khandoga, A. G., Khandoga, A., Wolburg-Buchholz, K., Wolburg, H., Marz, S., Krombach, F., and Vestweber, D. (2007). A CD99-related antigen on endothelial cells mediates neutrophil, but not lymphocyte extravasation *in vivo*. *Blood* **109,** 5327–5336.

Brown, D. C., and Larson, R. S. (2001). Improvements to parallel plate flow chambers to reduce reagent and cellular requirements. *BMC Immunol.* **2,** 9.

Butcher, E. C. (1991). Leukocyte-endothelial cell recognition: Three (or more) steps to specificity and diversity. *Cell* **67,** 1033–1036.

Carman, C. V., Sage, P. T., Sciuto, T. E., de la Fuente, M. A., Geha, R. S., Ochs, H. D., Dvorak, H. F., Dvorak, A. M., and Springer, T. A. (2007). Transcellular diapedesis is initiated by invasive podosomes. *Immunity* **26,** 784–797.

Cinamon, G., Shinder, V., and Alon, R. (2001). Shear forces promote lymphocyte migration across vascular endothelium bearing apical chemokines. *Nature Immunol.* **2,** 515–522.

Cuvelier, S. L., and Patel, K. D. (2001). Shear-dependent eosinophil transmigration on interleukin 4-stimulated endothelial cells: A role for endothelium-associated eotaxin-3. *J. Exp. Med.* **194,** 1699–1709.

Cuvelier, S. L., Paul, S., Shariat, N., Colarusso, P., and Patel, K. D. (2005). Eosinophil adhesion under flow conditions activates mechanosensitive signaling pathways in human endothelial cells. *J. Exp. Med.* **202,** 865–876.

Furie, M. B., Naprstek, B. L., and Silverstein, S. C. (1987). Migration of neutrophils across monolayers of cultured microvascular endothelial cells. An *in vitro* model of leucocyte extravasation. *J. Cell. Sci.* **88(Pt 2),** 161–175.

Goetz, D. J., el-Sabban, M. E., Pauli, B. U., and Hammer, D. A. (1994). Dynamics of neutrophil rolling over stimulated endothelium *in vitro*. *Biophys. J.* **66,** 2202–2209.

Goetz, D. J., Greif, D. M., Ding, H., Camphausen, R. T., Howes, S., Comess, K. M., Snapp, K. R., Kansas, G. S., and Luscinskas, F. W. (1997). Isolated P-selectin glycoprotein ligand-1 dynamic adhesion to P- and E-selectin. *J. Cell Biol.* **137,** 509–519.

Goetz, D. J., Greif, D. M., Shen, J., and Luscinskas, F. W. (1999). Cellcell adhesive interactions in an in vitro flow chamber. *Methods Mol. Biol.* **96,** 137–145.

Goldman, A., Cox, R., and Brenner, H. (1967). Slow viscous motion of a sphere parallel to a plane wall. II Couette Flow. *Chem. Eng. Sci.* **22,** 653–660.

Hammer, D. A., and Apte, S. M. (1992). Simulation of cell rolling and adhesion on surfaces in shear flow: General results and analysis of selectin-mediated neutrophil adhesion. *Biophys. J.* **63,** 35–57.

Kitayama, J., Hidemura, A., Saito, H., and Nagawa, H. (2000). Shear stress affects migration behavior of polymorphonuclear cells arrested on endothelium. *Cell. Immunol.* **203,** 39–46.

Lawrence, M. B., McIntire, L. V., and Eskin, S. G. (1987). Effect of flow on polymorphonuclear leukocyte/endothelial cell adhesion. *Blood* **70,** 1284–1290.

Liao, F., Ali, J., Greene, T., and Muller, W. A. (1997). Soluble domain 1 of platelet-endothelial cell adhesion molecule (PECAM) is sufficient to block transendothelial migration *in vitro* and *in vivo*. *J. Exp. Med.* **185,** 1349–1357.

Liao, F., Huynh, H. K., Eiroa, A., Greene, T., Polizzi, E., and Muller, W. A. (1995). Migration of monocytes across endothelium and passage through extracellular matrix involve separate molecular domains of PECAM-1. *J. Exp. Med.* **182,** 1337–1343.

Liao, F., Schenkel, A. R., and Muller, W. A. (1999). Transgenic mice expressing different levels of soluble platelet/endothelial cell adhesion molecule-IgG display distinct inflammatory phenotypes. *J. Immunol.* **163,** 5640–5648.

Lim, Y. C., Garcia-Cardena, G., Allport, J. R., Zervoglos, M., Connolly, A. J., Gimbrone, M. A., Jr., and Luscinskas, F. W. (2003). Heterogeneity of endothelial cells from different organ sites in T-cell subset recruitment. *Am. J. Pathol.* **162,** 1591–1601.

Lou, O., Alcaide, P., Luscinskas, F. W., and Muller, W. A. (2007). CD99 is a key mediator of the transendothelial migration of neutrophils. *J. Immunol.* **178**, 1136–1143.

Luscinskas, F. W., Kansas, G. S., Ding, H., Pizcueta, P., Schleiffenbaum, B. E., Tedder, T. F., and Gimbrone, M. A., Jr. (1994). Monocyte rolling, arrest and spreading on IL-4activated vascular endothelium under flow is mediated via sequential action of L-selectin, beta 1-integrins, and beta 2-integrins. *J. Cell Biol.* **125**, 1417–1427.

Muller, W. A. (1996). Transendothelial migration of leukocytes. *In* "Leukocyte Recruitment in Inflammatory Disease." (G. Peltz, ed.), pp. 3–18. R.G. Landis Company, Austin, TX.

Muller, W. A. (2001). Migration of leukocytes across endothelial junctions: Some concepts and controversies. *Microcirculation* **8**, 181–193.

Muller, W. A. (2003). Leukocyte-endothelial-cell interactions in leukocyte transmigration and the inflammatory response. *Trends Immunol.* **24**, 326–333.

Muller, W. A., Ratti, C. M., McDonnell, S. L., and Cohn, Z. A. (1989). A human endothelial cell-restricted, externally disposed plasmalemmal protein enriched in inter-cellular junctions. *J. Exp. Med.* **170**, 399–414.

Muller, W. A., and Weigl, S. (1992). Monocyte-selective transendothelial migration: Dissection of the binding and transmigration phases by an *in vitro* assay. *J. Exp. Med.* **176**, 819–828.

Muller, W. A., Weigl, S. A., Deng, X., and Phillips, D. M. (1993). PECAM-1 is required for transendothelial migration of leukocytes. *J. Exp. Med.* **178**, 449–460.

Noria, S., Cowan, D. B., Gotlieb, A. I., and Langille, B. L. (1999). Transient and steady-state effects of shear stress on endothelial cell adherens junctions. *Circ. Res.* **85**, 504–514.

Phillipson, M., Heit, B., Colarusso, P., Liu, L., Ballantyne, C. M., and Kubes, P. (2006). Intraluminal crawling of neutrophils to emigration sites: A molecularly distinct process from adhesion in the recruitment cascade. *J. Exp. Med.* **203**, 2569–2575.

Randolph, G. J., Beaulieu, S., Lebecque, S., Steinman, R. M., and Muller, W. A. (1998). Differentiation of monocytes into dendritic cells in a model of transendothelial traffick-ing. *Science* **282**, 480–483.

Randolph, G. J., Inaba, K., Robbiani, D. F., Steinman, R. M., and Muller, W. A. (1999). Differentiation of phagocytic monocytes into lymph node dendritic cells *in vivo*. *Immunity* **11**, 753–761.

Rao, R. M., Betz, T. V., Lamont, D. J., Kim, M. B., Shaw, S. K., Froio, R. M., Baleux, F., Arenzana-Seisdedos, F., Alon, R., and Luscinskas, F. W. (2004). Elastase release by transmigrating neutrophils deactivates endothelial-bound SDF-1alpha and attenuates subsequent T lymphocyte transendothelial migration. *J. Exp. Med.* **200**, 713–724.

Rao, R. M., Yang, L., Garcia-Cardena, G., and Luscinskas, F. W. (2007). Endothelial-depen-dent mechanisms of leukocyte recruitment to the vascular wall. *Circ. Res.* **101**, 234–247.

Schenkel, A. R., Mamdouh, Z., Chen, X., Liebman, R. M., and Muller, W. A. (2002). CD99 plays a major role in the migration of monocytes through endothelial junctions. *Nat. Immunol.* **3**, 143–150.

Schenkel, A. R., Mamdouh, Z., and Muller, W. A. (2004). Locomotion of monocytes on endothelium is a critical step during extravasation. *Nat. Immunol.* **5**, 393–400.

Shaw, S. K., Bamba, P. S., Perkins, B. N., and Luscinskas, F. W. (2001). Real-time imaging of vascular endothelial-cadherin during leukocyte transmigration across endothelium. *J. Immunol.* **167**, 2323–2330.

Shaw, S. K., Ma, S., Kim, M. B., Rao, R. M., Hartman, C. U., Froio, R. M., Yang, L., Jones, T., Liu, Y., Nusrat, A., Parkos, C. A., and Luscinskas, F. W. (2004). Coordinated redistribution of leukocyte LFA-1 and endothelial cell ICAM-1 accompany neutrophil transmigration. *J. Exp. Med.* **200**, 1571–1580.

Yang, L., Kowalski, J. R., Zhan, X., Thomas, S. M., and Luscinskas, F. W. (2006). Endothelial cell cortactin phosphorylation by Src contributes to polymorphonuclear leukocyte transmigration *in vitro*. *Circ. Res.* **98**, 394–402.

CHAPTER TEN

CELL MECHANICS AT MULTIPLE SCALES

Maxine Jonas,* Peter T. C. So,† *and* Hayden Huang‡

Contents

Abstract

The response of cells to mechanical stresses is a field of growing inquiry. It is well known that both the morphologic and molecular expression of cells depend, in part, on the local mechanical environment, especially for cells such as endothelial cells that experience shear stress, stretch, and pressures. To systematically study the large variety of responses of cells to physical forces

* BioTrove, Massachusetts
† Departments of Mechanical and Biological Engineering, Massachusetts Institute of Technology, Cambridge, Massachusetts
‡ Brigham and Women's Hospital, Cambridge, Massachusetts

Methods in Enzymology, Volume 443
ISSN 0076-6879, DOI: 10.1016/S0076-6879(08)02010-7

(e.g., signaling, adhesion, or stiffness changes), a number of techniques have been developed and used. Here we present methods for three types of cell mechanical studies, from the multicellular to the subcellular scales, and describe the basic principle and main use of each technique along with some design and setup considerations.

1. INTRODUCTION

Several cell types respond to mechanical stresses through changes in morphology and molecular signaling. The study of tissues, such as bone and heart that bear high mechanical stresses, clearly requires a good understanding of the mechanosensory pathways relevant to their cells and the local environment. However, mechanotransduction is also relevant in tissues where the mechanical stresses are comparatively lower. For example, the vasculature permeates the entire body and determines local blood perfusion, and the responses of endothelial cells to fluid shear flow are well known and subject to continuing investigation.

In this chapter, we present three techniques for studying cell mechanics at multiple scales. Although endothelial cells are emphasized in this volume, the methods described in this chapter can be used in a wide variety of cell types, and some of the sample data presented will involve other cell types. However, it should be noted that endothelial cells' responses to mechanical forces are not limited to those under shear stress. In general, an understanding of the behavior of endothelial cells requires an understanding of their responses under broad classes of mechanical stress. As a result, it is likely that angiogenesis, along with other endothelial-dependent processes (atherosclerosis, neutrophil adhesion, and paracellular migration, etc.) will remain incompletely characterized until mechanical influences are better understood.

The first section covers large-scale stretching, because flow-mediated assays will be covered in another chapter. This section will be brief, because this technique is relatively well established and well described in the literature (Brown, 2000; Dartsch *et al.,* 1986; Kanda *et al.,* 1992; Osol, 1995). The second section will focus on magnetic micromanipulation with an electromagnet, which allows assays of cell mechanics and adhesion on the scale of individual cells. The final section will focus on a recent advance in particle-tracking microrheology with fluorescent markers called fluorescence laser tracking microrheology (FLTM) to measure the mechanical properties of cells within a micron-scale vicinity of a tracer particle.

Finally it should be noted that this volume focuses mainly on methods as opposed to presenting an exhaustive review. Thus, the references cited mainly serve as illustrative examples.

2. Large-Scale Cell Stretching

Cost: ~$10,000 (not including incubator).

Experimental duration: seconds to days, typically at least 5 min to several hours.

Readouts: molecular changes by means of Western and Northern blotting or real-time PCR. Imaging of cell alignment and immunolabeling can also be performed.

Scale: Typically, signals (molecular or otherwise) from several hundreds to millions of cells are summed or averaged, depending on substrate size and readout. Morphology can be assayed on a single-cell basis.

Two broad categories of membrane-based stretching exist. Uniformly biaxial stretching can be performed with a piston on a fixed membrane and has the advantage that no preferred direction is induced. Uniaxial stretching can be performed with lateral stretching, but care must be taken to decide whether the free side is constrained or not. For endothelial cells, constrained stretch is a more physiologic approach. However, for generic mechanical response, uniformly biaxial strain may provide the lowest variance in readout. A piston-based uniformly biaxial stretching device developed in our laboratory is shown in Fig. 10.1.

Membrane stretch has been used to study molecular signaling in response to the applied strain in a variety of cells (Chien, 2006; Husse et al., 2007; Leung et al., 1976). It has also been used to assess the reorientation of cells in response to directional strain (Kanda and Matsuda, 1994; Kanda et al., 1992). Similar to endothelial elongation parallel to the direction of applied shear stress, studies have found that endothelial cell

Figure 10.1 A piston membrane stretch device is used to apply a uniformly biaxial strain to cells plated on flexible membranes. The pistons are coated with a lubricant to let the membrane slide during applied stretch.

cytoskeleton remodels perpendicular to the direction of applied (uniaxial) strain (Chien, 2006).

2.1. Instrument design

Thin silicone membranes (e.g., 0.005″) can be coated with desired adhesion molecules (for example, we dilute stock fibronectin to a final concentration of 1 to 4 μg/ml in buffer such as Hanks balanced salt solution (HBSS), and the silicone membrane is incubated overnight in this solution at 4 °C for coating) and then plated with cells (typically to attain full confluence within a day or two, depending on the experiment and cell type). The membrane is held in a holder ring that is slightly larger than the piston (care must be taken to ensure the diameters are not too different to avoid unstrained flexure of the membrane). To obtain uniform strain across the membrane, the piston and membrane are coated with some lubricant (such as Braycote-804 from Castrol). The stretching device is normally placed in a cell culture incubator to maintain physiologic temperatures and pH. To maintain sterility for long-term experiments, the membrane assembly can be covered; however, care must be taken to avoid the formation of a seal. If a seal is formed, the air in the chamber above the membrane can be evacuated when the piston pushes up; when the piston moves back down, a temporary vacuum may be formed that will distort the strain of the membrane. Commercial devices exist that use air pressure under or over the membrane to induce strains, but similar precautions should be used regardless of the flexing mechanism.

2.2. Experimental procedures

Typical strains induced are from 1 to 10%; some cells can withstand higher strains, but apoptosis will prevent the prolonged application of significantly higher strain in many cases. Typical frequencies range from 0.1 to 10 Hz. Although it is possible to go outside of this range, the mechanical response of the device, the potentially increased vibrations, and the contribution of incidental fluid flow will complicate analysis unless accounted for.

To harvest the cells after strain, the membrane is washed with either PBS or HBSS and then harvested with a kit or cell lysis buffer for Western analysis (2 ml of 1 M Tris-HCl, 1 ml Triton-X 100, 1 ml of 10% SDS, 3 ml of 5 M NaCl, 200 μl of 0.5 M EDTA, and then add water to 100 ml). Performing Northern or Western blotting can illuminate the responses of cells to applied stretch (Feng *et al.*, 1999; Yamamoto *et al.*, 2001). Cells can also be fixed immediately with 4% paraformaldehyde in PBS for immuno-histochemistry. The membrane can be sliced and placed on a microscope slide to minimize reagent use if desired. One simple way to do this is to place the slide under the membrane and excise the membrane with a razor, rather than attempting to cut out a section of the membrane and manipulate it in air onto a slide.

3. MAGNETIC MICROMANIPULATION

Cost: <$10,000 (not including a bright-field microscope).
Experimental duration: seconds to minutes.
Readouts: Cell stiffness, adhesion, and tether formation. With appropriate markers, it may be possible to induce or examine cell signaling, intracellular strain, etc.
Scale: Single-cell level, but can also examine a small groups of cells if desired.

Particle manipulation is useful for probing cells at an individual level. The motion of beads or other small probes adhered to the cell surface can reveal cellular mechanical strain in response to exerted stress. Two major methods for biologic particle manipulation are optical trapping and magnetic micromanipulation. These techniques are complementary, with optical trapping imposing fixed displacements on a particle and magnetic micromanipulator imposing fixed forces. With feedback control, it is possible to regulate either position or force level with either technique. Optical trapping currently generates lower forces compared with magnetic micromanipulation, because high laser power may induce cell damage by heating. However, optical traps allow the simultaneous and independent manipulation of multiple particles. Optical traps may also allow the manipulation of some organelles or unicellular organisms (Hormeno and Arias-Gonzalez, 2006; Nieminen *et al.*, 2007). Magnetic micromanipulation generates higher forces but requires seeding with magnetic beads. In addition, although configurations exist to apply similar forces to all beads in a small area, it is difficult to exert independent forces on separate beads within the same microscopy field of view.

Because many cellular mechanotransduction and mechanics studies require fairly high force, we will focus on magnetic micromanipulation. Several variants of magnetic micromanipulation exist. Typical magnetic micromanipulators use electromagnets for actuation. Although it is possible to use permanent magnets for experiments, electromagnets provide the ability to alter forces in very short time frames. Electromagnets also have no moving parts for periodic force generation. As a result, we will focus on electromagnet design in this section, but it should be noted that permanent magnets could be used for similar studies.

Among electromagnetic designs, there are single-pole, multiple-pole, and twisting configurations (Bausch *et al.*, 1998; Fabry *et al.*, 1999; Huang *et al.*, 2002; Wang and Ingber, 1995). Multiple poles can offer a more uniform force field, usually at the cost of force level. Twisting has been used for several studies on cell mechanical properties, and, similar to multiple-pole configurations, has the benefit of having a multiple-cell readout. However, because the single-pole configuration can usually provide higher forces and is more relevant for certain cell mechanical studies, we focus on the single-pole design in this section.

3.1. Instrument design

An electromagnet has three basic components: the power supply (which can be a simple on/off device or, preferably, can be programmed to control the current level conducting through the electromagnet), the magnet core, which can be a ferromagnetic material with high permeability, and the coils, which are wires wound around the core to induce magnetic fields.

The core should be shaped roughly like a pencil. There should be a long region of constant radius around which the coils can be wound (flanges at the two ends may help in the winding process). The tip should be sharp and, if desired, may be chiseled. Chiseling creates a slightly more uniform field near the tip region, whereas a sharp tip will have larger directional variances. The length of the tip should be short enough so that it is close to the coils but long enough that the generated field lines can concentrate at the tip. There is considerable leeway in determining this length; finite element simulations can help, but in general, for magnets of 6 inches (15 cm), a tip length of 1 to 2 inches (2 to 5 cm) is likely acceptable for a broad range of applications (Fig. 10.2).

Because the electromagnet is basically a solenoid, the larger the number of coils per unit length, the higher the magnetic field and the higher the force that can be generated, within an operating range. Increasing the

Figure 10.2 A custom-designed electromagnet for applying forces on magnetic beads. The ruler is in centimeters for reference; the actual size should depend on individual spatial and force requirements. The gold coating is for protection against corrosion from constant immersion into salt media. There are approximately 300 coils of wire over five layers.

number of coils per unit length can be accomplished by use of thinner wire or by overlapping the wires, creating several layers. Wiring more coils in a tighter space should be tempered against the heat generated by the wires. For short bursts of force, this is not likely to be an issue; however, for sustained readouts, heat dissipation may become a problem. The magnet should be placed in a micromanipulator for precise translational and angular control. Care should be taken to avoid exposed wires, because of the high current level, for both safety and instrument-preservation purposes. The wires can be connected to a variable power supply. If desired, resistance measurements may be taken to estimate the voltage or current required; for a typical-sized magnet shown in Fig. 10.2, with 300 coils, resistance is ~10 ohms for 20-gauge copper wire.

3.2. Experimental procedures

Magnetic beads can be purchased commercially from a number of places, or can be manufactured in-house. Coating the beads is typically accomplished with standard techniques (activated surface receptors, passive adsorption, or use of a precoated secondary antibody). One of the major advantages of magnetic micromanipulation is this ability to selectively coat the beads, which allows one to apply mechanical stresses to a specific type of receptor (e.g., beta-1 integrins) or class of receptor (e.g., collagen receptors) of interest. Use of adhesion molecules such as fibronectin or collagen results in high-affinity binding; however, some issues with bead clumping may reduce the efficiency of the bead seeding if single beads are desired. Antibodies can be used against specific receptors, but care should be taken to choose the correct antibody (e.g., targeted against an external epitope).

3.3. Calibration

Calibration of the electromagnet can be done with the magnetic beads being used. A high-viscosity fluid (e.g., dimethylpolysiloxane at 12,500 centistokes, or cS) is often used to calibrate the electromagnet for high-force applications. Resuspend an aliquot of uncoated magnetic beads (or BSA-coated beads) in 70% ethanol after a brief wash in buffer. Place a small volume (e.g., 1 to 2 μl) of beads solution, containing a few thousand beads, in a 30-mm dish. After air-drying, pour approximately 1 to 2 ml of the high-viscosity fluid onto the beads and then stir with a pipette tip to resuspend the beads. Place the dish on the microscope stage, lower the electromagnet into the fluid, and wait 5 to 10 min for the beads to settle. Once the bead is no longer visibly moving, engage the magnet at a fixed current and allow the bead to move through the region of interest. The bead centroid can be tracked over time, and its velocity can be calculated from the trajectory data. The velocity and force are related for low Reynolds number flow by Stokes' law:

$$F = 3\pi D \mu V \qquad (10.1)$$

where F is the force on the bead, D is the bead diameter, μ is the dynamic viscosity of the suspension fluid, and V is the velocity of the bead. To ensure this formula is applicable, the Reynolds number (Re) should be much lower than 1:

$$Re = \rho D V / \mu \qquad (10.2)$$

where ρ is the density of the suspension fluid. As an example, with 5 μm diameter magnetic beads with a specific gravity of 2 (noting that in a fluid, the effective density will be reduced because of buoyancy) in dimethylpolysiloxane at 12,500 cS viscosity, the bead will drop through the fluid at ~1 nm/s because of gravity. It is important to ensure that the calibration procedure takes less than a few minutes at this drop rate, so that gravity will not cause a significant vertical drift of the bead. Under a 10 nN force, the bead will have a (terminal) velocity of 17 μm/s and have a Reynolds number of ~7e-9, which is significantly lower than unity, and thus Stokes' law applies. Because the magnetic field strength is nonuniform but increases rapidly in the vicinity of the magnet tip, the velocity changes as the bead approaches the tip. Therefore, the bead is near terminal velocity for only slowly accelerating beads. Typically, distances of less than 10 to 20 μm away from the tip are not used, because the bead may undergo significant acceleration, and there may be a significant vertical force component.

3.4. Cell-based experiments

For cell culture experiments, cells can be plated and allowed to attach overnight in cell culture dishes (30-mm dishes, for example). The cells are typically seeded with a low density of beads (roughly 1 bead for every 1000 cells to 1 bead for every cell) for single-bead manipulation. Beads should be allowed to attach only as long as is necessary to ensure a good population of adhered beads. This generally lasts from 30 min to an hour. Prolonged incubation of beads on the cells may result in bead endocytosis (particularly for matrix-coated beads), which will interfere with adhesion measurements. After incubation, the cells can be gently washed once with buffer or media and then placed in fresh cell culture media for experiments. The cells should be maintained in a temperature-controlled environment wherever possible. For most experiments, a 20× or 40× objective is sufficient for beads sized 1- to 5-μm in diameter. After the cells are placed on the microscope stage, the magnetic micromanipulator is positioned near a bead. A bead should be chosen that is farther than 3 to 4 bead diameters away (preferably >10 diameters) from the next closest bead to minimize interference because of mutual attraction. To assess the binding of the bead, the microscope stage may be tapped (preferably, if the microscope is on a vibration isolation unit, that unit can be pressed and released). The bead will exhibit obvious rocking motion if it is not securely attached. This test is necessary, particularly for adhesion measurements, because no reasonable amount of bead washing with media or buffer will remove all unbound beads. If the bead is well attached, the magnet tip can be brought within close proximity to the bead (the distance should be determined by the force desired on the basis of a prior calibration) and be activated. The response of the bead to the applied magnetic force can be recorded by acquiring a movie.

For pure adhesion measurements, complete detachment of a bead qualifies as a loss of adhesion, as does the bead moving over one bead diameter, which typically suggests rolling, tethering, or reattachment after initial detachment. (This assumes a thin monolayer of cells or sparsely plated cells; if the cells or tissues of interest are very thick, the allowable range of motion before detachment can vary considerably). Tracking the magnetic bead with cross-correlation or centroid measurements allows the force response of the cell to be monitored, allowing quantitative modeling of cell stiffness (plain displacement, spring-and-dashpot or, if sine-wave forces were applied, rheologic power-law or other decomposition may be used, see Fig. 10.3 for sample bead tracking data). Note that for measuring cell stiffness, use of confluent monolayers of cells is advised because of the increased variability in cell thickness and local properties (such as actin density) associated with cell location (e.g., protrusions versus nuclei) in sparsely plated cells.

One common issue with magnetic micromanipulation is the residual magnetic field. Sometimes, the act of simply moving the magnet into the field of view on the microscope is sufficient to detach the bead of interest. For strong adhesion measurements, this is unlikely to be problematic, because the residual forces are typically much lower than the applied force under high currents. For low-force applications or experiments where the residual force significantly influences the results, this issue can be resolved with a gauss meter to provide feedback. Alternately, running the current backward through the electromagnet can significantly reduce the residual field, although the level of residual field must be determined empirically.

Results from magnetic micromanipulation can reveal fundamental roles for receptors and proteins. For example, it was found that disruption of the actin cytoskeleton results in a decrease in stiffness of the cell, in a receptor- and cell-type–dependent manner (Huang et al., 2001; 2006). Furthermore, in endothelial cells, it was recently discovered that cell–cell receptors involved in junctional regulation can influence not only cell migration but also adhesion (Huang et al., 2006). In the junctional study, it was also found that

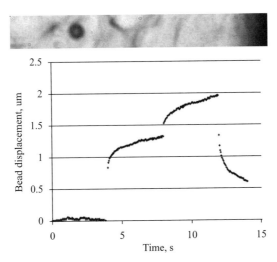

Figure 10.3 An image taken during a magnetic micromanipulation experiment is shown at top. The bead is the dark circle approximately one fourth of the way from the left edge. The cell monolayer (HEK 293 cells) can be seen throughout the background of the image. The shadow on the right side is the shadow caused by the electromagnet tip. The magnet was first left off for 4 s, then turned on half peak current (0.75 amps) for four sec, then at full peak current (1.5 amps) for 4 s, then turned off. Note the two upwards jumps in bead displacement at 4 and 8 s corresponding to the changes in force and the recoil after the force is removed.

confluence of the cells might influence the adhesion properties of the cells. The ability to obtain confluence-dependent results is one of the strengths of magnetic micromanipulation. Detachment of magnetic beads depends mainly on the properties of attachment of the bead to the cell surface. Other adhesion assays such as fluid shear assays or microplate-based assays can be more difficult to interpret because of the influence of cell–cell adhesion. For example, in fluid shear assays, if part of a confluent monolayer detaches, there is a chance that a portion of the detached sheet will "parachute" and peel away at adjacent cells, which will muddle the cell–matrix adhesion readouts.

4. FLUORESCENCE LASER TRACKING MICRORHEOLOGY

Cost: $50,000 (not including fluorescent microscope).
Experimental duration: seconds to minutes.
Readouts: Cell rheology (stiffness and viscosity) as a function of frequency.
Scale: Subcellular, one part of one cell at a time.

To probe cellular response at the nanoscale level, a laser tracking microrheometer was developed (Yamada *et al.*, 2000). Recently, fluorescence laser tracking microrheology (FLTM) was shown to allow quantitative studies of cytoskeletal mechanotransduction with fluorescent tracers (Jonas *et al.*, 2008b). FLTM monitors the brownian motion of a tracer particles in its local environment. The amplitude of a particle's displacement is related to the viscoelastic properties of its local environment, with increased viscosity or stiffness in the microenvironment leading to increased confinement of the probe, and decreased brownian motion. Importantly, FLTM allows the storage and the loss moduli of the cytoskeleton to be separated and quantified. In contrast to membrane stretch and magnetic micromanipulation (as well as micropipette aspiration and atomic force microscopy), FLTM is a passive method (similar to dynamic light scattering and interferometry). The main strength of passive microrheology is its broad bandwidth range, which takes advantage of the white spectrum of the thermal energy of the embedded probes. FLTM can examine cellular rheology at the subcellular level with nanometer resolution over a five-decade frequency range ($\sim$0.5 Hz to 50 kHz), allowing an in-depth investigation of the cellular cytoskeleton as a complex fluid.

4.1. Instrument design

For more technical details of the design described in the following sections, please see Jonas et al. (2008b). To characterize the rheology of cellular systems, fluorescence measurements have several advantages over existing bright-field configurations. First, because of the variety of fluorescent labels available, different tracers can target different cell components, and spectrally resolved fluorescence readouts allow these tracers to be distinguished within the same sample. Second, there is greatly diminished background, thus the localization of the fluorescent tracers is less susceptible to optical interference effects. Finally, FLTM studies may eventually be based on selective labeling of cellular organelles (e.g., endosomes and lysosomes), without requiring the use of exogenous tracer particles. However, fluorescence tracers have some shortcomings. For example, because of the relatively low signal level of fluorescent particles, localization of these tracers is limited by the photon shot noise. Furthermore, very long-term monitoring can be limited by photobleaching. Despite these issues, the unique advantages of FLTM make this technique an important extension of particle tracking rheometric measurement of cells.

Robust design of the FLTM's fluorescence detection requires consideration of two primary factors. First, to allow the detection of displacements in cytoskeletal networks with shear moduli up to several thousands of pascals (Pa) (a value encountered at high frequencies in many cell types), the spatial resolution of the fluorescent tracer's trajectory must be on the order of a few nanometers given the magnitude of the thermal driving force. Second, the detection bandwidth should be broad ($\sim$0.5 Hz to 50 kHz) to enable the probing of both molecular networks that deform at long-time scales and the fast bending and twisting fluctuations of single filaments in the cytoskeleton at shorter times. To meet these two requirements of spatial precision and speed, 10^4 photons should be detected within each 20-μs period. Tracer particles, such as fluorescent polystyrene spheres of 1 or 2 μm diameter, are often used experimentally, because they are resistant to photobleaching and can sustain a high fluorescent photon generation rate over several seconds. Quadrant photomultiplier tubes (PMT) are used in FLTM research work because of their low-noise high-gain properties (dark current of $\sim$50 pA for a gain of $G_{PMT} = 2 \times 10^5$ at a wavelength of 560 nm). Both the transimpedance circuitry at the output of the multianode PMT and the PMT supply voltage are then further optimized to achieve a data acquisition speed of $\sim$50 kHz without causing PMT saturation. The PMT signal is measured by a high sampling rate analog-to-digital converter (ADC) with linearity up to a photon detection rate of $\sim$10^9 photons/s. Taken together, these design parameters ensure that the FLTM can span the desired temporal and frequency ranges while maintaining required detection sensitivity.

4.2. Theoretical principles of FLTM

Because of the novelty of this technique, we provide a brief background into the principles of FLTM, culminating in the equations necessary to reduce the data to stiffness and viscous components (equations and explanation are also in Jonas *et al.* [2008b]). The viscoelastic properties of a material can be parameterized by the frequency-dependent complex shear modulus $G^*(\omega)$, defined as the ratio of the stress experienced by the material over the strain it undergoes at a frequency ω. Both the elastic and the viscous components of the material can be derived from $G^*(\omega) = G'(\omega) + iG''(\omega)$, where the storage (elastic) modulus $G'(\omega)$ is the real part of $G^*(\omega)$, and the loss (viscous) modulus $G''(\omega)$ is its imaginary part.

To empirically determine $G^*(\omega)$, the results obtained from FLTM can be described by the fluctuation-dissipation theorem (Mason *et al.*, 1997; Mason and Weitz, 1995; Yamada *et al.*, 2000). More specifically, the displacements of the probe bead can be specified by its mean squared displacement (MSD), a time-averaged autocorrelation function given by

$$\text{MSD} = \left\langle \Delta r^2(\tau) \right\rangle = \left\langle \left(r(t + \tau) - r(t) \right)^2 \right\rangle_t \qquad (10.3)$$

where $r(t)$ describes the particle's 2-dimensional (2D) trajectory, and τ corresponds to various lag times. Analytically, the motion of a single particle is expressed through a generalized Langevin equation. A generalized Stokes–Einstein relation (GSER) can then be derived:

$$G^*(\omega) = G'(\omega) + iG''(\omega) = \frac{2k_B T}{3\pi a i \omega \Im_u \left\{ \langle \Delta r^2(\tau) \rangle |_{\tau=1/\omega} \right\}} \quad (10.4)$$

where $\Im_u\{<\Delta r^2(\tau)>\}$ is the Fourier transform of the MSD, k_B is the Boltzmann constant, T the absolute temperature, a the radius of the probe particle, and $i = (-1)^{1/2}$. To avoid numerical transformation of the data or with functional forms to fit $G^*(\omega)$ in Fourier space, an algebraic form of the GSER can be yielded by estimating the MSD as a local power law:

$$\begin{aligned} G'(\omega) &= G^*(\omega)\cos(\pi\alpha(\omega)/2) \\ G''(\omega) &= G^*(\omega)\sin(\pi\alpha(\omega)/2) \end{aligned} \quad (10.5)$$

where

$$\alpha(\omega) \equiv \frac{\partial \ln\langle \Delta r^2(\tau) \rangle}{\partial \ln \tau} \bigg|_{\tau=1/\omega} \quad (10.6)$$

$$|G^*(\omega)| \approx \frac{2k_B T}{3\pi a \langle \Delta r^2(1/\omega) \rangle \Gamma[1 + \alpha(\omega)]} \quad (10.7)$$

where Γ represents the gamma function. Γ could be also be represented as:

$$\Gamma[1 + \alpha] \approx 0.457(1 + \alpha)^2 - 1.36(1 + \alpha) + 1.90 \quad (10.8)$$

for the range of values of $\alpha(\omega)$ spanned by thermally driven spheres: $\alpha(\omega)$ lies between zero, corresponding to elastic confinement, and one, corresponding to viscous diffusion.

FLTM conditions of operation meet the criteria for validity of the GSER: the fluorescent beads of diameter $a = 1$ μm tracked by the instrument are spherical and rigid, and they are embedded in materials whose typical mesh size is much smaller than a (for example, the mesh size is $\sim$10 to 100 nm in the cytoskeleton).

Only one particle is tracked at a time with the FLTM detection scheme. One constraint on the extent of the particle motion is the size of the field of view of the quadrant photodetector, (for our case 8 μm $\times$ 8 μm), which should be designed large enough for the particle to stay within view during the tracking time frame, but as small as possible otherwise.

4.3. FLTM hardware setup

FLTM measurements presented in this chapter were typically performed at 37° by means of a custom-made heating apparatus, and on a vibration-isolating floating optical table. All samples were ultimately prepared in cell culture dishes equipped with a glass bottom coverslip (e.g., number 1.0). A schematic of the FLTM system is presented in Fig. 10.4.

4.4. Cell sample preparation

Covalent coupling of antibodies to fluorescent polystyrene beads can be achieved with commercial kits using various surface properties or by passive adsorption. Cells (NIH 3T3 fibroblasts for the purposes of results presented

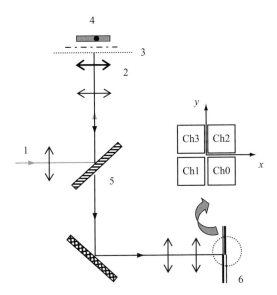

Figure 10.4   A laser beam (1) (532 nm from a Verdi Nd:YVO4, Coherent, Santa Clara, CA) is collimated through a custom light path and a 100× NA 1.30 oil objective (2) (Olympus, Melville, NY) and illuminates a 100-μm $\times$ 100-μm area of sample positioned on the stage of an Olympus IX71 inverted microscope and (3) mounted on an xy piezoelectric nanopositioning system. The photons emitted by the fluorescent beads (4) contained in the excitation volume are filtered by a dichroic mirror-barrier filter combination (5) (Q560LP and HQ585/40 m) and detected after beam expansion by a quadrant photomultiplier tube (6). Position signals are inferred at the PMT level from the difference in photocurrents between opposing pairs of quadrant elements, which are further amplified into digitized voltages by a 200-kHz 16-bit simultaneous 4-channel analog-to-digital converter.

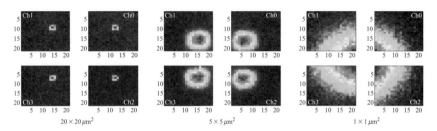

Figure 10.5 Sample data from bead centering. At progressively higher magnifications from left to right, the bead position is adjusted to be at the center of the quadrant detectors, which is achieved and best verified at the rightmost image. (See color insert.)

in this chapter) are cultured according to their type. The day before the fluorescence laser tracking experiments, cells are plated on 35-mm glass-bottom cell culture dishes coated with collagen I (1 $\mu g/cm^2 = 10$ $\mu g/ml$ for a 1-ml solution). On the day of the experiments, the cells should be at the desired confluence (for the experiments we describe, confluence reached 60%). The fluorescent beads are mixed with growth media (at a concentration of 5×10^5 microspheres/ml) and added to the plated cells for 24 h (for bead endocytosis). Before the microrheology experiments, nonadherent beads are washed off with culture media.

4.5. Bead centering

Single fluorescent tracer particles plated on coverslips without cells are positioned at the center of the quadrant detection zone in three steps (Fig. 10.5). The centering procedure is automated (e.g., by custom LabVIEW, C++, or Matlab programs) by the xy stage (for raster-scanning of three progressively smaller areas) and the ADC acquisition card (for fluorescent signal detection). Initial bead centering should be carried out at very low illumination power (<5 μW for a 100-μm $\times$ 100-μm excitation area) to avoid photobleaching the probes. This centering guarantees that the bead being tracked is located within a desired $\sim$1-μm $\times$ 1-μm-wide zone.

4.6. Calibration

Once the bead is centered, its brownian motion is monitored by the FLTM for 1.5 s, under high-illumination power ($\sim$5 $\times$ 10^5 W.m^{-2}), at an ADC sampling rate of 200 kS/sec (a total of $2^{18} = 262,144$ data points are collected). Calibration curves for the FLTM (bead x- or y-position vs ADC signals) can be obtained after each experiment for every bead. Calibration consists of translating the bead along both detection axes (x, then y)

for a total of approximately twenty 10-nm steps from the center position with the high-resolution (0.5 nm) xy stage. Results should show step patterns of bead displacement along either axis on the basis of quadrant detection output. Depending on optics used, the size of the steps may require adjustment to overcome noise. From the quadrant detector outputs (Ch0, Ch1, Ch2, Ch3), two quantities can be calculated: $RatioX = (Ch0 + Ch2)/(Ch1 + Ch3)$, and $RatioY = (Ch2 + Ch3)/(Ch0 + Ch1)$. Each calibration position can then be associated with specific RatioX and RatioY values. This matrix of values allows the position of the tracer particle to be determined by interpolation. With a 100× NA 1.30 objective lens, 5-nm steps were discernible, but the expected resolution would drop to approximately 10 nm with a 40× NA 1.30 objective. It should be noted that the resolution of the xy stage should be higher than the desired final resolution of displacement, because the final measurement resolution is ultimately limited by the resolution of the stage during calibration.

4.7. Determining the trajectory of the particle

At each time step, the position of tracer particle can be related to a pair of RatioX and RatioY values as defined in the calibration experiment. With the calibration matrix of RatioX and RatioY, the actual position of the tracer particle can be obtained from calibration (see Fig. 10.6 for an example). Repeating this procedure every 20 μs, one can establish the bead's 2D trajectory.

4.8. Determining rheology from the particle trajectory

Once the fluorescent particle's trajectory has been determined, the particle's mean squared displacement (MSD) can be calculated from its 2D trajectory as defined by Eq. (10.3). The storage and loss moduli $G'(\omega)$ and $G''(\omega)$ can then be computed from the probes' MSDs with Eqs. (10.5 to 10.8). Fig. 10.7 provides examples of MSD plots for a series of nine beads endocytosed by cells. At each time lag τ considered, or each frequency $\omega = 1/\tau$ studied, the parameter α defined in Eq. (10.6) is established by linearly fitting a logarithmic slope between two adjacent frequencies $\omega1 < \omega < \omega2$. Once α has been determined, both the magnitude $|G^*(\omega)|$ and Γ can be determined on the basis of Eqs. 10.7 and 10.8, and $G'(\omega)$ and $G''(\omega)$ can then be calculated with Eq. (10.5). Fig. 10.8 in the next section shows how MSD data translate into frequency-dependent microrheologic parameters.

4.9. Illustrative results

As an illustration of some results that may be obtained with FLTM, we present some data from cells that were treated with different compounds, on the basis of work under review (Jonas et al., 2008a) To assess the influence of

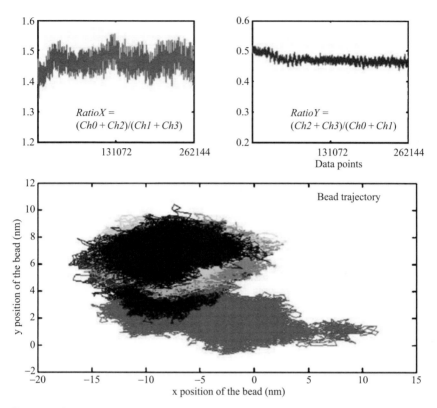

Figure 10.6 The *RatioX* and *RatioY* as a function of position is applied to the experimental particle's signals (upper plots), and the calculation from the calibration is reversed to obtain the displacements in x and y of the particle being tracked as a function of time (lower figure). (See color insert.)

cytoskeletal components on the rheology of living cells, fibroblasts were treated for 30 min with either 10 μM cytochalasin D or 17 μM nocodazole. Cells were washed with culture media immediately before the FLTM experiments.

From the brownian trajectories of 1-μm fluorescent beads endocytosed by the cells, the mean squared displacements (MSDs) of the probed particles were determined (Fig. 10.8A). The cells' local shear moduli $G^*(\omega)$ (Fig. 10.8B), storage moduli $G'(\omega)$ (Fig. 10.8C), and loss moduli $G''(\omega)$ (Fig. 10.8D) were then calculated. Three chemical conditions were examined: untreated cells (Fig. 10.8, *triangles*), cells treated with the actin-disrupting drug cytochalasin D (*circles*), and cells treated with the microtubule-disrupting drug nocodazole (*squares*). Our sample preparation led to most cells containing a single fluorescent microsphere. Untreated cells exhibit storage and loss moduli ranging from 1 to 10 Pa at 1 Hz to ~100 Pa

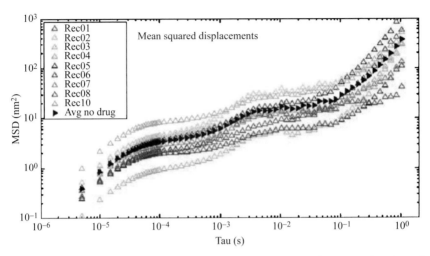

Figure 10.7   Mean squared displacements for nine endocytosed beads as a function of lag time Tau. (See color insert.)

at 1 kHz and ~500 to 1000 Pa at 50 kHz (Fig. 10.8, *triangles*), which concurs with results obtained in fibroblasts by other cell mechanical techniques.

Disrupting actin (with cytochalasin D) or depolymerizing microtubules (with nocodazole) results in altered cytoskeletal morphology (Fig. 10.8, E to J) and altered rheology (Fig. 10.8, A to D). Fluorescent cytoskeletal staining was performed to verify that cytochalasin D disrupted primarily actin without significantly changing microtubules (Fig. 10.8, F and I), and that nocodazole disrupted microtubules without significantly altering actin (Fig. 10.8, G and J). From the FLTM data, we found that disruption of the fibroblasts by either cytochalasin D or nocodazole resulted in statistically significantly lower stiffness ($G'_{drug} < G'_{untreated}$, Fig. 10.8C) and lower viscosity ($G''_{drug} < G''_{untreated}$, Fig. 10.8D) compared with untreated cells over the entire five-decade range of frequencies (0.5 Hz to 50 kHz). Cell stiffness decreased by a factor of 4 to 20 across this frequency spectrum and was accompanied by a decrease in viscosity by a factor 4 to 50 (Fig. 10.8 C and D, respectively).

Some experimental variables were examined in terms of their influence on FLTM measurements. First, we found that viscoelastic moduli depend on sample temperature. Specifically, tracers exhibited greater displacement at 37° than at 20°, suggesting the importance of conducting experiments at proper temperatures on the basis of the cell type. Second, we found that for beads larger than 1-μm diameter, FLTM is relatively independent of tracer size. These results suggest that FLTM measurements can be sensitive to cellular biologic environment and experimental design, and thus care must be taken to understand the basic principles behind this technique.

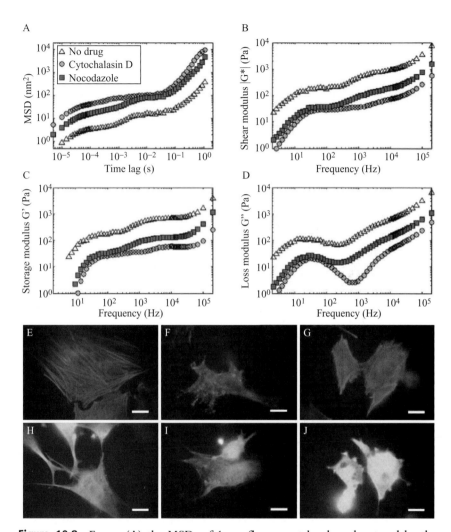

Figure 10.8 From (A) the MSDs of 1-μm fluorescent beads endocytosed by the cells, the fibroblasts' (B) shear moduli G^*, (C) storage moduli G', and (D) loss moduli G'' were calculated. Fibroblasts were left untreated (E and H, and *triangles* in A to D), treated with cytochalasin D (F and I, and *circles*), or nocodazole (G and J, and *squares*). Fluorescent labeling of cytoskeletal networks after FLTM measurements (Alexa Fluor® 568 phalloidin staining of F-actin: E to G, or FITC-conjugated β-tubulin antibodies: H to J) confirmed that the drug cytochalasin D disrupted the actin filaments only (F and I), and that the drug nocodazole promoted depolymerization of the microtubules only (G and J). Scale bars are 20 μm.

5. GENERAL SUMMARY AND CONCLUSION

In this chapter we describe three techniques for studying cell mechanics and mechanotransduction. The most commonly used membrane stretch technique can be used for molecular signaling or other large-scale studies. Membrane stretch represents one type of typical mechanical stress; other mechanical stresses such as fluid shear are described elsewhere in this volume. Although the membrane-stretch devices are commercially available, several groups choose to custom-machine the devices, for example, for better control over the range of motion and the size of the device. These devices permit a large-scale readout and can mimic strain fields found in tissues.

Magnetic micromanipulation is a more recent development and is complemented by several variants, including magnetic twisting and multiple-pole devices. As far as the authors are aware, there is no commercially available source for such devices; indeed, the magnetic microparticles used for manipulation are traditionally meant for cell sorting rather than force application. Electromagnetic forcing is an elegant, minimally invasive technique to apply fixed forces to cells by means of magnetic beads (in contrast to optical trapping, which typically applies fixed displacements to beads). This technique is useful for its range of forces (pN to hundreds of nN) and the ability for the force to be applied to a specific receptor.

FLTM is a novel method that can assess the local, frequency-dependent mechanical properties of living cells. As a passive technique, FLTM characterizes the local viscoelastic behavior of biomaterials in the absence of an active mechanical perturbation to the sample. FLTM data acquisition speeds (up to 50 kHz) and spatial resolution of the bead probe's trajectory ($\sim$4 nm) make this apparatus well suited to the study of heterogeneous and dynamic biologic systems. The extended frequency range ($\sim$0.5 Hz to 50 kHz) probed by FLTM allows appraisal of the macroscopic viscoelasticity of molecular networks at long time scales (on the order of seconds), as well as the fast fluctuations of single filaments in the cytoskeleton at shorter time scales, which are important for understanding fast transitions in cytoskeletal dynamics. Finally, the addition of fluorescence to a microrheometry system enables targeting of cellular structures with molecular specificity and overcomes the difficulty of finding large native particles to be tracked. Because of their spectral signatures, fluorescent probes linked to different cellular structures can be readily distinguished. Furthermore, tracking in fluorescent mode renders the detection process less sensitive to optical background effects than white light approaches.

These techniques span scales from multiple cells to subcellular and complement each other in terms of readouts and strengths. Indeed, because of the complementary nature of these techniques, different combinations can be used to further explore cell mechanics. For example, one can attach magnetic beads against different receptors of interest, pull on the beads, and assess the changes in cell stiffness and viscosity both near and far from the location of the bead with FLTM. Thus, the possibilities of scientific exploration are greatly amplified and can permit a better understanding of both cell biology and mechanics.

ACKNOWLEDGMENTS

M. J. and P. S. were supported by NIH P01 HL64858-01A1. H.H. was supported by an NIH NRSA Drs. Roger Kamm, Jan Lammerding, and Richard Lee assisted with design and initial methods. Michael Motion and Francisco Cruz assisted with the projects.

REFERENCES

Bausch, A. R., Ziemann, F., Boulbitch, A. A., Jacobson, K., and Sackmann, E. (1998). Local measurements of viscoelastic parameters of adherent cell surfaces by magnetic bead microrheometry. *Biophys. J.* **75,** 2038–2049.

Brown, T. D. (2000). Techniques for mechanical stimulation of cells *in vitro*: A review. *J. Biomech.* **33,** 3–14.

Chien, S (2006). Molecular basis of rheological modulation of endothelial functions: Importance of stress direction. *Biorheology* **43,** 95–116.

Dartsch, P. C., Hammerle, H., and Betz, E. (1986). Orientation of cultured arterial smooth muscle cells growing on cyclically stretched substrates. *Acta Anat.* **125,** 108–113.

Fabry, B. M. G., Hubmayr, R. D., Butler, J. P., and Fredberg, J. J. (1999). Implications of heterogeneous bead behavior on cell mechanical properties measured with magnetic twisting cytometry. *J. Magnetism Magn. Mater.* **194,** 120–125.

Feng, Y., Yang, J. H., Huang, H., Kennedy, S. P., Turi, T. G., Thompson, J. F., Libby, P., and Lee, R. T. (1999). Transcriptional profile of mechanically induced genes in human vascular smooth muscle cells. *Circ. Res.* **85,** 1118–1123.

Hormeno, S., and Arias-Gonzalez, J. R. (2006). Exploring mechanochemical processes in the cell with optical tweezers. *Biol. Cell* **98,** 679–695.

Huang, H., Cruz, F., and Bazzoni, G. (2006). Junctional adhesion molecule-A regulates cell migration and resistance to shear stress. *J. Cell. Physiol.* **209,** 122–130.

Huang, H., Dong, C. Y., Kwon, H. S., Sutin, J. D., Kamm, R. D., and So, P. T. (2002). Three-dimensional cellular deformation analysis with a two-photon magnetic manipulator workstation. *Biophys. J.* **82,** 2211–2223.

Huang, H., Kamm, R. D., So, P. T., and Lee, R. T. (2001). Receptor-based differences in human aortic smooth muscle cell membrane stiffness. *Hypertension* **38,** 1158–1161.

Husse, B., Briest, W., Homagk, L., Isenberg, G., and Gekle, M. (2007). Cyclical mechanical stretch modulates expression of collagen I and collagen III by PKC and tyrosine kinase in cardiac fibroblasts. *Am. J. Physiol. Regul. Integr. Comp. Physiol.* **293,** R1898–R1907.

Jonas, M., Huang, H., Kamm, R. D., and So, P. (2008a). Fast fluorescence laser tracking microrheometry II: Quantitative studies of cytoskeletal mechanotransduction. *Biophys. J.* **95,** 895–909.

Jonas, M., Huang, H., Kamm, R. D., and So, P. T. (2008b). Fast fluorescence laser tracking microrheometry I: Instrument development. *Biophys. J.* **94,** 1459–1469.

Kanda, K., and Matsuda, T. (1994). Mechanical stress–induced orientation and ultrastructural change of smooth muscle cells cultured in three-dimensional collagen lattices. *Cell Transplant* **3,** 481–492.

Kanda, K., Matsuda, T., and Oka, T. (1992). Two-dimensional orientational response of smooth muscle cells to cyclic stretching. *Asaio J.* **38,** M382–M385.

Leung, D. Y., Glagov, S., and Mathews, M. B. (1976). Cyclic stretching stimulates synthesis of matrix components by arterial smooth muscle cells *in vitro*. *Science* **191,** 475–477.

Mason, T. G., van Zanten, J., Wirtz, D., and C., K. S. (1997). Particle tracking microrheology of complex fluids. *Phys. Rev. Lett.* **79,** 3282–3285.

Mason, T. G., and Weitz, D. A. (1995). Optical measurements of frequency-dependent linear viscoelastic moduli of complex fluids. *Phys. Rev. Lett.* **74,** 1250–1253.

Nieminen, T. A., Knoner, G., Heckenberg, N. R., and Rubinsztein-Dunlop, H. (2007). Physics of optical tweezers. *Methods Cell. Biol.* **82,** 207–236.

Osol, G (1995). Mechanotransduction by vascular smooth muscle. *J. Vasc. Res.* **32,** 275–292.

Wang, N., and Ingber, D. E. (1995). Probing transmembrane mechanical coupling and cytomechanics using magnetic twisting cytometry. *Biochem. Cell. Biol.* **73,** 327–335.

Yamada, S., Wirtz, D., and Kuo, S. C. (2000). Mechanics of living cells measured by laser tracking microrheology. *Biophys. J.* **78,** 1736–1747.

Yamamoto, K., Dang, Q. N., Maeda, Y., Huang, H., Kelly, R. A., and Lee, R. T. (2001). Regulation of cardiomyocyte mechanotransduction by the cardiac cycle. *Circulation* **103,** 1459–1464.

CHAPTER ELEVEN

VASCULAR INTEGRIN SIGNALING

Ganapati H. Mahabeleshwar *and* Tatiana V. Byzova

Contents

1. INTRODUCTION

Integrins are a family of heterodimeric, transmembrane glycoproteins consisting of alpha (α) and beta (β) subunits (Hynes, 1987). The integrin heterodimeric complexes are expressed on various cell types in a tissue–specific

Department of Molecular Cardiology, JJ Jacobs Center for Thrombosis and Vascular Biology, The Cleveland Clinic Foundation, Cleveland, Ohio

Methods in Enzymology, Volume 443
ISSN 0076-6879, DOI: 10.1016/S0076-6879(08)02011-9

manner and are known to regulate cellular interactions with extracellular matrix proteins (Hynes, 2002). Studies from a number of laboratories have resulted in the identification and characterization of 18 types of α-subunits and 8 types of β-subunits of integrins in mammals. The combinations of these α- and β-subunits gave rise to 24 integrin heterodimers with well-defined ligand specificity (Burke, 1999; Plow et al., 2000). Integrins serve as adhesion receptors for extracellular matrix in animals from sponges to high vertebrates (Brower et al., 1997; Pancer et al., 1997). Integrins transmit information from the extracellular matrix environment across the cellular membrane to the cytoplasmic milieu. By integrating cells into surrounding extracellular matrix, this family of receptors controls various fundamental biologic processes, such as cellular adhesion and regulation of signaling events, which modulate cellular migration, proliferation, survival, and apoptosis (Alam et al., 2007; Berrier and Yamada, 2007; Ramsay et al., 2007; Stupack and Cheresh, 2003; Zutter, 2007). The relationship between matrix and cells is of a true reciprocal nature, and integrins, in turn, are responsible for extracellular matrix modifications by cells (Davis et al., 2002; Sepulveda et al., 2005; Wierzbicka-Patynowski and Schwarzbauer, 2003). At the structural level, integrins consist of large extracellular and relatively short transmembrane and cytoplasmic domains (Springer and Wang, 2004). The extracellular portions of α- and β-subunits are responsible for ligand binding, whereas transmembrane and cytoplasmic domains transmit signals from the extracellular matrix environment to the cytoplasm. The transmembrane domains of integrins are highly hydrophobic and are responsible for membrane localization of integrins. In most cases α-subunits have shorter cytoplasmic domains than β-subunits, thus it is believed that the β-subunit provides binding sites for cytoskeletal proteins, signaling intermediates and numerous cytoplasmic adapter molecules (Luo and Springer, 2006; Qin et al., 2004).

For many, but not all, integrins, high-affinity ligand binding requires integrin activation (Bazzoni and Hemler, 1998), which is achieved through conformational changes initiated from the cytoplasmic and transmembrane domains (Ginsberg et al., 2005; Ratnikov et al., 2005). These processes are generally known as "inside-out signaling" and on endothelial cells are triggered by cell stimulation with cytokines, growth factors, or shear conditions (Byzova et al., 2000; Tzima et al., 2005). Ligand binding to integrins results in further conformational changes in integrins and initiates a series of signaling events that are commonly referred to as "outside-in signaling" (Ginsberg et al., 2005). On endothelial cells, both inside-out and outside-in integrin signaling events are known to regulate the cellular responses underlying physiologic and pathologic angiogenesis (Fig. 11.1).

On many cells, stimulation with growth factors and integrin ligation lead to phosphorylation of tyrosine, serine, or threonine residues on the β-integrin cytoplasmic domain. These phosphorylation events influence

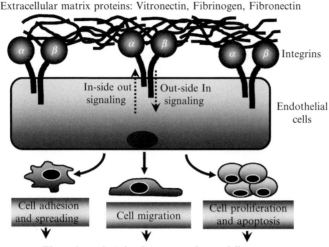

Extracellular matrix proteins: Vitronectin, Fibrinogen, Fibronectin

Integrins

In-side out signaling

Out-side In signaling

Endothelial cells

Cell adhesion and spreading

Cell migration

Cell proliferation and apoptosis

Tissue (vascular) development and remodeling
Physiological and Pathological neovascularization

Figure 11.1 Roles of integrin in endothelial cell physiology. Diagrammatic representation of integrin functions in endothelial cell physiology and pathophysiology. Integrin α- and β-heterodimers expressed on the endothelial cells recognize extracellular matrix proteins, and this interaction is responsible for "out-side in signaling." Out-side in signaling is responsible for the transmission of signals from the extracellular environment to the cytoplasm. A series of these signaling events leads to the transmission of signal to the nucleus and thence the regulation of gene expression. Simultaneously, integrin inside-out signaling events regulate conformational changes in integrin structure and thus regulate the affinity of integrin binding. Together, these signaling events regulate endothelial cell adhesion, migration, cell proliferation, apoptosis, angiogenic programming in endothelial cells, and, finally, regulate processes of angiogenesis.

conformational changes in integrins and their ability to interact with ligands (Cram and Schwarzbauer, 2004). Recent studies demonstrated that mutation of phosphorylation sites within β3-integrin leads to significant impairment of both inside-out and outside-in signaling (Mahabeleshwar et al., 2006; 2007; Phillips et al., 2001a,b).

Experimental evidence along with clinical data strongly demonstrates that integrins are a major type of cell adhesion molecules, involved in angiogenesis. Indeed, expression levels of $\alpha v \beta 3$-, $\alpha v \beta 5$-, and $\alpha 5 \beta 1$-integrin are upregulated in actively proliferating angiogenic endothelial cells (Stromblad and Cheresh, 1996). More importantly, the blockade of either $\alpha v \beta 3$- or $\alpha v \beta 5$-integrin alone with specific antagonists disrupted experimental angiogenesis and tumor growth in a variety of mouse tumor models (Felding-Habermann and Cheresh, 1993). Antagonists of these integrins, including function blocking monoclonal antibodies, are currently under various stages of clinical trials (Tucker, 2006). Furthermore, both $\alpha 5 \beta 1$-integrin and its ligand fibronectin have been shown to be overexpressed in angiogenic blood vessels of various species, and genetic deletions of both $\alpha5\beta1$ and fibronectin resulted in defective angiogenesis and vasculogenesis (Hynes, 2002; Tanjore *et al.*, 2008). Not surprisingly, antagonists of integrin $\alpha5\beta1$ blocked tumor-induced angiogenesis and are currently being evaluated in clinical trials (Tucker, 2002). Thus, in view of the pathophysiologic importance of integrins and their crucial role in angiogenesis, it is important to understand the molecular mechanisms involved in integrin regulation. In this review, we focus on methodologic strategies aimed at assessing the roles of individual integrins on endothelial cells and their functional activity during processes of angiogenesis. Because integrins mediate several cellular aspects of angiogenesis, this chapter addresses methods to investigate integrin-dependent cell adhesion to ECM proteins, cell migration, endothelial tube and sprout formation, integrin affinity modulation, integrin-dependent cell proliferation and apoptosis, and integrin interactions with tyrosine kinase receptors in endothelial cells.

2. INTEGRIN-MEDIATED CELL ADHESION

Cellular interaction with extracellular matrix (ECM) components through cell surface integrins is required for cell growth, differentiation, progression of cell cycle through G1 phase, and cell survival. Cellular adhesion to ECM induces cellular proliferation and the lack thereof triggers programmed cell death of nonadherent cells in suspension (Eliceiri and Cheresh, 2001). An increasing body of evidence suggests that a cooperative interaction between ECM substratum and cell surface integrin can affect cell cycle progression. Furthermore, ligation of integrins induces a wide variety of intracellular processes, including increased inositol lipid synthesis, tyrosine phosphorylation of FAK, and expression of several cell survival factors. Integrin-dependent cellular adhesion to ECM proteins is also crucial for ERK-1/2 activation, cyclin D1 expression, regulation of the cip/kip family of cdk inhibitors, and activation of cyclin-dependent kinases (Klein *et al.*, 2007). In our laboratory, we perform integrin-dependent cell adhesion assays with a variety of cell types, including, but not limited to, platelets, endothelial cells, fibroblasts, and epithelial cells. Here we have summarized general protocols for endothelial cell adhesion assays.

Endothelial cells express several integrins, whose expression level and identity are typically assessed by flow cytometry. Therefore, for complete analysis it is important to test cell adhesion to several ECM proteins. Fig. 11.2 shows a comparative analysis of adhesive activity of WT and DiYF (knock-in mutation of $\beta3$-integrin) microvascular endothelial cell primary cultures.

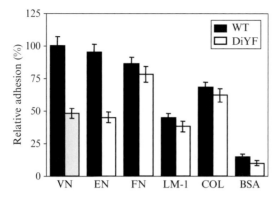

Figure 11.2   Cellular adhesion to extracellular matrix proteins. Integrins are major cell surface adhesion molecules that regulate cellular interaction with extracellular matrix proteins. Wild-type (WT) and DiYF endothelial cells were detached with 20 mM EDTA and plated on vitronectin (VN), entactin (EN), fibronectin (FN), laminin-1 (LM-1), collagen (COL), or bovine serum albumin (BSA)–coated plates. Cells were allowed to adhere for 45 min at 37° in a humidified CO_2 incubator. The number of cells attached per field was counted in several representative fields. The number of wild-type endothelial cells adhered to vitronectin was assigned a value of 100% and relative changes in adhesion are indicated (Reproduced from The Journal of Experimental Medicine, 2006; **203**, 2495–2507. Copyright 2006 The Rockefeller University Press.)

2.1. Materials

Complete tissue culture media suitable for the selected cell line (supplemented with antibiotics).
Sterile 1× phosphate-buffered saline (pH 7.4).
Bovine serum albumin, sterile, 1% solution in phosphate-buffered saline.
4% paraformaldehyde in 1× phosphate-buffered saline.
20 mM EDTA in sterile 1× phosphate-buffered saline (pH 7.4).
6-well or 12-well tissue culture plates coated with appropriate integrin ligand.

2.2. Equipment

Hemacytometer.
Inverted phase-contrast microscope.

2.3. Procedure

1. Coat the 6-well or 12-well tissue culture plates with the integrin ligand to which the adhesion of a specific cell type is to be tested. To coat the plates with integrin ligand, add sterile 10 μg/ml ligand solution to each

well of the culture plates. Incubate the coated plates at 4° for 24 h. Post-coat the plates with 1% bovine serum albumin at room temperature for 1 h. Wash these ligand-coated plates at least two times with sterile $1\times$ phosphate-buffered saline.
2. Gently wash cultured cells with sterile $1\times$ phosphate-buffered saline two times. Detach the cultured cells with a 20 mM EDTA solution in sterile 1x phosphate-buffered saline. Cellular detachment with 20 mM EDTA alone will take approximately 5 to 10 min. This time varies greatly, depending on cell type. It is advised not to use trypsin/EDTA solution to detach the cells to prevent any proteolytic damage created by trypsin. Alternately, it may be advisable to use commercially available enzyme-free cell dissociation buffer.
3. Collect the detached cells in 10 ml sterile $1\times$ phosphate-buffered saline. Centrifuge the cell suspension at 500g for 5 min. Gently wash this cell pellet by resuspending once with sterile $1\times$ phosphate-buffered saline then wash one more time with suitable serum-free tissue culture media.
4. Resuspend the cells in suitable serum-free tissue culture media. Count and adjust the cell number to 2×10^5 cells/ml with serum-free medium. Add an equal volume or 1 ml of cell suspension to the ligand-coated plates and incubate at 37° for 30 min in a humidified CO_2 incubator.
5. At the end of 30 min, remove unattached cells by aspirating the medium. Gently wash the culture plate wells to remove loosely attached cells. Adhered cells are fixed immediately with 4% paraformaldehyde in $1\times$ phosphate-buffered saline for 10 min.
6. Photographs of nonoverlapping fields can be taken with an inverted phase-contrast microscope.

2.4. Quantification

Average numbers of adhered cells per field are calculated from analysis of multiple nonoverlapping fields. The number of cells attached in the control group is assigned the value of 100%. Relative differences in attachment on different matrix proteins and differential attachment of various cell types toward different integrin ligands can be determined. Statistical significance in adhesion between two groups can be calculated through Student's t test.

3. Analysis of Integrin-Dependent Cell Migration

Extracellular matrix provides crucial support for proliferating vascular endothelium through adhesive interactions with endothelial cell surface integrins (Rose *et al.*, 2007; Stupack, 2007). Extracellular matrix also provides the scaffold essential for maintaining the organization of vascular endothelial cells in mature blood vessels. Endothelial cell adhesion to extracellular matrix is essential for endothelial cell proliferation, migration, and morphogenesis (Alghisi and Ruegg, 2006), whereas endothelial cell migration through extracellular matrix is one of the essential responses in angiogenesis. Integrins are known to mediate cellular migration by linking extracellular matrix proteins and cytoplasmic cytoskeleton machinery (Moissoglu and Schwartz, 2006). Active cell migration involves a series of well-defined steps such as directional protrusion of the cell, actin polymerization, integrin-dependent stable adhesion complex formation, and retraction of the trailing edge of the cell to complete forward movement. Each of these processes is integrin dependent, and functional abnormalities of integrins can severely impair these processes. The major integrins implicated in endothelial cell migration are $\alpha v\beta 3$-, $\alpha v\beta 5$-, and $\alpha 5\beta 1$-integrins (Byzova et al., 1998; Hood and Cheresh, 2002; Stupack and Cheresh, 2002). Here we describe general procedures used to measure endothelial cell migration on ECM proteins with a Boyden-type migration chamber. The later part of this section also describes a procedure to measure directional migration of endothelial cell on ECM proteins.

3.1. Boyden-type migration assay

3.1.1. Materials

Cell line or primary cell cultures whose migration on integrin ligands is to be tested by various stimuli.
Transwell tissue culture inserts (Boyden-type migration chambers), plates coated with suitable integrin ligand.
Complete tissue culture media suitable for cell type.
0.05% Trypsin and 0.53 mM EDTA cell detachment solution.
Chemoattractant such as VEGF-A or bFGF.
Sterile 1× phosphate-buffered saline (pH 7.4).
4% Paraformaldehyde in 1× phosphate-buffered saline.
0.3% Methylene blue stain.
Cotton swabs.

3.1.2. Equipment

Hemacytometer.
Inverted phase-contrast microscope.

3.1.3. Procedure

1. Coat both the upper and lower surfaces of Transwell tissue culture inserts with suitable integrin ligand on which the migration of a specific

cell type needs to be tested. To coat Transwell tissue culture inserts add 0.5 ml of sterile 10 μg/ml integrin ligand solution to the lower well of tissue culture plates and place the Transwell tissue culture inserts into the well. Then add 0.2 ml of 10 μg/ml integrin ligand solution into the upper well of the tissue culture insert. Incubate these tissue culture plates at 4° for 24 h. Gently wash the ligand-coated plates at least two times with sterile 1× phosphate-buffered saline before use.

2. Gently wash cultured cells with 1× phosphate-buffered saline and detach cells with trypsin–EDTA cell detachment solution. Collect detached cells in 10 ml sterile 1× phosphate-buffered saline. It is necessary to remove trace amounts of trypsin–EDTA after cellular detachment. Therefore, gently wash the resultant cell suspension once with sterile 1× phosphate-buffered saline one more time with suitable serum-free tissue culture media by centrifuging the cell suspension at 500g for 5 min.

3. Resuspend the cells in the appropriate media supplemented with 2% fetal bovine serum. Count the cells with a hemacytometer and adjust the cell number to 3 to 4 × 10^5 cells/0.2 ml with low serum culture media. Add an equal volume or 0.2 ml of cell suspension to the top wells of Transwell inserts. Fill the lower well with low serum culture media supplemented with appropriate concentrations of chemoattractant (20 ng/ml VEGF-A or bFGF). Incubate these culture plates further at 37° for an additional 6 to 12 h in a humidified CO_2 incubator.

4. At the end of incubation, aspirate the medium from the upper wells of the tissue culture inserts and remove unmigrated cells by gently but thoroughly rubbing the filter membrane with cotton swabs. Immerse the Transwell insert immediately in 4% paraformaldehyde (in 1× phosphate-buffered saline) for 10 min. Wash the Transwell inserts two times with 1× phosphate-buffered saline. The cells on the insert can then be stained with 0.3% methylene blue.

5. Photographs of nonoverlapping fields of migrated cells on the lower surface of the migration chamber can be taken with an inverted phase-contrast microscope.

3.1.4. Quantification

The cells in Boyden-type migration chambers are often not very homogeneously distributed. Both the center and the periphery of the chambers exhibit high cell density. Therefore, it is advisable to avoid these areas during quantification. Average numbers of cells migrated per field are calculated with multiple nonoverlapping fields of triplicates per experiment. The number of cells migrated in the control group is assigned the value of 100%, and relative differences in migration on different matrix proteins can be calculated (Fig. 11.3). Statistical significance in migration between two groups can be calculated by Student's t test.

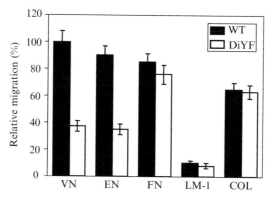

Figure 11.3 Cellular migration on integrin ligands. Integrins expressed on the cell surface are known to regulate the processes of migration. To examine the role of $\beta3$-integrin cytoplasmic tyrosine motifs in endothelial cell migration, wild-type and DiYF endothelial cells were used. Boyden-type migration chambers were coated with various integrin ligands. Wild-type and DiYF mouse lung ECs were detached and seeded into the top portion of the chamber. Cells were allowed to migrate, then were fixed and stained. The nonmigrated cells adhering to the top surface of migration chamber were removed. Three random fields were photographed, and the number of wild-type cells that migrated onto the vitronectin-coated insert was assigned a value of 100%. Relative differences in percentage of endothelial cell migration on various integrin ligands are indicated. (Reproduced from The Journal of Experimental Medicine, 2006; **203**, 2495–2507. Copyright 2006 The Rockefeller University Press.)

3.2. Endothelial wound healing assay

3.2.1. Materials

Cell line or primary cell cultures whose monolayer recovery on integrin ligand is to be tested after various stimuli.
6-Well or 12-well plates coated with suitable integrin ligand.
Complete tissue culture media appropriate for cell type.
0.05% Trypsin and 0.53 mM EDTA cell detachment solution.
Chemoattractant such as VEGF-A or bFGF.
Sterile 1× phosphate-buffered saline (pH 7.4).
4% Paraformaldehyde in 1× phosphate-buffered saline.

3.2.2. Equipment

Inverted phase-contrast microscope.

3.2.3. Procedure

1. Coat the wells of 6- or 12-well tissue culture plates with suitable integrin ligand on which the degree of wound recovery desired is to be tested. The tissue culture surface can be coated with integrin ligand as described

previously. After coating, gently wash the integrin-ligand coated surface two times with $1\times$ phosphate-buffered saline.

2. Detach subconfluent cultured cells grown in 25 or 75 cm^2 flasks with trypsin-EDTA cell detachment solution. Carefully remove the maximum possible amount of trypsin-EDTA to prevent integrin ligand degradation by trypsin activity. Collect the detached cells by resuspending in 10 ml sterile $1\times$ phosphate-buffered saline. Centrifuge cell suspensions at $500g$ for 5 min and carefully remove all of the supernatant. Resuspend the cell pellet in complete culture medium and seed on integrin-ligand coated tissue culture wells at the appropriate cell density.

3. Allow the cells to attach firmly; then, remove unattached cells by replacing the media. Incubate the tissue culture plate in a humidified CO$_2$ incubator at 37° until cells form a monolayer (this is cell-type dependent). When the cells are confluent, create a wound by removing a swath of cells across the monolayer with a 1000-μl pipette tip as a scraper.

4. Remove the dislodged cells by rinsing twice with sterile $1\times$ phosphate-buffered saline; then, further culture the cells in reduced serum medium (2% fetal bovine serum) supplemented with or without the appropriate chemoattractant. Photograph the denuded areas immediately after wounding (time zero). At the end of 12 h, fix the cells with 4% paraformaldehyde prepared in $1\times$ phosphate-buffered saline. Wash the fixed cells gently to remove any dislodged cells and photograph multiple wound areas with an inverted phase-contrast microscope.

3.2.4. Quantification

Cell migration is quantified by image analysis of five randomly selected fields of the denuded areas of triplicate experiments. The mean wound area is expressed as percent of wound recovery (%R) from three identically treated plates with the equation

$$\%R = [1 - (T_t/T_0)] \times 100,$$

where T_0 is the wounded area at 0 h and T_t is the wounded area after 12 h (Fig. 11.4, A and B).

3.3. Assessment of directional migration of cells

3.3.1. Materials

Cell line or primary cell cultures whose directional migration on integrin ligand after stimulation is to be assessed.

6-Well plates coated with the desired integrin ligand.

Complete tissue culture media appropriate for cell type.

0.05% Trypsin and 0.53 mM EDTA cell detachment solution.

Chemoattractant such as VEGF-A or bFGF.

Sterile $1\times$ phosphate-buffered saline (pH 7.4).

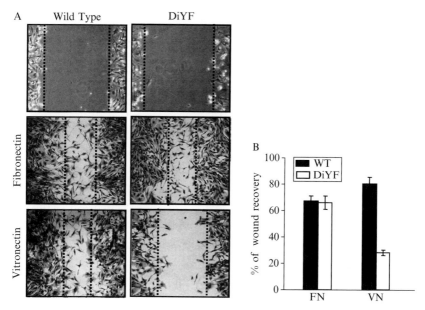

Figure 11.4   Endothelial cell monolayer recovery on extracellular matrix proteins. (A and B) An intrinsic property of endothelial cells is to form a monolayer both *in vitro* and *in vivo*. To examine the role of extracellular matrix and functional integrins in these processes, wild-type and DiYF mouse lung microvascular endothelial cells were grown in monolayer. These cells were serum starved then wounded across the cell monolayer by scraping away a swath of cells. Detached cells were removed by rinsing the wells two times with sterile $1\times$ PBS, then the monolayer was further cultured in DMEM medium containing 2% FBS. The wound sites were photographed immediately (T_0) and 12 h later (T_t). Percentage of wound recovery was calculated as described in the text. (Reproduced from The Journal of Experimental Medicine, 2006; **203**, 2495–2507. Copyright 2006 The Rockefeller University Press.)

3.3.2. Equipment

Leica DMIRB inverted phase-contrast microscope equipped with Cool-SNAP HQ cooled CCD camera (Roper Scientific). This microscope is further supported with a Prior motorized stage (Prior Scientific Instruments) and Pecon incubator. Image acquisition is controlled by computer by use of MetaMorph software (Universal Imaging, Downingtown, PA).

3.3.3. Procedure

1. Coat the wells of 6-well tissue culture plates with suitable integrin ligand for which the directional migration of a specific cell type needs to be tested.
2. Gently wash the integrin ligand coated wells with $1\times$ phosphate-buffered saline. Carefully detach semiconfluent cultured cells from their growth flasks with trypsin-EDTA cell detachment solution. Wash

these cells once with copious amounts of sterile $1\times$ phosphate-buffered saline. Resuspend the cells in complete culture medium and seed on integrin–ligand coated tissue culture wells at an appropriate cell density.
3. Incubate the tissue culture plate in a humidified CO_2 incubator at $37°$ for 12 h; then, replace the media to remove unattached cells. Allow this culture to reach a semiconfluent stage. When the cells are semiconfluent, create a wound by removing a strip of cells across the monolayer with a $1000\text{-}\mu l$ pipette tip as a scraper. Remove the dislodged cells by washing gently with sterile $1\times$ phosphate-buffered saline.
4. Culture with reduced serum medium (2% fetal bovine serum) with or without appropriate chemoattractant.
5. Warm up the stage incubator for at least 30 min before mounting the culture plate. Once the carbon dioxide level and temperature of the stage incubator have stabilized, mount the 6-well culture plate on the inverted phase-contrast microscope's motorized stage.
6. Select and save several fields for acquisition of photographs; record every 10 min for 12 h. Regularly examine for any changes in stage location. If necessary, refocus the selected field.

3.3.4. Quantification

Merge individually saved picture frames to create a motion picture with ImagePro Plus software. The paths of individual cells can be reconstructed by computer-assisted cell tracking. The individual track ball movement generates data related to distance of cell migration in every picture frame and angle with which it deviated from the center of origin (Fig. 11.5A). These data can be used to reconstruct individual migration paths by Microsoft Office Excel. The vector length between two x-y coordinates can be used as a direct measure of distance for the calculation of actual cell velocity. The degree of directional migration of cells on various integrin ligands can be determined by calculating the average direct distance between the point of origin and end points of migrated cells (Table 11.1).

4. Roles of Integrins in Endothelial Angiogenic Programming

During processes of angiogenesis, the proliferating and migrating endothelial cells organize to form three-dimensional capillary networks. These processes begin with the transition of endothelial cells into a spindle-shaped morphology. Previous studies indicated that integrins play crucial roles in each of these processes (Mousa, 2002; Ruegg et al., 2002). Therefore, integrins in coordination with ECM proteins are believed to serve as master regulators of angiogenic programming in endothelial cells.

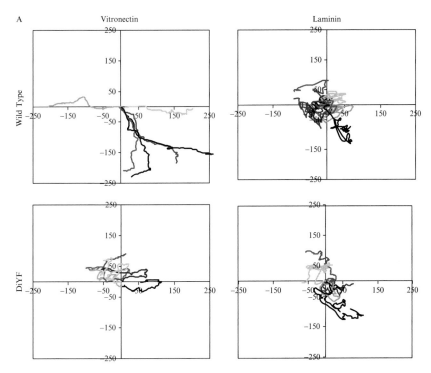

Figure 11.5 Assessment of integrin ligands in directional migration of endothelial cells. (A) Integrins in coordination with extracellular matrix proteins are known to regulate the processes of endothelial cell migration. Here we examined the roles of integrin and ECM proteins in directional migration of endothelial cells. Wild-type and DiYF mouse lung microvascular endothelial cells were grown to monolayers on vitronectin- and laminin-coated plates. A wound was created across the cell monolayer by scraping away a swath of cells, and detached cells were removed by washing two times with sterile 1× PBS. Migration of endothelial cells on matrix proteins were recorded with time-lapse video microscopy. Representative cell paths were generated as described in the text in the presence of 20 ng/ml VEGF over a period of 10 h. Migratory cell paths are reconstituted such that all paths start from the origin. Units of measure on x and y axes are micrometers. (Reproduced from Circulation Research, 2007; **101(6)**, 570–580 with permission.)

Integrins expressed on the endothelial cell surface are also known to regulate endothelium alignment and integration into solid, multicellular, precapillary cordlike structures that form an integrated polygonal network. During this vascular morphogenesis, extracellular matrix serves as an adhesive support and, through interaction with integrins, provides crucial signaling to regulate endothelial cell shape and contractility (Ingber, 2002). Here we describe methods to assess roles of integrins in endothelial capillary tube and cord formation in matrigel matrix. We also discuss methods for endothelial sprout formation with mouse aortic rings.

Table 11.1   Role of extracellular matrix in integrin-dependent directional migration of cells. Wild-type and DiYF endothelial cells were grown on vitronectin- and laminin-coated plates in monolayer. A wound was created by removing a layer of cells, and cellular migration was recorded by time-lapse video microscopy. Average total migration path length on vitronectin and laminin are indicated. The direct distance between the point of origin and the end of migration was used to assess the directional migration of cells.

	Total path length		Distance from origin	
	WT	*DiYF*	*WT*	*DiYF*
VITRONECTIN	308 μm	283 μm	229 μm	77 μm
LAMININ	532 μm	472 μm	71 μm	77 μm

4.1. Precapillary cord formation assay

4.1.1. Materials

Low passage endothelial cells.
6- or 12-Well tissue culture plates.
Complete endothelial culture media supplemented with antibiotics.
0.05% Trypsin and 0.53 mM EDTA cell detachment solution.
10 ng/μl VEGF-A.
Matrigel.
Sterile 1× phosphate-buffered saline (pH 7.4).

4.1.2. Equipment

Inverted phase-contrast microscope

4.1.3. Procedure

1. Grow the primary endothelial cells on gelatin- or fibronectin-coated tissue culture flasks. Allow these endothelial cells to reach a semiconfluent stage.
2. Thaw Matrigel overnight at 4°. Pipette out 1 ml of Matrigel into each well of 6-well plates with a cooled pipette. Incubate the Matrigel-coated plate at 37° for 15 min and confirm the solidification of Matrigel by tilting the tissue culture plates.
3. Detach primary endothelial cells with trypsin-EDTA cell detachment solution. Remove the trypsin-EDTA from the cells completely to prevent Matrigel digestion by trypsin enzymatic activity.
4. Gently wash these detached cells once with sterile 1× phosphate-buffered saline and one more time with complete endothelial cell culture media.

5. Count the cells with a hemacytometer and adjust the cell number to 2×10^5 cells/ml with complete culture media. Add 1 ml of cell culture suspension onto solidified Matrigel with or without 20 ng/ml VEGF-A. Incubate this tissue culture plate at 37° in a humidified CO_2 incubator for 12 h.

6. At the end of incubation, remove the culture media without disturbing the attached cells. Carefully overlay a layer of Matrigel, supplemented with or without 40 ng/ml VEGF, on top of the first layer. Incubate this culture plate at 37° for 20 min to solidify the fresh layer of Matrigel.

7. Confirm the solidification of the fresh Matrigel layer by tilting the tissue culture plate, then overlay the Matrigel layers with 2 ml of complete endothelial culture media. Incubate the culture plate at 37° in a humidified CO_2 incubator for at least 7 to 14 days. Observe this endothelial culture periodically for formation of endothelial cords with an inverted phase-contrast microscope.

4.1.4. Quantification

Usually, endothelial cords start to form approximately 7 days and continue until 14 to 18 days. Take photographs of nonoverlapping fields every alternate day. At the end of the experiment, count the number of cords formed in each field from several representative fields (Fig. 11.6A and B). Data can be represented as the number of cords formed per field or number of cords formed per square millimeter.

4.2. Tube formation assay

4.2.1. Materials

Low-passage endothelial cells.
12- or 24-Well tissue culture plates.
Complete endothelial culture media supplemented with antibiotics.
Serum-free culture media.
0.05% Trypsin and 0.53 mM EDTA cell detachment solution.
10 ng/μl VEGF-A.
Matrigel.
Sterile 1× phosphate-buffered saline (pH 7.4).

4.2.2. Equipment

Inverted phase-contrast microscope

4.2.3. Procedure

1. Maintain primary mouse or human endothelial cells on gelatin or fibronectin-coated tissue culture flasks. Coat 12- or 24-well plates with 0.5 ml or 0.25 ml of Matrigel as described earlier. Incubate the

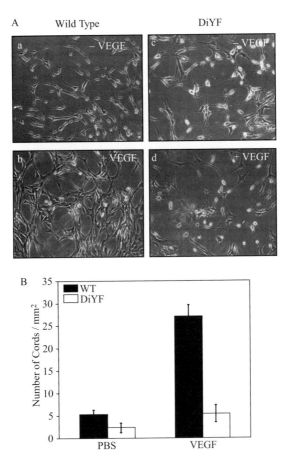

Figure 11.6 Endothelial cord formations in three-dimensional Matrigel matrix. Endothelial cell surface integrins are known to play very crucial roles in the morphogenesis of endothelial cells during processes of angiogenesis. Here we demonstrated the requirement of functional β3-integrin in these processes. (A and B) Functional β3-integrin is essential for endothelial cell organization into precapillary cords. WT and DiYF mouse lung endothelial cells were collected and resuspended in DMEM containing 10% FBS. Equal numbers of cells were seeded on Matrigel-coated plates, and cells were allowed to adhere. After 24 h, cells were overlaid with Matrigel, with or without 40 ng/mL VEGF, and maintained in culture for 6 to 8 days. (A) Three random fields were photographed periodically with a phase-contrast microscope (Leica). (B) Quantification of cord numbers per field ($P = 0.0004$). (Reproduced from The Journal of Experimental Medicine, 2006; **203**, 2495–2507. Copyright 2006 The Rockefeller University Press.)

Matrigel-coated plate at 37° for 20 min and confirm the solidification of Matrigel by tilting the plate.

2. Gently wash the cultured endothelial cells with sterile 1× phosphate-buffered saline. Detach the cultured endothelial cells with trypsin-EDTA

cell detachment solution. Carefully wash the detached cells two times with sterile 1× phosphate-buffered saline.

3. Resuspend the washed endothelial cells in reduced serum-containing medium (2% fetal bovine serum). Count the cells with a hemacytometer and adjust the cell numbers to 4×10^5 cells/ml by use of reduced serum medium. Add 1 ml or 0.5 ml of cell suspension to each well of 12-well or 24-well culture plates coated with Matrigel. Cell inoculations can be further supplemented with or without 20 ng/ml VEGF-A.

4. Incubate the cell-seeded tissue culture plate at 37° in a humidified CO_2 incubator for 8 to12 h. Generally, endothelial tubes start to appear at the end of 5 h and can be stable for up to 12 to 18 h. Cell suspensions supplemented with growth factors start to form endothelial tubes much earlier than cell suspensions devoid of growth factor.

5. Once a moderate number of endothelial tubes appears in the control group, take photographs of several representative fields.

4.2.4. Quantification

Bu use of ImagePro software, the degree of tube formation is quantified by measuring the length of tubes in control and growth factor–treated groups. Select at least five nonoverlapping random fields from each experiment to calculate average tube length. The degree of tube formation in various groups can be represented as the average tube length (μm)/mm^2 or average tube length (μm)/field (Fig. 11.7A and B).

4.3. Aortic ring assay

4.3.1. Materials

Aortas from 4- to 6-week-old healthy mice.
Anesthetic cocktail (ketamine, 1 ml; xylazine, 0.3 ml; saline, 1.5 ml).
Syringe (1 ml) fitted with 23-gauge needle.
Microdissection scissors and forceps.
Scalpel blade.
6-Well tissue culture plates.
Matrigel.
10 ng/μl VEGF-A.
Complete endothelial culture media supplemented with antibiotics.
Sterile 1× phosphate-buffered saline (pH 7.4).

4.3.2. Equipment

Inverted phase-contrast microscope.
Dissection microscope.
Flow cytometer (FACS).

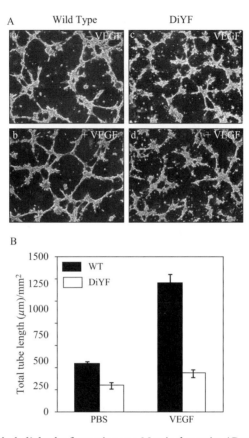

Figure 11.7 Endothelial tube formations on Matrigel matrix. (C and D) β3-integrin cytoplasmic tyrosine residues are critical for capillary tube formation. Single cell suspensions of WT and DiYF mouse lung ECs were transferred to Matrigel-coated plates and incubated at 37° for 8 h with or without 20 ng/mL VEGF. (C) Endothelial capillary tubes formed in Matrigel were photographed through an inverted phase-contrast microscope (Leica). (D) The lengths of tubes in random fields from each well were analyzed with ImagePro software. (Reproduced from The Journal of Experimental Medicine, 2006; **203**, 2495–2507. Copyright 2006 The Rockefeller University Press.)

4.3.3. Procedure

1. Anesthetize a pair of 4- to 6-week-old mice by subcutaneous injection of 50 μl of anesthetic cocktail. Quickly remove the thoracic aorta with microdissection forceps and collect them in ice-cold 1× sterile phosphate-buffered saline.

2. Place the aorta in a 60-mm petri dish and gently flush the aorta with sterile ice-cold 1× sterile phosphate-buffered saline, with a 1 ml syringe fitted with a 23-gauge needle to remove blood clots.

3. Carefully remove fibroadipose tissue and small lateral branch blood vessels with fine forceps under a dissection microscope. Cut the aorta into 1-mm rings with a sterile scalpel blade under laminar airflow. Wash these aortic rings with sterile $1\times$ phosphate-buffered saline and transfer them to a petri dish filled with complete endothelial growth medium.

4. Place a 6-well tissue culture plate on ice to cool it down (which prevents immediate solidification of Matrigel when poured into the wells). Homogeneously coat the tissue culture wells with 1 ml of Matrigel without introducing any air bubbles. If growth factors need to be added, aliquot the required amount of Matrigel into precooled tubes and mix with growth factor by pipetting up and down several times.

5. Select four to five freshly prepared aortic rings and insert them deep into the Matrigel layer. If necessary, pour a second layer of Matrigel on top of the aortic rings. Incubate these plates at $37°$ for 15 min and allow the Matrigel to solidify. Confirm Matrigel solidification by carefully tilting the culture plate.

6. Carefully add complete endothelial cell growth medium, with or without VEGF, and incubate the culture plate at $37°$ in a humidified CO_2 incubator. Endothelial cell sprouts will start to appear on the end of the second day and will grow rapidly for 5 to 12 days.

7. Observe the aortic ring sprouts periodically under the phase-contrast microscope; photographs should be taken between 4 and 7 days. Avoid taking photographs of overcrowded aortic sprouts. Alternately, each aortic ring can be numbered and photographs can be taken periodically. Analysis of these photographs demonstrates endothelial sprout formation from aortic rings over a period of time.

4.3.4. Quantification

Endothelial sprout growth from aortic rings can be quantified in several ways. When endothelial sprouts formed from aortic rings are low density, the number of sprouts formed from each aortic ring in each type of treatment can be counted and represented as the number of sprouts per aortic ring (Fig. 11.8A and B). When the aortic sprouts formed are dense, endothelial cells can be harvested by digesting Matrigel with collagenase/dispase solution. Fix these cells with 4% paraformaldehyde in $1\times$ PBS for 15 min. These cells can be stained with FITC-labeled anti-CD31 antibody and analyzed for CD31-positive cell populations by flow cytometry (FACS).

5. Integrin Affinity Modulation and Its Analysis

Integrins become "activated" to achieve a high-affinity state for their soluble ligands as a result of conformational modifications within the integrin heterodimers (Calderwood, 2004). Integrin activation actually precedes

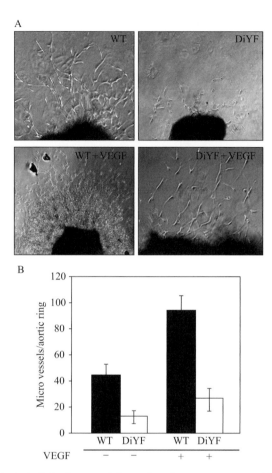

Figure 11.8   Endothelial microvessel formations from mouse aorta. (A and B) The $\beta3$-integrin cytoplasmic tyrosine motif is required for normal microvessel growth. (A) Wild-type (upper panel) and DiYF (lower panel) mouse aortic rings were embedded in Matrigel and maintained at 37° for 6 days. Microvessel outgrowths from aortic rings were observed periodically and photographed by use of a Leica DMIRB phase-contrast microscope. (B) Numbers of microvessel sprouts from aortic rings were counted and are represented by bar diagram. (Reproduced from The Journal of Experimental Medicine, 2006; **203**, 2495–2507. Copyright 2006 The Rockefeller University Press.)

ligand binding (Humphries *et al.*, 2003). Although integrin activation can be only detected by structural analyses, subsequent ligand binding can be measured biochemically. Here we describe the method to assess the ligand binding activities of $\alpha v \beta3$-integrin expressed on the endothelial cell surface with the natural ligand fibrinogen or the genetically engineered ligand WOW-1 Fab antibody.

5.1. Fibrinogen and WOW-1 binding assay

5.1.1. Materials

Cell line or primary cell culture in which integrin activation needs to be
   studied.
Complete tissue culture media supplemented with antibiotic.
FITC-labeled fibrinogen.
WOW-1 Fab fragments.
FITC-labeled anti-mouse IgG.
Growth factor such as VEGF-A.
4% Paraformaldehyde in $1\times$ PBS.
$1\times$ Phosphate-buffered saline (pH 7.4).

5.1.2. Equipment

Flow cytometer (fluorescence activated cell sorter, FACS).

5.1.3. Procedure

1. Grow the cell line or primary cell culture in 6-well tissue culture plates.
   Allow these cells to reach 70% confluence. Before the experiment gently
   wash these cells with copious amounts of $1\times$ phosphate-buffered saline.
2. Serum-starve the cells for 4 to 6 h in serum-free media. Wash the cells
   one more time with $1\times$ phosphate-buffered saline.
3. (a) Fibrinogen binding: Further incubate the serum-starved cells with or
   without growth factor (20 ng/ml VEGF or bFGF, depending on cell
   type) and 200 nM fluorescein isothiocyanate (FITC)–labeled fibrinogen
   in serum-free medium. Incubate the tissue culture plate in a humidified
   CO_2 incubator set to 37° for 30 min. At the end of the incubation, gently
   wash these cells with $1\times$ phosphate-buffered saline. Detach the cells by
   gently scraping the growth flask with a polyethylene cell lifter. Fix the
   detached cells immediately with 4% paraformaldehyde in phosphate-
   buffered saline. Wash these cells one more time with $1\times$ phosphate-
   buffered saline to remove trace amounts of fixative. (b) WOW-1 binding:
   To examine the $\alpha v\beta 3$ activation status, incubate serum-starved cells with
   appropriate amounts of growth factor (VEGF, bFGF, EGF, or PDGF
   depending on cell type) and 30 μg/ml WOW-1 Fab in serum-free
   medium. Incubate the tissue culture plate in a humidified CO_2 incubator
   at 37° for 30 min. At the end of the incubation, gently wash these cells
   with $1\times$ phosphate-buffered saline. Detach the cells by gently scraping
   with a polyethylene cell lifter. Fix the detached cells immediately with
   4% paraformaldehyde in phosphate-buffered saline. Wash the cells one
   more time with $1\times$ phosphate-buffered saline to remove trace amounts
   of fixative. Further incubate these cells with 10 μg/ml FITC-conjugated

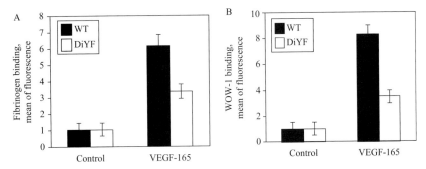

Figure 11.9 Integrin affinity modulations. Determination of integrin affinity modulation by natural and genetically engineered ligand binding assays. (A) Fibrinogen binding assay. Serum-starved WT and DiYF mouse lung ECs were incubated with FITC-labeled fibrinogen in the presence or absence of 20 ng/mL VEGF at 37°. These cells were fixed, washed, and analyzed by flow cytometry. Bars represent mean fluorescence intensity of several independent experiments performed in triplicate. (B) WOW-1 binding assay. Serum-starved WT and DiYF mouse lung endothelial cells were incubated with WOW-1 Fab fragments in the presence or absence of 20 ng/mL VEGF at 37°. These cells were washed and further incubated with FITC-conjugated goat anti-mouse IgG for 30 min. Cells were fixed, washed, and analyzed by flow cytometry. Bars represent mean fluorescence intensity of cells from three independent experiments performed in triplicate. (Reproduced from The Journal of Experimental Medicine, 2006; **203**, 2495–2507. Copyright 2006 The Rockefeller University Press.)

goat anti-mouse IgG in 2% bovine serum albumin freshly dissolved in 1× phosphate-buffered saline. Wash the cells two more times with 1× phosphate-buffered saline.

5.1.4. Quantification

Measure the mean florescence intensity (MFI) with a flow cytometer (FACS). Data can be represented as MFI from triplicate experiments (Fig. 11.9A and B). Generally, the mean florescence intensity of the control group is assigned the value of 1, and fold changes over control are indicated.

6. INTEGRIN AND ECM ROLES IN ENDOTHELIAL CELL PROLIFERATION AND APOPTOSIS

Integrin-mediated interactions with extracellular matrix proteins play important roles in many fundamental aspects of endothelial cell proliferation and apoptosis (Malik, 1997). When cells are completely devoid of integrin-dependent anchorage to extracellular matrix, they undergo a form of apoptosis that has been termed "anoikis" (Frisch and Screaton, 2001).

In addition, integrins coordinate signaling events underlying apoptosis (programmed cell death) (Reddig and Juliano, 2005). In endothelial cells, inhibition of $\alpha v \beta 3$-, $\alpha v \beta 5$-, or $\alpha 5 \beta 1$-integrin functions or their ligation could lead to apoptosis. Thus, several individual integrins have been implicated in cell protection from apoptosis and in the regulation of cellular proliferation induced by growth factors (Liu *et al.*, 2000). In the following subchapter, we summarize a general protocol to assess endothelial cell proliferation and apoptosis on various extracellular matrix proteins.

6.1. Endothelial cell proliferation assays

6.1.1. Materials

Mouse or human primary endothelial cell cultures.
Complete endothelial culture medium.
Integrin ligand-coated 96-well tissue culture plates.
Serum-free culture media.
Trypsin/EDTA.
Trypsin neutralizing solution.
$1\times$ Phosphate-buffered saline.
Hemacytometer.
^{3}H-thymidine.
Sodium dodecyl sulfate.
Sodium hydroxide.

6.1.2. Equipment

Orbital shaker.
β (scintillation) counter.

6.1.3. Procedure

1. Grow low-passage endothelial cells on gelatin-coated plates. Allow these cells to reach 70% confluence. It is advised to use actively proliferating cells for cell proliferation assays, avoiding overcrowded cell cultures for these assays.
2. Detach the cells with trypsin-EDTA cell dissociation solution. Gently wash the cells with $1\times$ phosphate-buffered saline to remove trypsin-EDTA.
3. Wash the detached cells one more time with trypsin-neutralizing solution to inactivate trypsin enzymatic activity. Resuspend the cells in complete endothelial cell culture media.
4. Count cells in suspension with a hemacytometer and adjust the cell number to 1×10^3 cells/0.2 ml. Add 0.2 ml of cell suspension to each well of a 96-well plate coated with integrin ligand.

5. Incubate the tissue culture plate in a humidified CO_2 incubator set at 37°
   for 6 h. Remove unattached/floating cells by washing two times with
   1× phosphate-buffered saline. Further incubate the attached cells with
   low serum containing media (2% fetal bovine serum), supplemented
   with or without an appropriate concentration of growth factor, for 24 h.
6. Add 1 μCi of ^{3}H-thymidine to each well of the 96-well plate and further
   incubate for an additional 24 h at 37° as before.
7. Carefully remove culture media containing ^{3}H-thymidine and discard it
   into a radioactive waste container. Wash the cells two times with 200 μl
   of 1× phosphate-buffered saline and discard the wash solution into the
   radioactive container.
8. Add 100 μl of 2% SDS with 0.2 M NaOH and incubate for 1 h on an
   orbital shaker. Collect the sample from each well, add to scintillation
   vials with scintillation fluid and count in a β-counter.

6.1.4. Quantification

The β-counter reading is generally obtained as radioactive counts per
minute (cpm). When cells are treated with proliferating agents or growth
factors, the data will be typically represented as fold increase in the rate of
proliferation over control cells. The rate of proliferation in the control
group is assigned the value of 100%, and relative increases or decreases in
the rate of proliferation are then indicated (Fig. 11.10A). Statistical significance in the rate of proliferation between two groups can be calculated by
Student's t test.

6.2. Endothelial cell apoptosis assays

6.2.1. Materials

Cultured mouse or human primary endothelial cells.
Complete endothelial culture medium.
Integrin ligand-coated 6-well tissue culture plate.
Serum-free culture media.
Trypsin/EDTA cell dissociation solution.
Annexin V–FITC.
Propidium iodide.
1× phosphate-buffered saline.
20 mM EDTA in 1× phosphate-buffered saline.
Binding buffer (10 mM HEPES/NaOH, pH 7.4, 140 mM NaCl, 2.5 mM
   $CaCl_2$).

6.2.2. Equipment

Hemacytometer.
Flow cytometer (FACS).

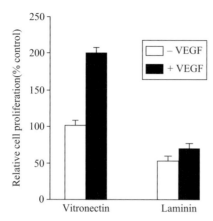

Figure 11.10 Cellular proliferation on integrin ligands. (A) Role of extracellular matrix in integrin-dependent cellular proliferation. HUVECs were grown on vitronectin- or laminin-coated plates. These cells were stimulated with 20 ng/ml VEGF-165 for 24 h. The cells were washed and further incubated with 1 μCi of ^{3}H-thymidine for an additional 24 h at 37° in a humidified incubator. Cells were lysed with alkaline lysis buffer and ^{3}H-thymidine incorporation was measured by β-scintillation counter. The rate of HUVEC proliferation on the vitronectin-coated plate was assigned the value of 100%, and the relative rate of proliferation over control is indicated (A).

6.2.3. Procedure

1. Detach actively proliferating low-passage endothelial cells grown on gelatin-coated plates with trypsin/EDTA cell dissociation solution. Wash the cells two times with copious amount of 1× phosphate-buffered saline.
2. Resuspend the cells in complete endothelial culture media supplemented with antibiotics. Allow the cells to attach and grow for an additional 12 h.
3. Remove unattached cells by washing two times with 1× phosphate-buffered saline. Replace the media with complete endothelial culture media or low serum containing media to incubate with apoptotic inducer.
4. Stimulate endothelial cells grown on different integrin ligands with an appropriate concentration of apoptotic inducer for a defined period of time.
5. Detach these cells after termination of treatment by incubating with 20 mM EDTA made in 1× phosphate-buffered saline. Wash these cells two times with binding buffer and resuspend in binding buffer.
6. Incubate 3×10^5 cells from each treatment in 140 nM annexin V-FITC and 5 μg/sample of propidium iodide for 15 min at room temperature in the dark.
7. At the end of the incubation, process immediately for flow cytometry analysis.

6.2.4. Quantification

Analyze the annexin V-FITC binding with a flow cytometer (FACS). Set the instrument to 488-nm excitation wavelength and detect emission at 530 nm. The FITC signal output is represented as FL1, and propidium iodide staining (detected by the phycoerythrin emission signal detector at ~600 nm) is represented as FL2. Cells stained positive both for annexin v and propidium iodide are considered as a true apoptotic cell population. The number of apoptotic cells plated in the control group is set as 100%, and fold changes over control are then indicated.

ACKNOWLEDGMENT

We acknowledge financial support from the U. S. National Institutes of Health (HL071625 and HL073311 to TVB) and the American Heart Association (0625271B to GHM). We thank Dr. Judith A. Drazba and scientists at the Image Core Facility at the Cleveland Clinic for help in image analysis. We thank The Rockefeller University Press and Lippincott Williams & Wilkins to provide permission to reuse previously published materials.

REFERENCES

Alam, N., Goel, H. L., Zarif, M. J., Butterfield, J. E., Perkins, H. M., Sansoucy, B. G., Sawyer, T. K., and Languino, L. R. (2007). The integrin-growth factor receptor duet. *J. Cell. Physiol.* **213,** 649–653.

Alghisi, G. C., and Ruegg, C. (2006). Vascular integrins in tumor angiogenesis: Mediators and therapeutic targets. *Endothelium* **13,** 113–135.

Bazzoni, G., and Hemler, M. E. (1998). Are changes in integrin affinity and conformation overemphasized? *Trends Biochem. Sci.* **23,** 30–34.

Berrier, A. L., and Yamada, K. M. (2007). Cell-matrix adhesion. *J. Cell. Physiol.* **213,** 565–573.

Brower, D. L., Brower, S. M., Hayward, D. C., and Ball, E. E. (1997). Molecular evolution of integrins: Genes encoding integrin beta subunits from a coral and a sponge. *Proc. Natl. Acad. Sci. USA* **94,** 9182–9187.

Burke, R. D. (1999). Invertebrate integrins: Structure, function, and evolution. *Int. Rev. Cytol.* **191,** 257–284.

Byzova, T. V., Goldman, C. K., Pampori, N., Thomas, K. A., Bett, A., Shattil, S. J., and Plow, E. F. (2000). A mechanism for modulation of cellular responses to VEGF: Activation of the integrins. *Mol. Cell* **6,** 851–860.

Byzova, T. V., Rabbani, R., D'Souza, S. E., and Plow, E. F. (1998). Role of integrin alpha (v)beta3 in vascular biology. *Thromb. Haemost.* **80,** 726–734.

Calderwood, D. A. (2004). Integrin activation. *J. Cell Sci.* **117,** 657–666.

Cram, E. J., and Schwarzbauer, J. E. (2004). The talin wags the dog: New insights into integrin activation. *Trends Cell Biol.* **14,** 55–57.

Davis, G. E., Bayless, K. J., and Mavila, A. (2002). Molecular basis of endothelial cell morphogenesis in three-dimensional extracellular matrices. *Anat. Rec.* **268,** 252–275.

Eliceiri, B. P., and Cheresh, D. A. (2001). Adhesion events in angiogenesis. *Curr. Opin. Cell Biol.* **13,** 563–568.

Felding-Habermann, B., and Cheresh, D. A. (1993). Vitronectin and its receptors. *Curr. Opin. Cell Biol.* **5,** 864–868.

Frisch, S. M., and Screaton, R. A. (2001). Anoikis mechanisms. *Curr. Opin. Cell Biol.* **13,** 555–562.

Ginsberg, M. H., Partridge, A., and Shattil, S. J. (2005). Integrin regulation. *Curr. Opin. Cell Biol.* **17,** 509–516.

Hood, J. D., and Cheresh, D. A. (2002). Role of integrins in cell invasion and migration. *Nat. Rev. Cancer* **2,** 91–100.

Humphries, M. J., McEwan, P. A., Barton, S. J., Buckley, P. A., Bella, J., and Mould, A. P. (2003). Integrin structure: Heady advances in ligand binding, but activation still makes the knees wobble. *Trends Biochem. Sci.* **28,** 313–320.

Hynes, R. O. (1987). Integrins: A family of cell surface receptors. *Cell* **48,** 549–554.

Hynes, R. O. (2002). Integrins: Bidirectional, allosteric signaling machines. *Cell* **110,** 673–687.

Ingber, D. E. (2002). Mechanical signaling and the cellular response to extracellular matrix in angiogenesis and cardiovascular physiology. *Circ. Res.* **91,** 877–887.

Klein, E. A., Yung, Y., Castagnino, P., Kothapalli, D., and Assoian, R. K. (2007). Cell adhesion, cellular tension, and cell cycle control. *Methods Enzymol.* **426,** 155–175.

Liu, W., Ahmad, S. A., Reinmuth, N., Shaheen, R. M., Jung, Y. D., Fan, F., and Ellis, L. M. (2000). Endothelial cell survival and apoptosis in the tumor vasculature. *Apoptosis* **5,** 323–328.

Luo, B. H., and Springer, T. A. (2006). Integrin structures and conformational signaling. *Curr. Opin. Cell Biol.* **18,** 579–586.

Mahabeleshwar, G. H., Feng, W., Phillips, D. R., and Byzova, T. V. (2006). Integrin signaling is critical for pathological angiogenesis. *J. Exp. Med.* **203,** 2495–2507.

Mahabeleshwar, G. H., Feng, W., Reddy, K., Plow, E. F., and Byzova, T. V. (2007). Mechanisms of integrin-vascular endothelial growth factor receptor cross-activation in angiogenesis. *Circ. Res.* **101,** 570–580.

Malik, R. K. (1997). Regulation of apoptosis by integrin receptors. *J. Pediatr. Hematol. Oncol.* **19,** 541–545.

Moissoglu, K., and Schwartz, M. A. (2006). Integrin signalling in directed cell migration. *Biol. Cell* **98,** 547–555.

Mousa, S. A. (2002). Vitronectin receptors in vascular disorders. *Curr. Opin. Invest. Drugs* **3,** 1191–1195.

Pancer, Z., Kruse, M., Muller, I., and Muller, W. E. (1997). On the origin of Metazoan adhesion receptors: Cloning of integrin alpha subunit from the sponge Geodia cydonium. *Mol. Biol. Evol.* **14,** 391–398.

Phillips, D. R., Nannizzi-Alaimo, L., and Prasad, K. S. (2001a). Beta3 tyrosine phosphorylation in alphaIIbbeta3 (platelet membrane GP IIb-IIIa) outside-in integrin signaling. *Thromb. Haemost.* **86,** 246–258.

Phillips, D. R., Prasad, K. S., Manganello, J., Bao, M., and Nannizzi-Alaimo, L. (2001b). Integrin tyrosine phosphorylation in platelet signaling. *Curr. Opin. Cell Biol.* **13,** 546–554.

Plow, E. F., Haas, T. A., Zhang, L., Loftus, J., and Smith, J. W. (2000). Ligand binding to integrins. *J. Biol. Chem.* **275,** 21785–21788.

Qin, J., Vinogradova, O., and Plow, E. F. (2004). Integrin bidirectional signaling: A molecular view. *PloS. Biol.* **2,** e169.

Ramsay, A. G., Marshall, J. F., and Hart, I. R. (2007). Integrin trafficking and its role in cancer metastasis. *Cancer Metastasis Rev.* **26,** 567–578.

Ratnikov, B. I., Partridge, A. W., and Ginsberg, M. H. (2005). Integrin activation by talin. *J. Thromb. Haemost.* **3,** 1783–1790.

Reddig, P. J., and Juliano, R. L. (2005). Clinging to life: Cell to matrix adhesion and cell survival. *Cancer Metastasis Rev.* **24,** 425–439.

Rose, D. M., Alon, R., and Ginsberg, M. H. (2007). Integrin modulation and signaling in leukocyte adhesion and migration. *Immunol. Rev.* **218,** 126–134.

Ruegg, C., Dormond, O., and Foletti, A. (2002). Suppression of tumor angiogenesis through the inhibition of integrin function and signaling in endothelial cells: Which side to target? *Endothelium* **9,** 151–160.

Sepulveda, J. L., Gkretsi, V., and Wu, C. (2005). Assembly and signaling of adhesion complexes. *Curr. Top. Dev. Biol.* **68,** 183–225.

Springer, T. A., and Wang, J. H. (2004). The three-dimensional structure of integrins and their ligands, and conformational regulation of cell adhesion. *Adv. Protein Chem.* **68,** 29–63.

Stromblad, S., and Cheresh, D. A. (1996). Integrins, angiogenesis and vascular cell survival. *Chem. Biol.* **3,** 881–885.

Stupack, D. G. (2007). The biology of integrins. *Oncology (Williston Park)* **21,** 6–12.

Stupack, D. G., and Cheresh, D. A. (2002). ECM remodeling regulates angiogenesis: Endothelial integrins look for new ligands. *Sci. STKE* **2002,** PE7.

Stupack, D. G., and Cheresh, D. A. (2003). Apoptotic cues from the extracellular matrix: Regulators of angiogenesis. *Oncogene* **22,** 9022–9029.

Tanjore, H., Zeisberg, E. M., Gerami-Naini, B., and Kalluri, R. (2008). Beta1 integrin expression on endothelial cells is required for angiogenesis but not for vasculogenesis. *Dev. Dyn.* **237,** 75–82.

Tucker, G. C. (2002). Inhibitors of integrins. *Curr. Opin. Pharmacol.* **2,** 394–402.

Tucker, G. C. (2006). Integrins: Molecular targets in cancer therapy. *Curr. Oncol. Rep.* **8,** 96–103.

Tzima, E., Irani-Tehrani, M., Kiosses, W. B., Dejana, E., Schultz, D. A., Engelhardt, B., Cao, G., DeLisser, H., and Schwartz, M. A. (2005). A mechanosensory complex that mediates the endothelial cell response to fluid shear stress. *Nature* **437,** 426–431.

Wierzbicka-Patynowski, I., and Schwarzbauer, J. E. (2003). The ins and outs of fibronectin matrix assembly. *J. Cell Sci.* **116,** 3269–3276.

Zutter, M. M. (2007). Integrin-mediated adhesion: Tipping the balance between chemosensitivity and chemoresistance. *Adv. Exp. Med. Biol.* **608,** 87–100.

CHAPTER TWELVE

METHODS FOR STUDYING MECHANICAL CONTROL OF ANGIOGENESIS BY THE CYTOSKELETON AND EXTRACELLULAR MATRIX

Akiko Mammoto, Julia E. Sero, Tadanori Mammoto, *and* Donald E. Ingber

Contents

Vascular Biology Program, Departments of Pathology & Surgery, Children's Hospital and Harvard Medical School, Boston, Massachusetts

Methods in Enzymology, Volume 443
ISSN 0076-6879, DOI: 10.1016/S0076-6879(08)02012-0

Abstract

Mechanical forces that capillary endothelial cells generate in their cytoskeleton and exert on their extracellular matrix adhesions feed back to modulate cell sensitivity to soluble angiogenic factors, and thereby control vascular development. Here we describe various genetic, biochemical, and engineering methods that can be used to study, manipulate, and probe this physical mechanism of developmental control. These techniques are useful as *in vitro* angiogenesis models and for analyzing the molecular and biophysical basis of vascular control.

1. INTRODUCTION

Most past work on control of angiogenesis focused on the role of soluble angiogenic factors that stimulate capillary endothelial (CE) cell growth and formation of new capillary blood vessels. Although these factors trigger the angiogenic response, cell sensitivity to these soluble cues and resulting vascular development are governed by mechanical forces that cells generate in their cytoskeleton and exert on their extracellular matrix (ECM) adhesions, which feed back to produce changes in cell shape and cytoskeletal structure (Huang and Ingber, 1999; Ingber and Folkman, 1989a; Ingber *et al.*, 1985). Vascular remodeling and directional outgrowth of capillaries also can be influenced by exogenous mechanical stress (e.g., ECM strain because of hemodynamic stresses) *in vitro* as well as *in vivo* (Matsumoto *et al.*, 2007; Pietramaggiori *et al.*, 2007; Tzima, 2006). Thus, there has been great interest in experimental systems that permit analysis of this mechanical form of angiogenic control.

In this chapter, we describe various methods that may be used to control, manipulate, and probe physical interactions between capillary cells and their ECM adhesions, while simultaneously measuring effects on vascular cell behavior and tissue development *in vitro* and *in vivo*. These methods range from use of microfabrication techniques to control the size, shape, and position of cell–ECM contacts to biochemical and genetic methods to manipulate cytoskeletal tension-generation mechanisms that are controlled by small Rho GTPases. Use of these techniques has led to recognition that physical distortion of cell shape and the actin cytoskeleton control cell cycle progression, as well as directional motility in CE cells, and that Rho signaling is critical for angiogenic control *in vitro* and *in vivo*.

2. Control of CE Cell Behavior with Different ECM Coating Densities

More than 20 years ago, we proposed that local variations in physical interactions between cells and their ECM might control the spatial differentials of cell growth and function that drive morphogenesis of tissues, such as branching capillary networks (Huang and Ingber, 1999; Ingber and Folkman, 1989a; Ingber *et al.*, 1985). To explore this hypothesis in the context of angiogenesis, it was necessary to develop methods that would allow us to vary physical interactions between cells and their ECM adhesions in a controlled manner. In early studies, we developed a method to control cell–ECM contact formation by varying the density of ECM molecules coated on otherwise nonadhesive plastic dishes. CE cells spread and proliferated on surfaces with high ECM coating densities, whereas they rounded and failed to grow on low-coating concentrations, even though all cells were stimulated with saturating amounts of soluble angiogenic mitogens (Ingber, 1990; Ingber and Folkman, 1989b; Ingber *et al.*, 1987). Importantly, when CE cells were cultured at a high density on a moderate coating density that only partially resisted cell traction forces, these cells collectively retracted and differentiated into branching capillary networks lined by hollow endothelial-lined tubular structures (Fig. 12.1) (Ingber and Folkman, 1989b). Moreover, the importance of mechanics for this form of differentiation control was shown by the fact that tube formation also could be induced on the highest ECM density that normally promotes cell growth by increasing the cell plating density and, thereby, increasing cumulative cell traction forces (Fig. 12.1). This same technique was shown to be useful for control of growth and differentiation of other cell types, including primary hepatocytes (Mooney *et al.*, 1992) and vascular smooth muscle cells (Kim *et al.*, 1998; 1999; Lee *et al.*, 1998). In fact, this is a popular technique, because it provides a method to exert fine control over cell shape

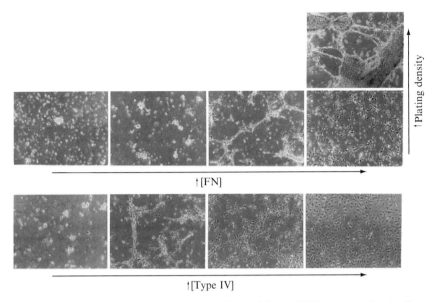

Figure 12.1 Control of angiogenesis *in vitro* by modulating ECM coating density. Bacteriologic dishes were precoated with 10, 50, 100, or 2500 ng/cm^2 (from left to right) of FN or type IV collagen (type IV). The highest concentration shown was saturating for both CE cell attachment and spreading. These phase-contrast views show that cell attachment and spreading increase in parallel with the number of ECM molecules available for cell attachment. Tube formation was only observed on dishes of intermediate adhesivity when cells were plated at a moderate density (4 × 10^4 cells/cm^2). CE cells formed extensive capillary networks on the highest FN coating density when higher numbers were plated (2 × 10^5 cells/cm^2; top right). Tube formation was observed on lower coating densities on type IV than on FN; type IV promoted more extensive cell attachment and spreading at all coating concentrations (original magnification of 50×; reprinted with permission from Ingber and Folkman, 1989b).

and function with commercially available substrates and ECM molecules that can be accomplished in any laboratory around the world.

2.1. Cell culture

CE cells isolated from bovine adrenal cortex (Folkman *et al.*, 1979) are maintained for 10 to 15 passages on gelatin-coated tissue culture dishes in low-glucose Dulbecco's modified Eagles medium (DMEM; Invitrogen) supplemented with 10% fetal calf serum (FCS) (Hyclone), 10 mM HEPES (JRH-Biosciences), and L-glutamine (0.292 mg/ml), penicillin (100 U/ml), streptomycin (100 μg/ml) at 37° under 10% CO$_2$ (Chen *et al.*, 1997; Ingber, 1990; Ingber and Folkman, 1989b; Matthews *et al.*, 2006; Numaguchi *et al.*, 2003).

2.2. Matrix coating procedures

To control cell–ECM contact formation, bacteriologic plastic substrates (35-mm petri dishes; Falcon, Lincoln Park, NJ) or multichamber glass culture slides (Lab-Tek; Miles) are precoated with fibronectin (FN; Cappel Laboratories, Malvern, PA) or type IV collagen (Calbiochem, San Diego, CA) at various densities (22 ng/cm^2 to 666 ng/cm^2) with a carbonate buffer-coating technique (Huang et al., 1998; Ingber, 1990; Ingber et al., 1987; Mammoto et al., 2004). Various amounts of human serum FN and collagen IV are dissolved in 0.1 M carbonate buffer (pH 9.4) and allowed to adsorb for 24 h at 4°. Dishes are washed with phosphate-buffered saline (PBS), DMEM, and blocked with 1% bovine serum albumin (BSA, Fraction V; Intergen, Purchase, NY) in DMEM for 1 h at 37° before use. CE cells are plated on the coated dishes in DMEM supplemented with 5 μg/ml transferrin (Collaborative Research, Lexington, MA), 10 μg/ml high-density lipoprotein (specific gravity, 1.063 to 1.21 g/cm^3; Perimmune, Rockville, MD), and 2 ng/ml recombinant basic FGF (Takeda Chemical Industries, Osaka, Japan). BSA (10 mg/ml) is also included as colloid in some studies without altering our results.

2.3. Analytical methods

CE cells are plated in defined medium on the ECM-coated substrates at a moderate (2 to 5 $\times$ 10^4 cells/cm^2) density and cultured for hours to days to analyze the effects on cell adhesion, spreading, cytoskeletal organization, growth, and differentiation with time-lapse cinematography, electron microscopic analysis, and measurement of DNA synthesis with [^{3}H] thymidine (Ingber and Folkman, 1989b). Extensive branching capillary networks consistently form within 24 to 48 h when CE cells are plated on dishes of intermediate adhesivity (100 to 500 ng FN or type IV collagen/cm^2), although tube formation also can be induced on higher ECM densities (>2500 ng FN or type IV collagen/cm^2) by plating cells at a higher density (1 to 3 $\times$ 10^5 cells/ cm^2) (Fig. 12.1) (Ingber and Folkman, 1989b). Phase-contrast images of living cells are recorded with in inverted microscope (Diaphot; Nikon Inc., Garden City, NY) with film (Plus–X-pan; Eastman Kodak Co., Rochester, NY). For electron microscopic analysis, reorganized capillary tube are fixed in 2.5% glutaraldehyde/ 2% paraformaldehyde in 0.1 M sodium cacodylate buffer, pH7.4, post-fixed in 1% osmium tetroxide, dehydrated in a graded series of alcohols, and embedded in Epon. Thin sections (800 A thick) are cut on an UltraCut microtome (Reicher Scientific Instruments, Buffalo, NY), counterstained with uranyl acetate and lead citrate, and studied under an electron microscope (No. 100B; JOEL USA, Cranford, NJ). To measure of DNA synthesis, light microscopic autoradiography is carried out by adding [^{3}H] thymidine

(1 mCi/ml final concentration; New England Nuclear, Boston, MA) to the defined medium during the first 24 h of culture (Ingber, 1990; Ingber and Folkman, 1989b). Radiolabelled CE cells are fixed with Karnovsky's solution, dehydrated with methanol, and overlaid with nuclear track emulsion (NTB-2; Eastman Kodak Co.). Autoradiographic grains are developed with D-19 developer (Eastman Kodak Co.).

3. CONTROL OF CE CELL FATE SWITCHING BY USE OF MICROFABRICATED ECM ISLANDS

Although the ECM coating-density method for cell shape control is powerful, it may be complicated by the fact that altering ECM-coating densities also may change the degree of integrin receptor clustering on the cell surface, which can alter intracellular biochemical signaling responses independent of cell shape (Schwartz *et al.*, 1991). We, therefore, developed another technique that permits one to vary cell shape and cytoskeletal distortion while maintaining the ECM coating density and concentration of soluble angiogenic factors constant. This approach involved use of a microcontact printing technique (Kumar and Whitesides, 1994; Prime and Whitesides, 1991) to microfabricate ECM islands of defined size, shape, and position on the micrometer scale that are the same size as individual cultured cells, surrounded by nonadhesive barrier regions (Fig. 12.2A) (Chen *et al.*, 1997; Singhvi *et al.*, 1994). By use of this method, we showed that single adherent, growth factor–stimulated CE cells increase their ability to pass through the G1/S transition and proliferate when they are cultured on large ECM islands (>2500 μm^2) that promote cell spreading, whereas cells turn off growth and become quiescent on intermediate sized islands (1000 to 2000 μm^2); the same CE cells switch on apoptosis (programmed cell death) when spreading is completely inhibited by plating cells on tiny (<500 μm^2) islands (Chen *et al.*, 1997). Furthermore, when CE cells are cultured on long thin (10 to 30 μm) ECM islands in the same growth factor–containing medium, they differentiate into linear hollow capillary tubes (Dike *et al.*, 1999). The same micropatterned ECM substrates have been shown to exert similar control over cell fate switching in many other kinds of cells as well, including vascular smooth muscle cells, hepatocytes, and fibroblasts (Chen *et al.*, 2000; Dike *et al.*, 1999; Thakar *et al.*, 2003).

Directional cell motility is critical for normal tissue development and the immune response, angiogenesis, and cancer metastasis. Most work on directional control of cell movement has focused on gradients of soluble chemokines (Ridley *et al.*, 2003) that, in turn, generate intracellular gradients of signal transduction (Ridley *et al.*, 2003; Van Haastert and Devreotes, 2004). However, cell movement also can be influenced by physical

A

B

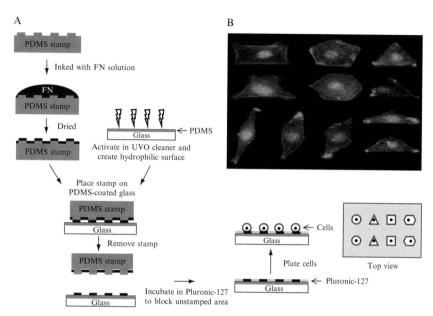

Figure 12.2 Method for controlling cell shape and function with micrometer-scale ECM islands microfabricated with microcontact printing. (A) Schematic of microcontact printing technique with soft lithography and direct printing of protein onto the surface of a thick layer of PDMS. See text for a detailed description. (B) Fluorescence microscopic images of CE cells cultured on microscale FN islands (900 μm^2) of different geometric shapes and stained with fluorescent-phalloidin to visualize F-actin and DAPI to visualize nuclei. Note that cells preferentially extend new cell processes from their corners and that they prefer acute rather than obtuse angles.

interactions between cells and their ECM adhesions. For example, cells move from regions of high ECM compliance to more stiff regions, a process as known as "durotaxis" (Discher *et al.*, 2005; Lo *et al.*, 2000). As described previously, when plated on single-cell–sized ECM islands of different geometric shapes created with microfabrication techniques, various cells preferentially extend new motile processes (e.g., lamellipodia, filopodia, microspikes) from their corners rather than along their edges (Brock *et al.*, 2003; Parker *et al.*, 2002). Focal adhesions are also preferentially formed in their corners near sites where new lamellipodia will form when cells are stimulated with soluble chemokines, and cells on circular islands (i.e., without corners) do not exhibit any directional bias (Parker *et al.*, 2002).

Importantly, this microcontact printing technique also can be used to study directional cell motility. CE cells, fibroblasts, and muscle cells cultured on angulated polygonal ECM islands (e.g., squares, rectangles, hexagons, pentagons) preferentially extend new motile processes, including lamellipodia, filopodia, and microspikes, from the corners of these islands (Fig. 12.2B), whereas cells display no directional bias when they are cultured on circular ECM islands (Brock *et al.*, 2003; Parker *et al.*, 2002). These cells preferentially exert greatest traction force, form focal adhesions, and deposit new ECM fibrils in their corners, and thus this technique is especially well suited for studying spatial control of motility signaling and oriented lamellipodia formation in adherent cells.

Microcontact printing also can be used to engineer substrates containing microarrays of much smaller (1 to 8 μm diameter) ECM islands that are on the same scale as individual focal adhesions (Fig. 12.3A) (Chen *et al.*, 2003; Tan *et al.*, 2003). Because focal adhesions seem to preferentially form directly behind the leading edge of cells migrating in the direction of increasing ECM stiffness (Beningo *et al.*, 2001; Pelham and Wang, 1997) and changes in mechanical force transfer across integrins can modulate focal

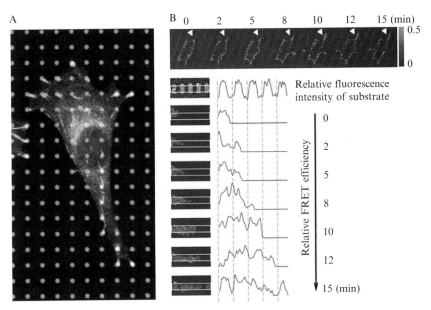

Figure 12.3 Method for analyzing spatial control of motility signaling pathways by cell–ECM interactions. (A) Immunofluorescence microscopic image of a cell cultured for 8 h on microarrays of circular 1-μm diameter rhodamine–FN–coated islands spaced 3 μm from neighbors and stained with anti-vinculin antibodies; white indicates formation of focal adhesions directly above the ECM islands. (B) Rac-FRET analysis of a cell spreading over an array of multiple FN islands (1 × 3 μm). *Upper;* Low-magnification view showing the same cell during a 15-min period of spreading showing increased Rac activity (relative FRET efficiency shown at right) concentrated in FAs directly above periphery FN islands. *Lower;* Higher power view of 5 FN islands showing the leading edge of the cell moving progressively from FN island to island (left to right) and activating Rac activity locally within minutes after forming an adhesive contact (FA) with the new ECM substrate (black line indicates relative FRET efficiency).

adhesion assembly (Riveline *et al.*, 2001), the spatial positioning of focal adhesions may determine the location in which new motile processes form and, hence, govern the direction in which cells move. Recently, we used microarrays of focal adhesion–sized ECM islands created with this micro-contact printing method to demonstrate that we can influence the direction of cell spreading and movement over a period of multiple hours by altering focal adhesion shape or position (Xia *et al.*, 2008). These findings are physiologically important, because cells also preferentially extend and migrate along ECMs with specialized shapes (e.g., fibrils) during tissue development (Nakatsuji and Johnson, 1984) and tumor angiogenesis (Folkman *et al.*, 1989). Again, this method may be useful for any adherent cell type that exhibits directional movement.

3.1. Cell culture

These studies are carried out with bovine CE cells cultured as described previously; with primary human pulmonary CE cells (Lonza Inc.) cultured for 3 to 5 passages in EBM medium (Lonza) supplemented with 10 ng/ml human recombinant epidermal growth factor (EGF), 12 mg/ml bovine brain extract, 1 mg/ml hydrocortisone, and 10% FBS at 37° under 5 % CO_2; or with NIH 3T3 cells cultured in DMEM/F12 nutrient media containing 10% bovine calf serum, 2 mM glutamine, penicillin (100 U/ml), streptomycin (100 mg/ml), 250 mg/ml amphotericin B, and 205 mg/ml of sodium deoxycholate (Life Technologies) under 5% CO_2 (Yu *et al.*, 2001).

3.2. Microcontact printing

Micropatterning techniques include four main categories of techniques: photolithography (Bhatia *et al.*, 1993), soft lithography (Whitesides, *et al.*, 2001), direct writing (Odde and Ren, 1999; Roth, *et al.*, 2004; Veiseh, *et al.*, 2004), and laser patterning (Corey, *et al.*, 1991; Duncan, *et al.*, 2002; Vaidya, *et al.*, 1998). Among them, soft lithography has drawn the largest attention and found the most extensive applications. We mainly use two methods of microcontact printing to study shape regulation of cell behavior that involved use of either self-assembled monolayers (SAMs) of alkane thiols on gold-plated substrates (Chen *et al.*, 1997; Dike *et al.*, 1999; Parker *et al.*, 2002; Singhvi *et al.*, 1994) or micropatterning proteins directly on poly (dimethyl-siloxane) (PDMS) (Lele *et al.*, 2007; Xia *et al.*, 2008) (Fig. 12.2A).

In both methods, the desired pattern of ECM islands is designed with computer software (e.g., AutoCAD) and transferred to a mask with high-resolution laser printing or by electron-beam etching of chromium on glass (for smaller features). PDMS stamps are fabricated by casting the polymer onto silicon wafers that have been etched by photolithography with micro-scale features (see Whitesides [2001] for more details). To micropattern

SAMs on substrates sputter-coated with a thin (12-nm) layer of gold, the PDMS stamp is inked with an ethanol solution of hexadecanethiol, dried thoroughly with filtered air or nitrogen, and then gently brought into conformal contact with the gold substrate, such that only the raised features of the stamp make contact for at least 30 sec. This prints the gold with SAMs that will allow protein adsorption; the remainder of the substrate is blocked by incubating in a solution of $(EG)_3OH$-terminated alkane thiols, forming SAMs that will resist protein and cell adsorption. Finally, the substrates are rinsed and incubated with ECM proteins, such as FN, type IV collagen, or laminin, which adsorbs exclusively to the printed areas. After washing off excess protein, taking care not to allow the substrates to dry out, the microcontact printed surface is ready for cell plating. Even in the presence of serum, cells will adhere only to the printed, adhesive areas. We have published details methods for this technique previously (Chen *et al.*, 2000).

More recently, we have used a more simple, cost-effective method of microcontact printing that allows direct stamping of protein onto the surface of a thin layer of PDMS (Sylgard-184, Dow Corning) (Tan *et al.*, 2002) (Fig. 12.2A). We typically use a 9:1 ratio of elastomer base to curing agent, which forms a moderately stiff substrate (Young's modulus greater than 1 MPa) when cured (Gray *et al.*, 2003). The base and curing agent should be well mixed and degassed before use with a vacuum or quick spin (1 to 2 min) at 1000 rpm in a tabletop centrifuge. Coverslips are coated with a spin coater: a drop of PDMS (200 μl for a 25-mm $\times$ 25-mm coverslip, Corning) is applied to the center of the coverslip and spun at 4000 rpm for 4 min. The polymer can be cured by incubation at $60°$ for 1 h or at room temperature for at least 24 h. Coated coverslips can then be cut with a diamond pen to desired size. The resulting layer of PDMS is only tens of microns thick and optically clear, allowing high magnification imaging of cells through the coated coverslip. Coverslip-bottomed petri dishes (Mat-Tek) can be spin-coated in the same manner to perform high-magnification imaging of live cells on microcontact printed substrates.

The PDMS stamps are cleaned in 70% ethanol in a sonicating water bath for 30 min, rinsed with water, and dried with filtered compressed air or nitrogen gas. The surface of the clean stamps containing the raised micro-patterned features are incubated with 50 μg/mL FN (or other ECM protein) in aqueous solution for 1 h and dried thoroughly with filtered nitrogen gas or compressed air. To ink the hydrophobic stamps, it is helpful to "paint" the protein solution onto the patterned surface with a pipet tip. Before stamping, the hydrophilic PDMS-coated coverslips are made hydrophilic by treatment with plasma, which causes the surface to become temporarily negatively charged and adsorptive for proteins (Whitesides *et al.*, 2001). We plasma treat the PDMS-coated coverslips in a Jelight UVO cleaner (Specialty Coating Systems G3P-8, Cookson Electronics) for 8 min, while the inked PDMS stamps are rinsed in water or PBS and dried. The stamps are pressed gently against the plasma-treated PDMS surface with the fingertips or forceps within 15 min of plasma treatment to ensure complete contact of stamp with substrate. By use of rhodamine-conjugated fibronectin (Cytoskeleton, Denver, CO), we have found that 1 to 3 min of contact is sufficient to efficiently transfer protein to the PDMS-coated coverslip, but less than 30 sec contact may result in incomplete protein transfer. Unstamped areas are then made nonadhesive by incubating the substrates in 1% Pluronic 127 in PBS for 1 h at room temperature or overnight at 4°. Pluronic surfactants are nontoxic triblock polymers whose polypropylene segments adsorb to the hydrophobic PDMS surface, whereas their polyethylene glycol segments block protein adsorption (Amiji and Park, 1992; Chen et al., 2004). Before use, substrates are washed three times with PBS. Microstamped PDMS substrates should be used within a few days of preparation, and they are most useful for short-term experiments (less than 1 week).

3.3. Cell culture on microfabricated ECM islands

Cells that are grown to confluence in serum-containing medium (Chen et al., 1997; Huang et al., 1998; Yu et al., 2001) are serum-deprived for 1 to 2 days before use in experiments. These quiescent cells are then trypsinized and plated sparsely (3×10^3 cells/cm^2) on the micropatterned ECM substrates to ensure that individual islands are seeded with single cells. For morphologic and cell cycle analysis, bovine CE cells are cultured in serum-free DMEM supplemented with bFGF (5 ng/ml) and primary human pulmonary CE cells are cultured with EGM2 supplemented with 2% FBS, 10 μg/ml high-density lipoprotein and 5 μg/ml; transferring for 18 to 24 h. When bovine CE and NIH3T3 cells are cultured for extended times (up to 30 min) on the microscope stage, these studies are carried out in bicarbonate-free minimum essential medium (MEM) containing Hank's balanced salts lacking phenol red and bicarbonate (Sigma Chemical Co., St. Louis, MO), MEM amino acids (Sigma), MEM vitamins (Sigma), 2 mM L-glutamine, 1 mM sodium pyruvate, 20 mM N-2-hydroxyethylpiperazine-N#-2-ethanesulfonic acid, D-glucose (1 g/L), hydrocortisone (1 μg/ml), and 1% BSA. This medium is supplemented with 20 μg/ml high-density lipoprotein and 5 μg/ml transferrin for studies with endothelial cells; studies with fibroblasts used the same medium with high glucose (5 g/L).

3.4. Analysis of cell cycle progression

For cell cycle analysis, human CE cells are synchronized by treatment with 40 mM lovastatin (Merck, Rahway, NJ) in standard culture medium for 32 h (Huang et al., 1998) or serum deprivation for 48 h (Mammoto et al., 2004). The cells are trypsinized, washed, and replated on the micropatterned islands in culture medium described previously. The ability of cells to enter S phase is measured by quantitating the percentage of cells that exhibited nuclear incorporation of 5-bromo-29-deoxyuridine (BrdU), as detected with a commercial assay (Amersham, Arlington Heights, IL). Cells are fixed in 5% acetic acid/90% ethanol for 30 min at room temperature. BrdU-positive cells are identified by incubating cells with anti–BrdU antibody and nuclease (RPN202, Amersham Biosciences) for 60 min and stained with biotinylated anti-mouse Ig antibodies and Texas red avidin (Vector Laboratories, Burlingame, California). BrdU-positive cells are visualized and scored with a Nikon epifluorescence microscope with 20× objectives; all nuclei are counterstained with DNA binding dye, 4,6-diamino-2-phenylindole (DAPI; 1 μg/ml). At least 12 random fields with a total of 500 cells are counted per sample (Huang *et al.*, 1998; Numaguchi *et al.*, 2003).

The extent of pRb hyperphosphorylation is also a marker of G1/S transition, which is measured indirectly with an *in situ* nuclear labeling technique (Huang *et al.*, 1998; Latham *et al.*, 1996; Mittnacht and Weinberg, 1991). This *in situ* technique is based on the finding that hyperphosphorylated pRb easily dissociates from nuclei when treated with a nuclear extraction buffer (Mittnacht and Weinberg, 1991). Cells are washed once in PBS after 18 h of culture, incubated in nuclear extraction buffer (10 mM HEPES-KOH, pH 7.9, 10 mM KCl, 1.5 mM MgCl$_2$, 0.1% Triton X-100, 1 mM dithiothreitol) for 15 min at room temperature, fixed for 20 min in 4% PFA/PBS, and washed with 0.1% BSA/PBS. pRb is visualized by indirect immunostaining with anti-human pRb antibody LM95.1 (2 g/ml; Calbiochem); cells are also counterstained with DAPI to facilitate quantitation of the percentage of pRb-negative nuclei. Negative pRb staining indicates cells that contain hyperphosphorylated pRb that dissociated from nuclear matrix and, hence, cells that successfully passed through the late G1 restriction point (Huang *et al.*, 1998; Latham *et al.*, 1996; Mittnacht and Weinberg, 1991).

3.5. Analysis of directional lamellipodia extension

For studies on directional extension of new motile processes on single cell–sized ECM islands, lamellipodia formation is synchronously activated in NIH 3T3s by addition of human PDGF-BB (5 ng/mL, BioVision, Mountain View, CA) and in bovine CE cells by addition of bFGF (5 ng/mL) or 10% calf serum. Living cells are visualized with a Hamamatsu CCD camera on a Nikon Diaphot 300 inverted microscope equipped with phase–contrast optics and epifluorescence illumination. Temperature is controlled by a stage mount (Micro Video Instruments, Avon, MA) equipped with a temperature controller (Omega technologies Co., Stamford, CT). The total projected area of lamellipodia per cell is quantitated with the computerized image acquisition and analysis tools of IP Lab Spectrum and RatioPlus software (Scanalytics, Fairfax, VA). F–actin, vinculin, fibronectin, and DNA (nuclei) are visualized in cells fixed with 4% PFA with phalloidin (300 ng/ml), mouse anti-vinculin antibody, rabbit anti-fibronectin antibody, and DAPI staining (all from Sigma), respectively (Parker *et al.*, 2002) (Fig. 12.2B). We have found that $40 \times 40 \ \mu m$ and $30 \times 30 \ \mu m$ square ECM islands are highly effective for this analysis in bovine CE cells and NIH3T3 fibroblasts, respectively. However, any shape and size island can be used for these experiments, and the optimal island geometry needs to be empirically determined for each cell type and set of experimental conditions.

Any cell projection that extends over the nonadhesive region surrounding the ECM island and stains positively for F–actin is considered a lamellipodium. But to account for registration errors, we only count projections that are greater than $1 \ \mu m^2$ in area and have a pixel intensity greater than background. To determine relative changes in lamellipodia length in different regions of the cell, corners are defined as parts of the square perimeter within $6 \ \mu m$ from the intersection of the two sides; sides are defined as the 18-μm interval between these corner regions. To control for bias in morphometric calculations because of the geometry of the orthogonal corner regions relative to the linear sides and the large range of lamellipodia morphology, the normalized lamellipodia length is determined by transforming the total lamellipodia area measured in each region into a similar shaped region (corner or side) composed of a lamellipodium that extended equally from all points along its perimeter. Total cumulative data may be presented by overlaying 20 to 40 images of cells cultured on FN islands and stained for F–actin with fluoresceinated-phalloidin. The pixel occupancy at each position relative to the FN island is determined with IP Lab software, and the pixel distribution is color coded for frequency. Immunofluorescence microscopy is carried out with the epifluorescence optics of the Nikon Diaphot microscope.

3.6. Microcontact printing of microarrays of focal adhesions sized ECM islands

Various different planar microarrays of focal adhesion-sized circular or linear FN islands, separated by nonadhesive regions, are created by use of microcontact printing (Chen *et al.*, 2003; Tan *et al.*, 2003; Xia *et al.*, 2008). As displayed in Fig. 12.3, these include patterns of 1-μm diameter circular FN islands that are either separated by $3 \ \mu m$ in both the X and Y directions (1C-3,3) or by $1.5 \ \mu m$ in the X direction and $3 \ \mu m$ in the Y direction (1C-1.5,3); linear FN islands $1 \ \mu m$ high (Y direction) and 3 or $8 \ \mu m$ wide (X direction) separated by $3 \ \mu m$ in both directions (3L-3,3 and 8L-3,3 respectively); linear FN islands $3 \ \mu m$ wide $\times 1 \ \mu m$ high separated by 4.5 and $1 \ \mu m$ in

X and Y directions, respectively (3L-4.5,1); and linear 8 μm wide $\times$ 1 μm high FN lines equally spaced by 3 μm in both directions, but staggered along a 60-degree offset (8L-3,3st). Subconfluent NIH 3T3 cell monolayers are serum-starved for 1 day, trypsinized, and plated on micropatterned substrates in DMEM containing high-density lipoprotein (10 μg/ml;), transferrin (5 μg/ml;), and 1% BSA. The plating density is low to minimize cell–cell contacts.

3.7. Morphometric analysis of cell spreading, orientation, and directional movement

Live cell images are carried out as described previously (Brock *et al.*, 2003; Parker *et al.*, 2002), and fluorescence images are acquired on a Leica TCS SP2 confocal laser scanning microscope with a 63 $\times$ /1.4 NA oil immersion objective and processed with Leica software or Adobe Photoshop. Morphometric analysis of cell shape, orientation, and migration is carried out by outlining the borders of individual cells within differential interference contract images with the computerized image acquisition and analysis tools described previously; 25 to 50 cells are analyzed for each experimental condition. Cell elongation is defined as the ratio of the maximal to minimal cell length. In the migration assays, cells are stimulated with PDGF-BB (50 ng/ml) and the centroids of the migrating cell recorded at 20-min intervals over 8 h are plotted and connected to generate the migration path, which is then used to calculate the speed and direction of cell migration.

Cell movement is analyzed by its speed, direction, and pattern of the migration path. For example, cell migration direction is random (45 degrees with a wide distribution) on the isotropic FN circles and more oriented on the anisotropic patterns and linear FN islands. Importantly, even cells migrated in a random Brownian walk on the isotropic array of FN circles, cells could be made to preferentially migrate in either the X or Y direction with great efficiency by decreasing interisland spacing in the X or Y direction with the 1C-1.5, 3 and 3L-4.5,1 substrates, respectively.

 ## 4. CONTROL OF CELL FATE SWITCHING BY MODULATING THE CYTOSKELETON

Cell fate switching also can be controlled in CE cells by altering cytoskeletal structure from within. For example, CE cell growth can be inhibited by treating cells with pharmacologic agents that disrupt the actin cytoskeleton, such as cytochalasin D (Cyto D) or latrunculin B (LatB), or with drugs that inhibit cytoskeletal tension generation (e.g., 2,3-butanedione 2-monoxime; BDM) (Huang *et al.*, 1998; Mammoto *et al.*, 2004).

Treatment of cells with Cyto D or with nocodazole (Noc) to depolymerize the microtubule system also promotes apoptosis in these cells (Flusberg *et al.*, 2001). Alternately, overexpression of nonmuscle caldesmon (CaD), a protein component of the contractile actomyosin filament apparatus, which binds to actin, myosin, tropomyosin, and calmodulin (Huber, 1997; Matsumura and Yamashiro, 1993), can be used to disrupt actin stress fibers, disassemble focal adhesions, and induce cell retraction in cultured cells (Helfman *et al.*, 1999; Numaguchi *et al.*, 2003). When CaD binds to actin, it inhibits the ATPase activity of actomyosin in a calcium- and calmodulin-dependent manner (Chalovich *et al.*, 1998; Marston *et al.*, 1994) and thereby suppresses formation of actin fiber bundles and focal adhesions (Helfman *et al.*, 1999). We used an adenovirus-mediated expression system under control of a tetracycline (Tet)-inducer (AdTet-Off) to achieve efficient, synchronous, and tunable expression of CaD-GFP in CE cells and found that it is involved in angiogenesis (Numaguchi *et al.*, 2003).

4.1. Pharmacologic modifiers of cytoskeletal structure

To disrupt actin microfilaments and microtubules, cells are incubated with Cyto D (1 μg/ml; Sigma), Lat B (0.1 μg/ml; Calbiochem, La Jolla, CA) or Noc (10 μg/ml; Sigma) for 15 min before cells are plated on ECM-coated substrates in the continued presence of the drugs. These doses have been shown to fully disrupt the integrity of actin microfilaments and microtubules in CE cells (Flusberg *et al.*, 2001; Huang *et al.*, 1998; Ingber *et al.*, 1995; Mammoto *et al.*, 2004). To inhibit cytoskeletal tension generation and prevent the formation of focal adhesion and actin bundles in CE cells, cells are treated with 5 mM BDM (Sigma), a dose that does not significantly alter intracellular calcium concentration (Blanchard *et al.*, 1990; Chicurel *et al.*, 1998; Chrzanowska-Wodnicka and Burridge, 1996). The specificity of these cytoskeletal-disrupting agents in CE cells has been demonstrated experimentally (Ingber *et al.*, 1995; Wang *et al.*, 1993).

4.2. Control of cell shape and function with adenoviral CaD

To create an adenoviral form of CaD, the plasmid containing GFP-tagged cDNA encoding the full-length rat nonmuscle CaD gene (kindly supplied by Dr David M. Helfman, Cold Spring Harbor Laboratory) is used as the template plasmid. The shuttle vector plasmid pTRE2-shuttle containing the minimal cytomegalovirus promoter and tet-operator sequences cloned upstream of the cDNA to be expressed and the Tet-Off system is purchased from BD Biosciences Clontech (Palo Alto, CA). The GFP-CaD gene is excised from the template plasmid at *XbaI/BamHI* sites, and both intruding ends are blunted with a DNA blunting kit (TAKARA, Japan) and ligated at the *EcoRV* site of the pTRE2-shuttle vector. Human kidney–derived 293 epithelial cells (QBIOgene, Carlsbad, CA) are cultured in 10% FBS/ DMEM to a subconfluent density before transfection. The pTRE2 shuttle vector containing GFP-CaD gene is ligated with the adenoviral genome DNA (Clontech) and transfected into the 293 cells diluted in Optimem (Invitrogen, Carlsbad, CA) with transfection reagent, Lipofectamine 2000 (Invitrogen). After 2 or 3 days, cells that become round and detach from the substrate because of the cytopathic effect of the adenoviral infection float in the medium until 7 days after transfection. The cells are then collected in a tube, repeatedly frozen, thawed five times, and collected by centrifugation (1500 rpm). The supernatants containing recombinant adenovirus particles encoding GFP-CaD (AdGFP-CaD; first seed) are aliquoted and stored at $-80°$. To obtain higher titers of the adenoviral vectors, this process is repeated two additional times. The final (third) round produces viral titers ranging from 10^9 to 10^{10} pfu/ml, as determined by plaque assay in 293 cells. Induction of expression of the integrated target gene is controlled solely by removing Tet from the culture medium. The presence of the cDNA insert is confirmed by direct observation of disruption of actin fibers in GFP-labeled cells transduced with AdGFP-CaD and AdTet-Off in Tet-free medium and by Western blot analysis of cell lysates.

To determine the effects of expressing AdGFP-CaD, 1 day after bovine CE cells are plated, the DMEM/10%FBS is replaced 0.4% FBS/DMEM containing 10^8 pfu/ml of AdGFP-CaD, AdTet-Off, and 10 μg/ml of Tet, and cells are cultured for 36 to 48 additional hours (Numaguchi et al., 2003). After CE cells synchronize in G0 by culturing in low serum for 2 to 3 days, GFP-CaD expression is induced by culturing the cells in the same low-serum medium without Tet for 24 additional hours. Then, cells are induced to reenter into the cell cycle by replacing the medium with 10% FBS/ DMEM containing BrdU, in the continued absence of Tet. Control cultures are treated identically except in the medium with Tet throughout the entire experiment. Progression of quiescent CE cells through the cell cycle and into S phase is measured with BrdU analysis as described previously (Huang et al., 1998; Numaguchi et al., 2003).

To measure effects on apoptosis under similar conditions, CE cells cultured for 24 h under various level of induction of GFP-CaD are fixed with 4% PFA/ PBS for 15 min at room temperature. The fixed cells are permeabilized with 0.1% sodium citrate/0.1% Triton-X 100 in PBS and stained with terminal deoxynucleotidyl transferase dUTP nick end labeling (TUNEL) enzyme reagent (In situ Cell Death Detection kit; Roche Molecular Biochemicals, Indianapolis, IN). The apoptosis index is calculated as the percentage of DAPI-labeled nuclei that exhibit positive TUNEL staining as detected with fluorescence microscopy (Flusberg et al., 2001; Huang et al., 1998; Numaguchi et al., 2003).

We also use a caspase assay as another way to detect apoptosis (Flusberg et al., 2001). Caspase proteases play an essential role in apoptosis by degrading specific structural, regulatory, and DNA repair proteins within a cell (Casciola-Rosen et al., 1994; Lazebnik et al., 1994). Caspase-9 and -8 are activated early in the apoptotic cascade, which is released from the mitochondria in response to apoptotic stimuli (Li et al., 1997). Activated caspase then initiates the proteolytic activity of other downstream caspases, including caspase-3 and others (Muzio et al., 1997; Srinivasula et al., 1996). For caspase activity analysis, the cells are lysed in ice-cold lysis buffer (ApoA-lert Fluorometric Caspase-3 Activity Assay; Clontech, Palo Alto, CA), centrifuged, and the supernatants are transferred to a 96-well plate. Lysates are incubated for 1 h at 37° with the caspase-3–specific fluorescent substrate DEVD-AFC (50 μM; CLONTECH) or the caspase-8–specific substrate IETD-AFC, and fluorescence is measured with the use of a fluorometric plate reader (Bio-Rad, Hercules, CA) at 380-nm excitation and 460-nm emission. For caspase inhibition, cell monolayers are pretreated with the caspase inhibitor z-VAD.fmk (100 μM; Calbiochem, San Diego, CA) for 1 h at 37° before the start of the experiment and after replating.

5. ANALYSIS OF THE MECHANISM OF CELL SHAPE–DEPENDENT GROWTH CONTROL

Adhesion-dependent control of cell growth requires joint regulation of the ERK/MAPK pathway by integrins and growth-factor receptors during G1-phase of the cell cycle (Meloche et al., 1992; Weber et al., 1997; Zhu and Assoian, 1995). However, activation of ERK/MAPK is not sufficient for passage through the late-G1 checkpoint. For example, when CE cells are prevented from spreading by culturing them in the presence of soluble mitogens on dishes coated with a low density of FN or on small, micrometer-sized, high-density FN islands as described previously, the G1/S transition is similarly inhibited, despite normal activation of the canonical ERK/MAPK pathway (Huang et al., 1998). Cytoskeletal disruption also can prevent G1 progression in many cell types, including CE cells (Bohmer et al., 1996; Ingber et al., 1995; Iwig et al., 1995; Reshetnikova et al., 2000). Thus, additional signals that emanate from the intact cytoskeleton of spread cells in mid-G1 seem to be critical for the successful passage through late G1 and entry into S phase (Assoian, 1997; Huang and Ingber, 2002). Here, we describe how the techniques previously described can be used to analyze the molecular basis of cell shape–dependent control of cell cycle progression that characterizes virtually all normal anchorage-dependent cells.

Cell-cycle progression through the late G1/S restriction point, which represents the "point of no return" in the cell cycle, is associated with the hyperphosphorylation of retinoblastoma protein (pRb) by cyclin-dependent kinases (cdks) (Assoian, 1997; Sherr and Roberts, 1999; Weinberg, 1995). Notably, the cdk inhibitor p27*kip1* (p27) that binds and inactivates the cyclin D1/cdk4 and cyclin E/cdk2 complexes is a major target for many physiologic growth regulatory signals (Pagano *et al.*, 1995). Importantly, when CE cell spreading is prevented by either altering the ECM substrate or disrupting the actin cytoskeleton with various cytoskeletal modulators, p27 levels remain high and cell-cycle progression is blocked in mid to late G1 (Huang *et al.*, 1998; Huang and Ingber, 2002; Mammoto *et al.*, 2004). The F-box protein Skp2 that is required for ubiquitination-dependent degradation of p27 restores G1 progression in these cells (Mammoto *et al.*, 2004). The effects of cell shape on cell-cycle progression are similarly mediated by p27 in other cell types (Carrano and Pagano, 2001; Zhu *et al.*, 1996). Thus, we measure phosphorylation level of pRb and the expression levels of Skp2 and p27 as markers for the G1/S transition. Any laboratory can use these quantitative methods, because they only require conventional molecular biologic and biochemical reagents, along with the ECM density modulation technique previously described; however, similar types of cell cycle analysis can be carried out with micropatterned ECM islands (Huang *et al.*, 1998).

5.1. Cell culture

Human microvascular endothelial (HMVE) cells from neonatal dermis and lung (Lonza) are cultured in EBM-2 (Lonza), supplemented with 5% fetal bovine serum (FBS) and growth factors (bFGF, insulin-like growth factor, vascular endothelial growth factor) according to the manufacturer's instructions and maintained at 37° in 5% CO_2 (Mammoto *et al.*, 2004). Cells are synchronized at the G0/G1 border by serum starvation (0.3% FBS/EBM-2) for 40 to 42 h and then released into G1 by trypsinizing the cells and replating them on ECM-coated (e.g., FN) bacteriologic dishes in EBM-2 containing 1% FBS and growth factors.

5.2. Cell-cycle analysis

We use various ways to detect G1/S transition, including BrdU incorporation and *in situ* labeling of pRb, as described previously, as well as the use of Western blotting and reverse transcription (RT)-PCR to detect cell cycle–associated proteins (Huang *et al.*, 1998; Mammoto *et al.*, 2004). For these latter forms of analysis, HMVE cells cultured on 60-mm dishes coated with various densities of FN (Ingber and Folkman, 1989b) are lysed with 0.3 ml of boiling lysis buffer (1% SDS, 50 mM Tris-HCl, pH 7.4), scraped, and lysates are collected at various time points after replating (4 to 24 h). Homogenized total cell lysates (10 μg protein) are subjected to SDS-PAGE, transferred to nitrocellulose membranes, and immunoblotted with specific primary antibodies. The primary antibodies are detected with horseradish peroxidase–conjugated secondary antibodies (Vector Laboratories, Burlingame, CA) and Super Signal Ultra (Pierce) as a chemiluminescence substrate. Monoclonal antibody against pRb (LM95.1) is from Calbiochem (San Diego, CA), Skp2 (SKP2–8D9) is from Zymed Laboratories Inc. (San Francisco, CA), actin (AC-15) is from Sigma, and polyclonal antibody against p27 (clone 57) is purchased from Santa Cruz Biotechnologies (Santa Cruz, CA). Results are quantified by densitometric analysis with image J software.

RT-PCR is used to determine the expression of Skp2 mRNA. HMVE cells are lysed and total RNA was isolated with the RNeasy RNA extraction kit (Qiagen, Valencia, CA) according to instruction manual. The RNA (500 ng/sample) is treated for 1 h at $37°$ with reverse transcriptase with OMNI Script reverse-transcriptase assay kit (Qiagen). The PCR is carried out with a series of 1:3 dilutions of the RT product (2 μl). Only reactions in the log-linear range (product quantity versus input template quantity) are used. The forward and reverse PCR primers for Skp2 are 5′- CAAC TACCTCCAACACCTATC-3′ and 5′-TCCTGCCTATTTTCCCTGT TCT-3′, respectively. PCR cycling conditions are 3 min at $94°$, then 26 cycles of 30 sec at $94°$, 30 sec at $55°$, and 1 min at $72°$. For internal control, we use actin mRNA, whose forward and reverse primers are 5′- TGACGGGGTCACCCACACTGTGCC-3′ and 5′-TAGAAGCATT TGCGGTGGACGATG-3′, respectively. PCR products are analyzed by agarose gel electrophoresis. Primers are designed with Oligo version 4.0 software (National Biosciences, Plymouth, MN) and synthesized by Sigma Genosys (Biotechnologies Industries, The Woodlands, TX).

5.3. Determination of cell cycle position

To determine their position in the cell cycle, the levels of the critical G1 proteins, p27, Skp2, and pRb phosphorylation are measured in total cell lysates, with the phosphorylation status of pRb being used as a readout of successful progression through the late G1 restriction point. The point at which the levels of Skp2 and hyperphosphorylated pRb start to increase, and p27 begins to decrease, represents the G1 restriction point. This usually occurs between 12 and 18 h after replating synchronized CE cells on dishes coated with a high density (666 ng/cm^2) of FN that allows maximal cell spreading (Fig. 12.4). Round cells on a low density (22 ng/cm^2) of FN do not down-regulate p27, increase Skp2, or hyperphosphorylate pRb over the same time course, even though they are stimulated by the same soluble mitogens (bFGF, insulin-like growth factor, and vascular endothelial growth factor in 1% serum) (Fig. 12.4).

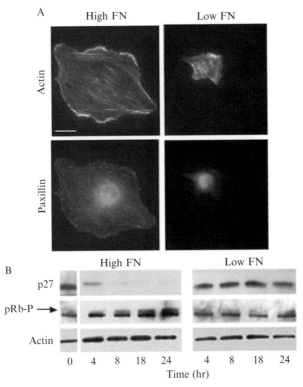

Figure 12.4 Variable ECM coating density method for analyzing cell shape–dependent control of cell cycle progression. (A) Fluorescence micrographs showing control of cell shape, stress fiber formation visualized with Alexa-488-phallodin and focal adhesion formation stained with anti-paxillin antibody in spread versus round HMVE cells on high versus low FN density (bar, 5 μm). Well-defined stress fibers and focal adhesions appear in highly spread cells on high FN, whereas round cells on low FN fail to exhibit either stress fibers or detectable focal adhesions. (B) Immunoblots showing changes in the level of expression of p27 protein relative to actin, as well as Rb protein phosphorylation status, in total cell lysates from spread and round cells cultured on high and low FN, respectively, for the indicated times after release from G0. For pRb, the slower-migrating (top) band represents the hyperphosphorylated form of the protein. The levels of hyperphosphorylated pRb increase and p27 decrease as the spread cells on high FN progress from G0 (time 0) to late G1 phase (from 8 to 18 h after release).

6. METHODS FOR ANALYZING RHO-DEPENDENT CONTROL OF CELL GROWTH AND MOVEMENT

The studies and methods previously described revealed that vascular cell growth, differentiation, and directional motility could vary greatly, depending on local variations in physical interactions between cells and their ECM adhesions that alter cytoskeletal structure. Members of the Rho family of small GTPases, including Rho, Rac, and Cdc42, mediate the effects of cytokine and ECM binding on actin polymerization and cytoskeletal tension generation, which regulate stress fiber and focal adhesion assembly (Jaffe and Hall, 2005). However, recent findings show that physical changes in cytoskeletal structure produced by mechanical cues can also feed back to alter Rho activity (Mammoto et al., 2004; 2007a). In fact, by use of protein transfection and FRET analysis techniques, we have found that Rho and Rac mediate cell shape–dependent control CE cell growth (Mammoto et al., 2004) and directional lamellipodia formation (Brock and Ingber, 2005; Xia et al., 2008). Thus, we present methods here that describe how control of cell growth and motility by Rho GTPases can be studied.

6.1. Preparation and delivery of recombinant Rho proteins

The constitutively active form of RhoA (RhoA14V) and the Rho activator, cytotoxic necrotizing factor 1 (CNF1), are expressed and purified from *Escherichia coli* expression plasmid pGEX4T-RhoA14V and pCNF24-CNF1 (kindly provided by Alan Hall, Memorial Sloan-Kettering Cancer Center, NY, and Melody Mills, Uniformed Services University of the Health Sciences, Bethesda, MD, respectively). GST-tagged RhoA14V and GST recombinant proteins are purified from *E. coli* (Mammoto et al., 2004). For RhoA14V, the GST tag is removed by proteolytic cleavage with thrombin (10 units/ml, Sigma) at 4° for 8 to 10 h; thrombin is removed by incubating the supernatant with p-amino-benzamidine-agarose (Sigma). The His-tagged CNF1 is purified with Ni-agarose nitrilotriacetic acid beads following the manufacturer's instructions (Qiagen). C3 exoenzyme is purchased from Cytoskeleton Inc. (Denver, CO), and ROCK inhibitor Y27632 is purchased from Calbiochem.

For protein transfection (proteofection) of living cells, BioPORTER protein delivery reagent (Gene Therapy Systems, San Diego, CA) is used according to the manufacturer's instructions. In brief, 0.5 to 5 μg of recombinant protein in 200 μl of PBS is incubated in a tube containing a film of 15 μl of BioPORTER that is formed by drying for 5 min. The complexes are then added to the cells (400,000 cells/60-mm dish) in 2.5 ml of serum-free EBM-2; after 4 h of incubation at 37°, cells are replated onto the experimental FN-coated dishes with experimental medium. In the case of RhoA14V and C3, we use the BioPORTER reagent alone without added protein cargo as a proteofection control. CNF1 (100 ng/ml) and Y27632 (10 μM) are added to the medium when cells are replated. The samples are collected at the indicated time points after replating for cell cycle experiments and cell staining.

6.2. Rac FRET analysis

The small GTPase, Rac1, controls lamellipodia formation and cell motility, and it can be activated by cell–ECM adhesion in an integrin-dependent manner (Clark *et al.*, 1998; del Pozo *et al.*, 2000; Price *et al.*, 1998). An intramolecular Rac1-FRET reporter analysis allows us to characterize the spatial distribution of Rac activity in cells cultured on the different FN microarrays previously described. The FRET probe, Raichu-Rac1, was kindly provided by Dr. Michiyuki Matsuda (Osaka University, Osaka, Japan) (Itoh *et al.*, 2002). NIH 3T3 cells are transfected with the FRET reporter of Rac activity, Raichu-Rac1 with effectene transfection reagent (Qiagen, Valencia, CA), and cultured for 24 h on the different micropatterned substrates in serum-free medium. We studied cells cultured on the 3L-4.5,1 pattern (Fig. 12.3B) in studies analyzing spatial control of Rac activation dynamics at high resolution, because this pattern has the largest difference in island spacing between the X and Y directions. For FRET analysis, fluorescence images are acquired every minute on a Leica TCS SP2 confocal laser scanning microscope with a 63×/1.4 NA oil immersion objective with Leica FRET sensitization software (CFP and FRET excitation at 458 nm; YFP excitation at 514 nm). Calibration images are acquired from the samples containing only Donor (CFP), Acceptor (YFP) and FRET (Raichu-Rac1), respectively, and the software automatically calculates the apparent FRET efficiency.

7. Rho-Dependent Control of Angiogenesis in Whole Organ Culture

Tissue morphogenesis is controlled in embryonic tissues by altering the cellular mechanical force balance (Ingber, 2006; Moore *et al.*, 2005; Sanchez-Esteban *et al.*, 2006), and Rho-dependent control of cytoskeletal tension generation seems to play a central role in this process (Moore *et al.*, 2005). For example, when cytoskeletal tension generation is suppressed in whole lung organ rudiments cultured for 48 h after isolation from embryonic mice on day E12 with the Rho-associated kinase (ROCK) inhibitor Y27632 or other drugs (e.g., BDM, Cyto D or the MLCK inhibitor, ML-9) that inhibit cytoskeletal tension generation, epithelial budding is inhibited and the local thinning of the basement membrane that is normally observed in regions of new epithelial bud formation is lost (Fig. 12.5) (Moore *et al.*, 2005). By contrast, when cytoskeletal tension is increased by activating Rho with CNF-1, lung branching is accelerated. Importantly, increasing and decreasing cell tension, respectively, promotes and inhibits angiogenesis (capillary elongation) within the neighboring connective tissue (Fig. 12.5). Therefore, changes in cytoskeletal tension play an important role in the establishment of the regional variations in cell growth and ECM remodeling

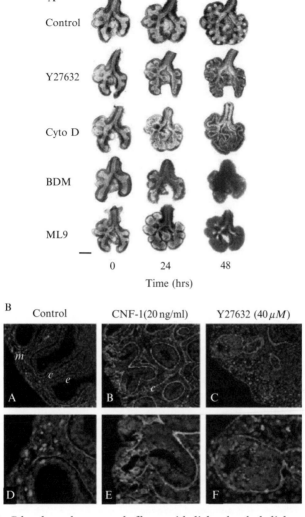

Figure 12.5 Rho–dependent control of lung epithelial and endothelial morphogenesis in whole embryonic organs. (A) Low-magnification bright-field views showing E12 mouse lung rudiment explants at time of isolation (time 0) and 24 and 48 h after treatment with agents that disrupt the cytoskeleton or suppress tension generation by various mechanisms. Lungs treated with Y27632 (40 μM) for 24 h or more exhibit enlarged bud ends and fail to form normal clefts or tight symmetric bud formation as seen in control lungs. Treatment with cyto D (100 ng/ml), BDM (20 mM), or ML9 (20 μM) completely prevents increases in epithelial bud number and size over a similar time course (scale bar, 500 μm). (B) Effects of the Rho stimulator CNF-1 and ROCK inhibitor Y27632 on epithelial and vascular development in developing lung rudiments. Immunofluorescence microscopic images of histologic sections through control lungs after 48 h culture in the absence (A and D) or presence of CNF-1 (20 ng/ml; B and E) or Y27632 (40 μM; C and F) and viewed at low (A to C) and high (D to F) magnification, showing BrdU

that drive epitheliogenesis and angiogenesis during embryonic organ development. In the developing lung, tensional forces exerted on capillary blood vessels, in part as a result of expansion of neighboring epithelial buds, apparently promote the elongation of these microvessels specifically in regions of lobular expansion.

7.1. Whole embryonic lung rudiment culture

Embryos from day 12 (E12) timed pregnant Swiss Webster mice (Taconic Farms, MA) are removed aseptically and placed into bacteriologic dishes containing Waymouth's MB medium (Invitrogen). Lung rudiments are microdissected en bloc (with all lobes still attached to the trachea), washed in serum-free medium, and transferred to a semipermeable membrane (Falcon cell inserts, 0.4-μm pore size) that is placed over 2.5 ml of serum-free BGJb medium (Fitton-Jackson modification; Invitrogen) supplemented with penicillin, streptomycin, and ascorbic acid (0.2 mg/ml) in a 6-well plate. Three lungs are placed in each well and subjected to the same dose of pharmacologic agent, which is added to individual wells at 0 h and again at 24 h with fresh BGJb medium. The agents included cyto D (100 ng/ml; Sigma), BDM (20 mM; Sigma), ML9 (20 μM; Sigma), Y27632 (10, 20, or 40 μM; Calbiochem), or CNF-1 (2, 20, or 200 ng/ml). *In vitro* development is monitored within whole organs by morphologic observations and serial measurements of branch points (number of buds) at 12-h intervals from 0 to 48 h with light microscopy. Results are expressed as percentage increase in number of terminal lung buds formed at each branch point relative to Time 0 baseline controls ($n = 9$ lungs/condition). Data are analyzed with an analysis of variance (ANOVA) single factor test and the two-sample independent t test.

---

incorporation (green), laminin (red), and DAPI-stained nuclei (blue). Note that control lung contains the greatest number of growing cells along the periphery of the growing epithelium in regions where the basement membrane thins. Well-developed capillaries (*c*) surrounded by their own basement membrane can been seen within the mesenchyme (*m*) between adjacent epithelial buds (*e*). Also, treatment with CNF-1 at 20 ng/ml results in increased numbers of epithelial buds and cell proliferation, although the proliferating cells remain localized to the periphery of the gland near regions of basement membrane thinning. CNF-1 treatment also promotes further extension and elongation of these capillary networks (*c*) in parallel with extension of the epithelium. In contrast, treatment with Y27632 inhibits basement membrane thinning and results in piled up clusters of disorganized cells that fill much of the epithelial lumen. Proliferative cells are still detected; however, their normal spatial localization at the periphery is lost. ROCK inhibition also results in disorganization of vascular architecture and disruption of the capillary network formation, because CE cells appear to be bunched together within isolated cell clusters.

7.2. Quantification of cell proliferation *in vivo*

Cell proliferation in lung rudiments is measured by quantifying the percentage of cells that exhibited nuclear incorporation of BrdU in control and drug-treated lungs that are pulsed with BrdU (10 μM, Amersham, Arlington Heights, IL) from hours 42 to 48 in culture. Lungs are fixed in 4% PFA for 1 h at room temperature, dehydrated, and paraffin-embedded. Three micrometer-thick sections are cut, deparaffinized, rehydrated, treated with proteinase K (10 μg/ml; Sigma) for 20 min at room temperature, blocked in TNB buffer (NEL-700A, NEN Life Sciences Products, Boston, MA), probed with a monoclonal mouse antibody against BrdU (RPN-202, Amersham) for 90 min at room temperature, detected with a biotinylated goat anti-mouse IgG antibody (BA-9200, Vector) and Texas Red–avidin (A2006, Vector), and counterstained with Hoescht (1:1,000). BrdU-positive fluorescent cells are visualized and scored with a Nikon Diaphot microscope with 25 and 63 objectives and the IPLab image acquisition and processing computer program (Vaytek). At least five random fields are counted per sample. Results are presented as percentage of cells incorporating BrdU. Data are analyzed by use of an ANOVA single-factor test and the two-sample independent *t* test.

7.3. Immunohistochemistry

Lungs are fixed in 4% PFA, paraffin-embedded, sectioned (3 μm), and stained with hematoxylin and eosin for light microscopic analysis. To visualize laminin in basement membrane, paraffin sections are treated with proteinase K (10 μg/ml) for 20 min at room temperature, blocked in TNB buffer, probed with a rabbit polyclonal anti-laminin antibody (L9393, Sigma; 1:100), detected by use of a biotinylated goat anti-rabbit antibody (BA-1000, Vector; 1:400) and Texas Red avidin (A2006, Vector; 1:250), counterstained with Hoechst (1:1,000), and visualized by use of immuno-fluorescence microscopy. Actin is visualized in the cytoskeleton of cells with a fluorescein isothiocyanate–conjugated monoclonal anti-actin antibody (F3046, Sigma, 1:50) in parallel sections.

8. RHO-DEPENDENT CONTROL OF VASCULAR PERMEABILITY

In addition to regulating angiogenesis, Rho and Rac are also central modulators of vascular permeability (Birukova *et al.*, 2004; Mammoto *et al.*, 2007b; Wojciak-Stothard *et al.*, 2001). We recently showed that the Rho inhibitor, p190RhoGAP, mediates the vessel sealing effects of angiopoietin-1 (Ang-1) by balancing Rho and Rac activities (Mammoto *et al.*, 2007b).

Given that p190RhoGAP mediates cytoskeleton-dependent inactivation of Rho (Mammoto *et al.*, 2007a), resulting changes in the cytoskeleton may also feed back to further modulate Rho and its vessel sealing effects. In any case, completion of these studies required lentiviral transduction of Rho proteins into CE cells and *in vivo* delivery of siRNA into vascular endothelial cells. Thus, we describe these methods here because they may be useful to other investigators as well.

8.1. Lentiviral transduction of Rho proteins

The dominant negative form of Rac1 (Rac1T17N) and the constitutively active form of RhoA (RhoAG14V) are constructed by PCR with pcDNA-Rac1T17N or -RhoAG14V (University of Missouri, Rolla, cDNA Resource Center) as a template and subcloned into the pHAGE lentiviral backbone vector at the *NotI/BamHI* sites. Generation of lentiviral vectors is accomplished by a five-plasmid transfection procedure (Mostoslavsky *et al.*, 2005). 293T cells are transfected with TransIT 293 (Mirus, Madison, WI) according to the manufacturer's instructions with the backbone pHAGE vector together with four expression vectors encoding the packaging proteins Gag-pol, Rev, Tat, and the G protein of the vesicular stomatitis virus (VSV). Viral supernatants are collected starting 48 h after transfection, for four consecutive times every 12 h, pooled, and filtered through a 0.45-μm filter. Viral supernatants are then concentrated 100-fold by ultracentrifugation in a Beckman centrifuge, for 1.5 h at 16,500 rpm. By use of these protocols, titers of 5×10^8 to 1×10^9/ml are achieved. HMVE cells are incubated with viral stocks in the presence of 5 μg/ml polybrene (sigma) and 90 to 100% infection is achieved 3 days later (Mammoto *et al.*, 2007b).

8.2. *In vitro* transendothelial permeability assays

HMVE cells from lung (1×10^5 cells in 100 μl of the medium) are seeded onto Coster Transwell membranes (6.5-mm diameter, 0.4-μm pore size) coated with 1% gelatin or fibronectin. After 1-day incubation (the cells reach confluent), the abluminal and luminal medium is carefully aspirated and replaced with the medium (100 μl medium supplemented with FITC-dextran at the final concentration of 1 mg/ml for luminal chamber and 600 μl medium for abluminal chamber). At the desired time points, samples are taken from both the luminal and abluminal chamber for fluorometry analysis (λex 485 nm; λem 525 nm). The readings are converted with the use of a standard curve to albumin concentration. These concentrations are then used in the following equation to determine the permeability coefficient of albumin (Pa): Pa = [A]/t $\times$ 1/A $\times$ V/[L], ([A]; abluminal concentration, where *t* is time in seconds, *A* is the area of membrane in cm^2, *V* is the volume of abluminal chamber, and *[L]* is the luminal concentration). Data are analyzed from a representative of at least three experiments.

8.3. *In vivo* vascular permeability assay

FVB mice (6- to 8-weeks of age) are anesthetized with intraperitoneal Avertin (2,2,2 Tribromoethanol) (125 to 240 mg/kg). 2% Evans blue (50 μl) is then injected into the retroorbital sinus or tail vein. Ten minutes after the Evans blue injection, mice are sacrificed and perfused with PBS with 2 mM EDTA for 10 min through a canula placed in the right ventricle. Blood and PBS are vented through an incision in the vena cava, thus allowing perfusate to pass through the pulmonary and systemic circulations. After 10 min of perfusion, the outflow from the vena cava is observed to be clear, confirming that blood (and intravascular Evans blue) had been flushed out the circulation. Organs (e.g., lungs, liver) are then harvested and homogenized in 1.5 ml of formamide. Evans blue is eluted by incubating the samples at 70° for 24 h, and the concentration of Evans blue is estimated by dual-wavelength spectrophotometer (620 nm and 740 nm). The following formula is used to correct optical densities (E) for contamination with heme pigments: E620 (corrected) = E620 (raw) − (1.426 × E740 (raw) + 0.03).

In vivo delivery of siRNA is used to knock down the gene. Delivery of siRNA into mice is performed with TransIT hydrodynamic delivery solution (Mirus, Madison, WI) according to the manufacturer's instructions. Mice are injected with either 10 μg of control siRNA or interested-gene siRNA in 1 ml of delivery solution into the tail vein over 7 sec. The method of siRNA delivery is known to suppress 80 to 90% gene expression in multiple organs, including the lung (Mammoto *et al.*, 2007b). High-pressure hydrodynamic delivery does not adversely affect serum chemistries or result in end-organ injury (Liu *et al.*, 1999). Three days after injection, a lung permeability assay is performed. Gene knockdown is confirmed by lysing organs in RIPA buffer and immunoblotting with specific antibody. Mouse lung wet-to-dry weight ratio (W/D ratio) is used to measure lung water accumulation. Lung wet weight is determined immediately after removal of the lung. Lung dry weight is determined after the lung is dried in an oven at 50° for 24 h. The W/D ratio is calculated by dividing the wet weight by the dry weight.

9. CONCLUSION

Vascular cell behavior is controlled through interplay between mechanical and chemical stimuli that regulate intracellular biochemistry. Understanding the molecular biophysical basis of mechanotransduction, therefore, represents a critically important challenge in vascular biology. As described in this chapter, we have developed various hybrid techniques that combine approaches from materials science and engineering with more conventional cell and molecular biologic tools to meet this challenge. These methods may be used to help better understand the fundamental mechanisms that underlie vascular control. In the future, these techniques may be further modified for more extensive use *in vivo*, which may greatly facilitate development of novel therapeutic approaches for various angiogenesis-dependent diseases, including cancer, arthritis, and various forms of blindness.

ACKNOWLEDGMENTS

This work was supported by grants from NIH, NASA, DARPA ,and DoD; D. E. I. is also a recipient of a DoD Breast Cancer Innovator Award.

REFERENCES

Amiji, M., and Park, K. (1992). Prevention of protein adsorption and platelet adhesion on surfaces by PEO/PPO/PEO triblock copolymers. *Biomaterials* **13**, 682–962.

Assoian, R. K. (1997). Anchorage-dependent cell cycle progression. *J. Cell Biol.* **136**, 1–4.

Beningo, K. A., Dembo, M., Kaverina, I., Small, J. V., and Wang, Y. L. (2001). Nascent focal adhesions are responsible for the generation of strong propulsive forces in migrating fibroblasts. *J. Cell Biol.* **153**, 881–888.

Bhatia, S. K., Teixeira, J. L., Anderson, M., Shriver-Lake, L. C., Calvert, J. M., Georger, J. H., Hickman, J. J., Dulcey, C. S., Schoen, P. E., and Ligler, F. S. (1993). Fabrication of surfaces resistant to protein adsorption and application to two-dimensional protein patterning. *Anal. Biochem.* **208**, 197–205.

Birukova, A. A., Smurova, K., Birukov, K. G., Kaibuchi, K., Garcia, J. G., and Verin, A. D. (2004). Role of Rho GTPases in thrombin-induced lung vascular endothelial cells barrier dysfunction. *Microvasc. Res.* **67**, 64–77.

Blanchard, E. M., Smith, G. L., Allen, D. G., and Alpert, N. R. (1990). The effects of 2,3-butanedione monoxime on initial heat, tension, and aequorin light output of ferret papillary muscles. *Pflüger's Arch.* **416**, 219–221.

Bohmer, R. M., Scharf, E., and Assoian, R. K. (1996). Cytoskeletal integrity is required throughout the mitogen stimulation phase of the cell cycle and mediates the anchorage-dependent expression of cyclin D1. *Mol. Biol. Cell* **7**, 101–111.

Brock, A., Chang, E., Ho, C. C., LeDuc, P., Jiang, X., Whitesides, G. M., and Ingber, D. E. (2003). Geometric determinants of directional cell motility revealed using microcontact printing. *Langmuir* **19**, 1611–1617.

Brock, A. L., and Ingber, D. E. (2005). Control of the direction of lamellipodia extension through changes in the balance between Rac and Rho activities. *Mol. Cell Biomech.* **2**, 135–143.

Carrano, A. C., and Pagano, M. (2001). Role of the F-box protein Skp2 in adhesion-dependent cell cycle progression. *J. Cell Biol.* **153**, 1381–1390.

Casciola-Rosen, L. A., Miller, D. K., Anhalt, G. J., and Rosen, A. (1994). Specific cleavage of the 70-kDa protein component of the U1 small nuclear ribonucleoprotein is a characteristic biochemical feature of apoptotic cell death. *J. Biol. Chem.* **269,** 30757–30760.

Chalovich, J. M., Sen, A., Resetar, A., Leinweber, B., Fredricksen, R. S., Lu, F., and Chen, Y. D. (1998). Caldesmon: Binding to actin and myosin and effects on elementary steps in the ATPase cycle. *Acta Physiol. Scand.* **164,** 427–435.

Chen, C. S., Alonso, J. L., Ostuni, E., Whitesides, G. M., and Ingber, D. E. (2003). Cell shape provides global control of focal adhesion assembly. *Biochem. Biophys. Res. Commun.* **307,** 355–361.

Chen, C. S., Mrksich, M., Huang, S., Whitesides, G. M., and Ingber, D. E. (1997). Geometric control of cell life and death. *Science* **276,** 1425–1428.

Chen, C. S., Ostuni, E., Whitesides, G. M., and Ingber, D. E. (2000). Using self-assembled monolayers to pattern ECM proteins and cells on substrates. *Methods Mol. Biol.* **139,** 209–219.

Chen, C. S., Tan, J., and Tien, J. (2004). Mechanotransduction at cell-matrix and cell-cell contacts. *Annu. Rev. Biomed. Eng.* **6,** 275–302.

Chicurel, M. E., Singer, R. H., Meyer, C. J., and Ingber, D. E. (1998). Integrin binding and mechanical tension induce movement of mRNA and ribosomes to focal adhesions. *Nature* **392,** 730–733.

Chrzanowska-Wodnicka, M., and Burridge, K. (1996). Rho-stimulated contractility drives the formation of stress fibers and focal adhesions. *J. Cell Biol.* **133,** 1403–1415.

Clark, E. A., King, W. G., Brugge, J. S., Symons, M., and Hynes, R. O. (1998). Integrin-mediated signals regulated by members of the rho family of GTPases. *J. Cell Biol.* **142,** 573–586.

del Pozo, M. A., Price, L. S., Alderson, N. B., Ren, X. D., and Schwartz, M. A. (2000). Adhesion to the extracellular matrix regulates the coupling of the small GTPase Rac to its effector PAK. *EMBO J.* **19,** 2008–2014.

Dike, L. E., Chen, C. S., Mrksich, M., Tien, J., Whitesides, G. M., and Ingber, D. E. (1999). Geometric control of switching between growth, apoptosis, and differentiation during angiogenesis using micropatterned substrates. *In Vitro Cell Dev. Biol. Anim.* **35,** 441–448.

Discher, D. E., Janmey, P., and Wang, Y. L. (2005). Tissue cells feel and respond to the stiffness of their substrate. *Science* **310,** 1139–1143.

Flusberg, D. A., Numaguchi, Y., and Ingber, D. E. (2001). Cooperative control of Akt phosphorylation, bcl-2 expression, and apoptosis by cytoskeletal microfilaments and microtubules in capillary endothelial cells. *Mol. Biol. Cell* **12,** 3087–3094.

Folkman, J., Haudenschild, C. C., and Zetter, B. R. (1979). Long-term culture of capillary endothelial cells. *Proc. Natl. Acad. Sci. USA* **76,** 5217–5221.

Folkman, J., Watson, K., Ingber, D., and Hanahan, D. (1989). Induction of angiogenesis during the transition from hyperplasia to neoplasia. *Nature* **339,** 58–61.

Gray, D. S., Tien, J., and Chen, C. S. (2003). Repositioning of cells by mechanotaxis on surfaces with micropatterned Young's modulus. *J. Biomed. Mater. Res. A* **66,** 605–614.

Helfman, D. M., Levy, E. T., Berthier, C., Shtutman, M., Riveline, D., Grosheva, I., Lachish-Zalait, A., Elbaum, M., and Bershadsky, A. D. (1999). Caldesmon inhibits nonmuscle cell contractility and interferes with the formation of focal adhesions. *Mol. Biol. Cell* **10,** 3097–3112.

Huang, S., Chen, C. S., and Ingber, D. E. (1998). Control of cyclin D1, p27(Kip1), and cell cycle progression in human capillary endothelial cells by cell shape and cytoskeletal tension. *Mol. Biol. Cell* **9,** 3179–3193.

Huang, S., and Ingber, D. E. (1999). The structural and mechanical complexity of cell-growth control. *Nat. Cell Biol.* **1,** E131–E138.

Huang, S., and Ingber, D. E. (2002). A discrete cell cycle checkpoint in late G(1) that is cytoskeleton-dependent and MAP kinase (Erk)–independent. *Exp. Cell Res.* **275,** 255–264.

Huber, P. A. (1997). Caldesmon. *Int. J. Biochem. Cell Biol.* **29**, 1047–1051.

Ingber, D. E. (1990). Fibronectin controls capillary endothelial cell growth by modulating cell shape. *Proc. Natl. Acad. Sci. USA* **87**, 3579–3583.

Ingber, D. E. (2006). Mechanical control of tissue morphogenesis during embryological development. *Int. J. Dev. Biol.* **50**, 255–266.

Ingber, D. E., and Folkman, J. (1989a). How does extracellular matrix control capillary morphogenesis? *Cell* **58**, 803–805.

Ingber, D. E., and Folkman, J. (1989b). Mechanochemical switching between growth and differentiation during fibroblast growth factor–stimulated angiogenesis *in vitro*: Role of extracellular matrix. *J. Cell Biol.* **109**, 317–330.

Ingber, D. E., Madri, J. A., and Folkman, J. (1987). Endothelial growth factors and extracellular matrix regulate DNA synthesis through modulation of cell and nuclear expansion. *In Vitro Cell Dev. Biol.* **23**, 387–394.

Ingber, D. E., Madri, J. A., and Jamieson, J. D. (1985). Neoplastic disorganization of pancreatic epithelial cell–cell relations. Role of basement membrane. *Am. J. Pathol.* **121**, 248–260.

Ingber, D. E., Prusty, D., Sun, Z., Betensky, H., and Wang, N. (1995). Cell shape, cytoskeletal mechanics, and cell cycle control in angiogenesis. *J. Biomech.* **28**, 1471–1484.

Itoh, R. E., Kurokawa, K., Ohba, Y., Yoshizaki, H., Mochizuki, N., and Matsuda, M. (2002). Activation of rac and cdc42 video imaged by fluorescent resonance energy transfer–based single-molecule probes in the membrane of living cells. *Mol. Cell Biol.* **22**, 6582–6591.

Iwig, M., Czeslick, E., Muller, A., Gruner, M., Spindler, M., and Glaesser, D. (1995). Growth regulation by cell shape alteration and organization of the cytoskeleton. *Eur. J. Cell Biol.* **67**, 145–157.

Jaffe, A. B., and Hall, A. (2005). Rho GTPases: Biochemistry and biology. *Annu. Rev. Cell Dev. Biol.* **21**, 247–269.

Kim, B. S., Nikolovski, J., Bonadio, J., Smiley, E., and Mooney, D. J. (1999). Engineered smooth muscle tissues: Regulating cell phenotype with the scaffold. *Exp. Cell Res.* **251**, 318–328.

Kim, B. S., Putnam, A. J., Kulik, T. J., and Mooney, D. J. (1998). Optimizing seeding and culture methods to engineer smooth muscle tissue on biodegradable polymer matrices. *Biotechnol. Bioeng.* **57**, 46–54.

Kumar, A., and Whitesides, G. M. (1994). Patterned condensation figures as optical diffraction gratings. *Science* **263**, 60–62.

Latham, K. M., Eastman, S. W., Wong, A., and Hinds, P. W. (1996). Inhibition of p53-mediated growth arrest by overexpression of cyclin-dependent kinases. *Mol. Cell Biol.* **16**, 4445–4455.

Lazebnik, Y. A., Kaufmann, S. H., Desnoyers, S., Poirier, G. G., and Earnshaw, W. C. (1994). Cleavage of poly(ADP-ribose) polymerase by a proteinase with properties like ICE. *Nature* **371**, 346–347.

Lee, K. M., Tsai, K. Y., Wang, N., and Ingber, D. E. (1998). Extracellular matrix and pulmonary hypertension: Control of vascular smooth muscle cell contractility. *Am. J. Physiol.* **274**, H76–H82.

Lele, T. P., Sero, J. E., Matthews, B. D., Kumar, S., Xia, S., Montoya-Zavala, M., Polte, T., Overby, D., Wang, N., and Ingber, D. E. (2007). Tools to study cell mechanics and mechanotransduction. *Methods Cell Biol.* **83**, 443–472.

Li, P., Nijhawan, D., Budihardjo, I., Srinivasula, S. M., Ahmad, M., Alnemri, E. S., and Wang, X. (1997). Cytochrome c and dATP-dependent formation of Apaf-1/caspase-9 complex initiates an apoptotic protease cascade. *Cell* **91**, 479–489.

Liu, F., Song, Y., and Liu, D. (1999). Hydrodynamics-based transfection in animals by systemic administration of plasmid DNA. *Gene Ther.* **6**, 1258–1266.

Lo, C. M., Wang, H. B., Dembo, M., and Wang, Y. L. (2000). Cell movement is guided by the rigidity of the substrate. *Biophys. J.* **79,** 144–152.

Mammoto, A., Huang, S., and Ingber, D. E. (2007a). Filamin links cell shape and cytoskeletal structure to Rho regulation by controlling accumulation of p190RhoGAP in lipid rafts. *J. Cell Sci.* **120,** 456–467.

Mammoto, A., Huang, S., Moore, K., Oh, P., and Ingber, D. E. (2004). Role of RhoA, mDia, and ROCK in cell shape–dependent control of the Skp2-p27kip1 pathway and the G1/S transition. *J. Biol. Chem.* **279,** 26323–26330.

Mammoto, T., Parikh, S. M., Mammoto, A., Gallagher, D., Chan, B., Mostoslavsky, G., Ingber, D. E., and Sukhatme, V. P. (2007). Angiopoietin-1 requires p190RhoGAP to protect against vascular leakage *in vivo. J. Biol. Chem* **282,** 23910–23918.

Marston, S. B., Fraser, I. D., and Huber, P. A. (1994). Smooth muscle caldesmon controls the strong binding interaction between actin-tropomyosin and myosin. *J. Biol. Chem.* **269,** 32104–32109.

Matsumoto, T., Yung, Y. C., Fischbach, C., Kong, H. J., Nakaoka, R., and Mooney, D. J. (2007). Mechanical strain regulates endothelial cell patterning *in vitro. Tissue Eng.* **13,** 207–217.

Matsumura, F., and Yamashiro, S. (1993). Caldesmon. *Curr. Opin. Cell Biol.* **5,** 70–76.

Matthews, B. D., Overby, D. R., Mannix, R., and Ingber, D. E. (2006). Cellular adaptation to mechanical stress: Role of integrins, Rho, cytoskeletal tension and mechanosensitive ion channels. *J. Cell Sci.* **119,** 508–518.

Meloche, S., Pages, G., and Pouyssegur, J. (1992). Functional expression and growth factor activation of an epitope-tagged p44 mitogen-activated protein kinase, p44mapk. *Mol. Biol. Cell* **3,** 63–71.

Mittnacht, S., and Weinberg, R. A. (1991). G1/S phosphorylation of the retinoblastoma protein is associated with an altered affinity for the nuclear compartment. *Cell* **65,** 381–393.

Mooney, D., Hansen, L., Vacanti, J., Langer, R., Farmer, S., and Ingber, D. (1992). Switching from differentiation to growth in hepatocytes: Control by extracellular matrix. *J. Cell Physiol.* **151,** 497–505.

Moore, K. A., Polte, T., Huang, S., Shi, B., Alsberg, E., Sunday, M. E., and Ingber, D. E. (2005). Control of basement membrane remodeling and epithelial branching morphogenesis in embryonic lung by Rho and cytoskeletal tension. *Dev. Dyn.* **232,** 268–281.

Mostoslavsky, G., Kotton, D. N., Fabian, A. J., Gray, J. T., Lee, J. S., and Mulligan, R. C. (2005). Efficiency of transduction of highly purified murine hematopoietic stem cells by lentiviral and oncoretroviral vectors under conditions of minimal *in vitro* manipulation. *Mol. Ther.* **11,** 932–940.

Muzio, M., Salvesen, G. S., and Dixit, V. M. (1997). FLICE induced apoptosis in a cell-free system. Cleavage of caspase zymogens. *J. Biol. Chem.* **272,** 2952–2956.

Nakatsuji, N., and Johnson, K. E. (1984). Experimental manipulation of a contact guidance system in amphibian gastrulation by mechanical tension. *Nature* **307,** 453–455.

Numaguchi, Y., Huang, S., Polte, T. R., Eichler, G. S., Wang, N., and Ingber, D. E. (2003). Caldesmon-dependent switching between capillary endothelial cell growth and apoptosis through modulation of cell shape and contractility. *Angiogenesis* **6,** 55–64.

Pagano, M., Tam, S. W., Theodoras, A. M., Beer-Romero, P., Del Sal, G., Chau, V., Yew, P. R., Draetta, G. F., and Rolfe, M. (1995). Role of the ubiquitin-proteasome pathway in regulating abundance of the cyclin-dependent kinase inhibitor p27. *Science* **269,** 682–685.

Parker, K. K., Brock, A. L., Brangwynne, C., Mannix, R. J., Wang, N., Ostuni, E., Geisse, N. A., Adams, J. C., Whitesides, G. M., and Ingber, D. E. (2002). Directional control of lamellipodia extension by constraining cell shape and orienting cell tractional forces. *FASEB J.* **16,** 1195–1204.

Pelham, R. J., and Wang, Y. (1997). Cell locomotion and focal adhesions are regulated by substrate flexibility. *Proc. Natl. Acad. Sci. USA* **94**, 13661–13665.

Pietramaggiori, G., Liu, P., Scherer, S. S., Kaipainen, A., Prsa, M. J., Mayer, H., Newalder, J., Alperovich, M., Mentzer, S. J., Konerding, M. A., Huang, S., Ingber, D. E., and Orgill, D. P. (2007). Tensile forces stimulate vascular remodeling and epidermal cell proliferation in living skin. *Ann. Surg.* **246**, 896–902.

Price, L. S., Leng, J., Schwartz, M. A., and Bokoch, G. M. (1998). Activation of Rac and Cdc42 by integrins mediates cell spreading. *Mol. Biol. Cell* **9**, 1863–1871.

Prime, K., and Whitesides, G. (1991). Self-assembled organic monolayers: Model systems for studying adsorption of proteins at surfaces. *Science* **252**, 1164–1167.

Reshetnikova, G., Barkan, R., Popov, B., Nikolsky, N., and Chang, L. S. (2000). Disruption of the actin cytoskeleton leads to inhibition of mitogen-induced cyclin E expression, Cdk2 phosphorylation, and nuclear accumulation of the retinoblastoma protein-related p107 protein. *Exp. Cell Res.* **259**, 35–53.

Ridley, A. J., Schwartz, M. A., Burridge, K., Firtel, R. A., Ginsberg, M. H., Borisy, G., Parsons, J. T., and Horwitz, A. R. (2003). Cell migration: Integrating signals from front to back. *Science* **302**, 1704–1709.

Riveline, D., Zamir, E., Balaban, N. Q., Schwarz, U. S., Ishizaki, T., Narumiya, S., Kam, Z., Geiger, B., and Bershadsky, A. D. (2001). Focal contacts as mechanosensors: Externally applied local mechanical force induces growth of focal contacts by an mDia1-dependent and ROCK-independent mechanism. *J. Cell Biol.* **153**, 1175–1186.

Sanchez-Esteban, J., Wang, Y., Filardo, E. J., Rubin, L. P., and Ingber, D. E. (2006). Integrins beta1, alpha6, and alpha3 contribute to mechanical strain-induced differentiation of fetal lung type II epithelial cells via distinct mechanisms. *Am. J. Physiol. Lung Cell Mol. Physiol.* **290**, L343–L350.

Schwartz, M. A., Lechene, C., and Ingber, D. E. (1991). Insoluble fibronectin activates the Na/H antiporter by clustering and immobilizing integrin alpha 5 beta 1, independent of cell shape. *Proc. Natl. Acad. Sci. USA* **88**, 7849–7853.

Sherr, C. J., and Roberts, J. M. (1999). CDK inhibitors: Positive and negative regulators of G1-phase progression. *Genes Dev.* **13**, 1501–1512.

Singhvi, R., Kumar, A., Lopez, G. P., Stephanopoulos, G. N., Wang, D. I., Whitesides, G. M., and Ingber, D. E. (1994). Engineering cell shape and function. *Science* **264**, 696–698.

Srinivasula, S. M., Ahmad, M., Fernandes-Alnemri, T., Litwack, G., and Alnemri, E. S. (1996). Molecular ordering of the Fas-apoptotic pathway: The Fas/APO-1 protease Mch5 is a CrmA-inhibitable protease that activates multiple Ced-3/ICE-like cysteine proteases. *Proc. Natl. Acad. Sci. USA* **93**, 14486–14491.

Tan, J. L., Tien, J., and Chen, C. S. (2002). Microcontact printing of proteins on mixed self-assembled monolayers. *Langmuir* **18**, 519–523.

Tan, J. L., Tien, J., Pirone, D. M., Gray, D. S., Bhadriraju, K., and Chen, C. S. (2003). Cells lying on a bed of microneedles: An approach to isolate mechanical force. *Proc. Natl. Acad Sci. USA* **100**, 1484–1489.

Thakar, R. G., Ho, F., Huang, N. F., Liepmann, D., and Li, S. (2003). Regulation of vascular smooth muscle cells by micropatterning. *Biochem. Biophys. Res. Commun.* **307**, 883–890.

Tzima, E. (2006). Role of small GTPases in endothelial cytoskeletal dynamics and the shear stress response. *Circ. Res.* **98**, 176–185.

Van Haastert, P. J., and Devreotes, P. N. (2004). Chemotaxis: Signalling the way forward. *Nat. Rev. Mol. Cell Biol.* **5**, 626–634.

Wang, N., Butler, J. P., and Ingber, D. E. (1993). Mechanotransduction across the cell surface and through the cytoskeleton. *Science* **260**, 1124–1127.

Weber, J. D., Raben, D. M., Phillips, P. J., and Baldassare, J. J. (1997). Sustained activation of extracellular-signal-regulated kinase 1 (ERK1) is required for the continued expression of cyclin D1 in G1 phase. *Biochem. J.* **326**(Pt 1), 61–68.

Weinberg, R. A. (1995). The retinoblastoma protein and cell cycle control. *Cell* **81**, 323–330.

Whitesides, G. M., Ostuni, E., Takayama, S., Jiang, X., and Ingber, D. E. (2001). Soft lithography in biology and biochemistry. *Annu. Rev. Biomed. Eng.* **3**, 335–373.

Wojciak-Stothard, B., Potempa, S., Eichholtz, T., and Ridley, A. J. (2001). Rho and Rac but not Cdc42 regulate endothelial cell permeability. *J. Cell Sci.* **114**, 1343–1355.

Xia, N., Thodeti, C. K., Hunt T. P., Xu Q., Ho, M., Whitesides, G. M., Westervelt, R., Ingber, D. E. (2008) Directional control of cell motility through focal adhesion positioning and spatial control of Rac activation. *FASEB J.* **22**, 1649–1659.

Yu, J., Moon, A., and Kim, H. R. (2001). Both platelet-derived growth factor receptor (PDGFR)-alpha and PDGFR-beta promote murine fibroblast cell migration. *Biochem. Biophys. Res. Commun.* **282**, 697–700.

Zhu, X., and Assoian, R. K. (1995). Integrin-dependent activation of MAP kinase: A link to shape-dependent cell proliferation. *Mol. Biol. Cell* **6**, 273–282.

Zhu, X., Ohtsubo, M., Bohmer, R. M., Roberts, J. M., and Assoian, R. K. (1996). Adhesion-dependent cell cycle progression linked to the expression of cyclin D1, activation of cyclin E-cdk2, and phosphorylation of the retinoblastoma protein. *J. Cell Biol.* **133**, 391–403.

VEGF RECEPTOR SIGNAL TRANSDUCTION

Xiujuan Li,* Lena Claesson-Welsh,* *and* Masabumi Shibuya[†]

Contents

Abstract

Signal transduction by vascular endothelial growth factors (VEGFs) through their cognate VEGF receptor tyrosine kinases follows the consensus scheme for receptor tyrosine kinases. Thus, binding of ligand induces receptor dimerization and activation of the tyrosine kinase through transphosphorylation between receptor molecules, leading to initiation of intracellular signal transduction pathways. Certain signal transduction pathways are shared with most, if not all, receptor tyrosine kinases, whereas some may be unique (e.g., transduced only by VEGF receptors). Indications that such unique signaling pathways may be discerned only when VEGF receptors are expressed in their proper

* Uppsala University, Department of Genetics and Pathology, Rudbeck Laboratory, Uppsala, Sweden
† Department of Molecular Oncology, Graduate School of Medicine and Dentistry, Tokyo Medical and Dental University, Tokyo, Japan

Methods in Enzymology, Volume 443
ISSN 0076-6879, DOI: 10.1016/S0076-6879(08)02013-2

context (i.e., in endothelial cells of microcapillary origin). In this chapter, we describe a number of methods for the study of signal transduction in endothelial cells. We describe how to isolate and examine endothelial cell lines. We also describe the embryoid body model representing vasculogenesis and angiogenesis, the procedure for subcutaneous Matrigel plugs, and, finally, how to construct gene-targeted mouse models. We emphasize the need for validation of *in vitro* data in more complex models, where endothelial cells reside in their proper three-dimensional context.

 ## 1. INTRODUCTION

Regulation of angiogenesis has a wide number of clinical applications. Therefore, drugs are developed that regulate angiogenesis (e.g., by neutralizing the effects of angiogenic growth factors) (Ferrara *et al.*, 2005). The interest in understanding the details of the angiogenic response is, to a large extent, motivated by potential clinical applications. It is, therefore, important that methods applied to study and measure angiogenesis are meaningful in the sense that they represent *in vivo* biology. It is also important to standardize methods to compare results both within and between different laboratories. It is not surprising that a single assay cannot fully represent all aspects of the angiogenic response. We describe a number of complementing strategies to study the effects of growth factors *in vitro* and *in vivo*.

The isolation of vascular endothelial cells (EC) and their culture *in vitro* paved the way for a breakthrough in the vascular biology field (Folkman and Haudenschild, 1980). Indeed, primary ECs remain very useful and may be the optimal model for functional screening of critical growth, differentiation, or survival factors for the vasculature. However, it is important to ensure that the ECs chosen as a model retain tissue- or organ-specific aspects of endothelial cells. This is particularly critical when studying signal transduction that may be EC specific. In this aspect, a comparative study on the signaling induced by angiogenic growth factors in primary ECs and artificial cell lines overexpressing EC-related receptors such as VEGFR is quite relevant. After identifying major signal transduction pathway(s) in cultured ECs, it is essential to validate these results in more complex models. The *in vitro* cultures can obviously not represent all aspect of angiogenesis *in vivo*. One vital missing aspect is the tubular organization of cells into three-dimensional (3D) lumen–containing vessel structures in a proper microenvironment providing perivascular cells and a vascular basement membrane. The embryonic stem (ES) cell culture offers a very good complement in this regard. Early steps in the differentiation of endothelial cell precursors can be studied. ES cells differentiating as defined aggregates (embryoid bodies; EBs) allow the study of later stages in the organization of endothelial cells. In a manner mimicking the *in vivo* development, angiogenic growth factors induce formation of endothelial precursors organizing in crude vascular structures (vasculogenesis) and, later, angiogenic remodeling and sprouting (Jakobsson *et al.*, 2006 and 2007a). Angiogenic sprouts present with a lumen and are surrounded by perivascular cells and a vascular basement membrane (Jakobsson *et al.*, 2007b). These cultures allow the study of growth factor–induced signal transduction, both by conventional biochemical analyses and also by use of phospho site–specific antibodies now available against a growing number of signal transduction molecules. In this chapter we give the details of the EB method.

Although the EB model has the clear advantages of allowing biochemical analyses of signal transduction in a 3D *in vivo*-like environment, it has certain inherent restrictions. Thus, the EBs lack blood flow, which is known to affect vascular remodeling. Moreover, processes occurring later in development such as lymphangiogenesis are difficult to control (Kreuger *et al.*, 2006), perhaps because of the many preceding steps that need to develop with proper timing.

The Matrigel plug assay has become commonly used as a model for *in vivo* testing of growth factor effects. The exact composition of Matrigel has not been determined, but it is clear that it contains extracellular matrix components such as laminin, collagen IV, heparan sulfate proteoglycans, entactin, and growth factors such as epidermal growth factor, fibroblast growth factors, and platelet-derived growth factors. In this assay, a liquid Matrigel mixture including growth factors, such as VEGF or FGFs, is injected subcutaneously into mice. At the body temperature of the mouse, the Matrigel solidifies. Angiogenic growth factors promote vascularization of the plug during the course of approximately a week. In contrast, vehicle-containing Matrigel will not be vascularized. Subsequently, invading inflammatory cells and fibroblasts promote reorganization of the plug, which looses its flexibility and translucency. The extent of angiogenesis can be quantified by immunohistochemistry with wholemount or histologic sections. Alternately, the total hemoglobin content can be determined, although this is considered to give less consistent results. The challenge of this assay includes a considerable variability, primarily because it is difficult to generate identical 3D plugs, even though the total Matrigel volume is kept constant. Moreover, different batches of Matrigel can give slightly variable results. Despite such challenges, this assay is regarded as one of the best assays for rapid screening of potential pro-and anti-angiogenic compounds. Eventually, it is critical to provide genetic evidence for the role played by the protein of interest by creating *in vivo* mouse models with various strategies such as gene inactivation, knockout of a particular domain, or amino acid substitution (knock in).

2. Methods

2.1. Preparation of rat liver sinusoidal endothelial cells (SEC)

Fresh preparations of vascular endothelial cells (VECs) are important when analyzing the characteristics of these cells and their dependency on growth factors such as VEGF. In rats, the liver is one of the best sources of primary VECs in terms of absolute numbers obtained and relative ease of preparation (Shibuya, 2000). Most of the VECs from adult rodents proliferative poorly, and the dependency of their growth on cytokines is not clear *in vitro*. However, rat sinusoidal endothelial cells (SECs) from the liver proliferate for a certain period of time, approximately 1 week, in the presence of VEGF.

2.1.1. Preparation of rat SECs

2.1.1.1. Preparation of the preperfusion solution (Sodium phosphate buffer with EDTA) and collagenase solution (Sodium phosphate buffer with collagenase), Each 1 L Prepare sodium phosphate buffer (basal solution).

NaCl	16 g
KCl	0.8 g
NaH_2PO_4-2 H_2O	0.156 g
Na_2HPO_4-12 H_2O	0.302 g
HEPES	4.76 g
Phenol red	0.012 g
$NaHCO_3$	0.70 g
Deionized H_2O	Adjust to 2 L

Add a few milliliters of 2 *N* NaOH to adjust the pH to 7.4. Distribute the basal solution into two bottles (each 1 L).

To prepare the preperfusion solution (1 L), add 0.19 g of Na–EDTA and 0.9 g of glucose to 1 L of basal solution, dissolve well, and distribute into two 500-ml bottles. Autoclave and store at 4 °C.

To prepare the collagenase solution, add 0.73 g of $CaCl_2$-2 H_2O and 0.5 g of collagenase (from Wako, Japan, or from other companies), and 0.05 g of trypsin-inhibitor (Sigma, T9128) to 1 L of basal solution, and mix well with a magnetic stirrer (longer than 30 min) to dissolve the collagenase. Filter with a 0.45-μm filter to sterilize the solution and to remove insoluble materials. Distribute into two 500-ml bottles, and store at 4 °C.

2.1.1.2. Collagenase perfusion

Anesthetize an adult rat (for example, Fisher rat, 200 to 250 g body weight), put the animal back in the cage, close the cover, and check whether the animal is anesthetized.

When the rat is well anesthetized, fix it, abdomen-up, with its four legs extended on a flat plate with pins. Sterilize the skin of the abdomen with 70% ethanol-containing cotton wool.

Cut the abdominal skin at the central line and extend the cut in both right and left directions at the top and bottom with sterile scissors, and fix the skin at both sides with pins to open the abdominal wall.

Cut and open the abdominal wall with sterile scissors, with as little bleeding as possible.

With 70% ethanol-dipped cotton wool, shift the intestine and colon to the right side, lift the liver lobes up, and check the position of the portal vein. Put a sterile thread around the portal vein.

Put a sterile catheter (or a small wing-type needle) into the portal vein, and take out half of the inner needle. Confirm that the blood is coming into the catheter.

Remove the inner needle from the catheter, and immediately connect the catheter to the silicon tube from the bottle of warmed preperfusion solution that has been kept in a 38 °C waterbath. With a peristaltic pump, deliver the preperfusion solution into the liver for approximately 5 to 10 sec (flow-rate, approximately 5 ml/min), and when the liver has changed from a dark brownish to light brownish color, which indicates that the solution has moved into the portal vein network, immediately cut the inferior vena cava (abdominal central vein) to release the blood outside. Fix the catheter to the portal vein with thread.

Next, mildly open the thorax with scissors, and cut the inferior vena cava in between the liver and heart. If the perfusion has been successful, a clear solution (preperfusion solution) flows up into the thoracic cavity.

After approximately 10 min, stop the peristaltic pump, and shift the origin of the silicon tube to the collagenase solution (100 ml, 38 °C). Perfuse the liver with approximately 80 ml of collagenase solution set at 38 °C in a waterbath.

After perfusion, pick up the liver with a sterile pincette and cut it away from the connective tissues of the abdomen. Transfer the liver into a sterile dish with collagenase solution.

Disperse the liver cells as a single cell suspension with a pincette and scissors in the solution. Pipette the cell suspension in a 50-ml plastic tube with a 20-ml pipette (wide-tip type) to separate the cells to create a single cell suspension.

Spin down the cells at 300 rpm for 1 min at 4 °C. Transfer the supernatant to another tube, and repeat the spinning at 300 rpm for 1 min. Transfer the supernatant to a fresh tube, and spin the cells down at 500 rpm for 2 min. Most hepatocytes (large cells) are removed by these spinning steps.

Transfer the supernatant to a fresh tube, and spin small cells down at 1000 rpm for 3 min. Resuspend the cell pellet in DMEM (Dulbecco's modified Eagle's Medium) containing 5% fetal bovine serum (FBS). Usually, a few cell aggregates (dead cells) can be observed. Let the tube stand for 1 min to allow cell aggregates to sediment. Transfer the single cell suspension to another tube, and spin again at 1000 rpm for 3 min. Approximately 80 to 90% of the cells in the pellet from this supernatant will be rat liver SECs.

Resuspend the SEC-containing cell pellet in an endothelial cell–compatible medium (for example, HuMedia-DG2, Kurabo, Japan) and seed them into collagen-coated dishes (SEC numbers from one rat liver are 3 to 10×10^6). Use 8 to 10 6-cm dishes or 2 10-cm dishes. Three to 10 h after incubation in a 5% CO_2/37 °C incubator, wash the attached cells by flushing the medium by pipetting, and remove small debris from the cells. Change the medium to fresh HuMedia and distinguish the cultured cells from the debris. Cells including SECs are now easily recognized under a phase-contrast microscope.

When changing medium (3 to 10 h after seeding), add a growth factor such as VEGF-A to examine effects on SEC proliferation and survival. Incubate the cells in 5% CO_2 in a 37 °C incubator.

POINTS: During perfusion, continuously check the catheter and the flow of solution to confirm that the perfusion is successful. Particularly carefully check the position of the catheter at the portal vein.

Some 80 to 90% of the attached cells on the collagen dish are SECs. The rest are macrophage-like Kupffer cells, epithelial cells, and fibroblastic cells including lipid storage cells (Ito cells). Depending on the technique, some hepatocytes carrying several nuclei per cell may be present in the final culture.

2.1.2. Effects of VEGF on the proliferation of SECs

Usually, VEGF-A$_{165}$ at 10 ng/ml stimulates the proliferation of SECs threefold to fourfold during a 5-day culture (Fig. 13.1). On the other hand, without VEGF-A, SECs rapidly die from apoptosis within 2 or 3 days (Yamane *et al.*, 1994). When adding 100 to 200 ng/ml of soluble VEGFR1 (sFlt-1) together with VEGF-A, the proliferation rate of SECs significantly decreases because of the trapping and neutralization of VEGF-A by soluble VEGFR1. In the presence of VEGF-A, SECs proliferate for 5 to 6 days, but thereafter, they start to die of apoptosis even in the presence of VEGF-A (Takahashi and Shibuya, 2001). Rat SECs are highly dependent on VEGF-A for their growth and survival. Even 10% or higher concentrations of FBS cannot sustain proliferation or survival of SECs. The early SEC cultures are useful for the analysis of intracellular signaling and morphologic changes and for the establishment of stable endothelial cell lines.

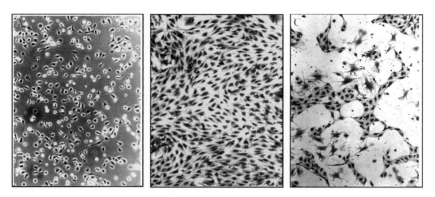

Figure 13.1 VEGF-dependent growth of rat SECs. (A) Cultured SECs 5 h after seeding (day 0). (B) Stimulation of SECs with VEGF-A165 (10 ng/ml) for 5 days. (C) SECs cultured with both VEGF-A165 (10 ng/ml) and soluble VEGFR1 (100 ng/ml) for 5 days.

2.1.3. VEGF-induced VEGFR2 signaling in rat SECs

Autophosphorylation of VEGFR2 and phosphorylation of other signaling molecules on rat SECs with VEGF.

Prepare the SECs as described previously and culture them at 5% CO_2/ 37 °C for 3 to 4 days in the presence of 10 ng/ml of VEGF-A$_{165}$.

Starve the cells of VEGF by changing the medium to a VEGF-minus medium (small volume, 1.5 ml/6-cm dish). Incubate for 6 h at 37 °C.

Stimulate the cells with VEGF-A (final concentration, 50 to 200 ng/ml) by adding a highly concentrated VEGF-A into the medium. Immediately mix the VEGF solution well, and incubate the cells at 37 °C for appropriate time periods such as 5, 10, or 30 min.

To stop the reaction, wash the cells with ice-cold phosphate-buffered saline (PBS) containing 0.1 mM Na$_3$VO$_4$, twice.

Lyse the cells in 1% Triton X-100 lysis buffer (50 mM HEPES, pH 7.4, 150 mM NaCl, 10% glycerol, 1% Triton X-100, 1.5 mM MgCl$_2$, 2% aprotinin, 1 mM phenylmethylsulfonyl fluoride [PMSF], 50 mM NaF, 10 mM Na$_4$P$_2$O$_7$, and 2 mM Na$_3$VO$_4$).

Spin the lysates at 15,000 rpm for 10 min, and use the clarified lysates for regular SDS-PAGE and immunoblotting analysis with antibodies against VEGFR1, VEGFR2, phospho-MAPK, or phospho-tyrosine (PY20).

For immunoprecipitation, incubate the cell lysates with the primary antibody against VEGFR1, VEGFR2, or phospholipase-Cγ. Then, precipitate the immune complex with Protein-A or -G beads, separate the precipitates by SDS-PAGE, and immunoblot with antibody against phosphotyrosine or other targets.

2.1.3. Establishment of permanent SEC lines

To establish an immortalized SEC line, SECs growing in the presence of VEGF may be transfected with a cell-immortalizing gene-containing plasmid DNA such as the Simian virus 40 (SV40) T-antigen (Maru *et al.*, 1998).

Essentially all the SECs without the SV40 T-antigen plasmid die during a 10-day culture. Pick the surviving and growing colonies with an endothelial-like morphology for further analysis. So far, all immortalized SEC lines show a downregulation of VEGFR gene expression at the protein level, and cell growth becomes independent of exogenous VEGF, although VEGFR mRNA is detected by RT-PCR.

2.2. Establishment of VEGFR1- or VEGFR2-expressing cell lines with NIH3T3 fibroblasts

2.2.1. Preparation of VEGFR1 or VEGFR2 cDNA-containing expression plasmid vectors

Prepare full-length cDNAs (approximately 4.5 kbp each) for VEGFR1 or VEGFR2, which should carry approximately 200 bp 5'-noncoding sequence.

Insert the cDNA (for example, the 4.7 kbp *SalI-NotI* fragment of the VEGFR1 cDNA) into the *XhoI-NotI* site of bovine papilloma viral (BPV) genome-based plasmid vector DNA, BCMGSneo (Karasuyama *et al.*, 1990), which contains 69% of the BPV genome and the CMV promoter, for gene expression. The BCMGSneo vector replicates in the host nucleus as a multicopy episome.

2.2.2. Transfection of DNA into NIH3T3 Cells and selection of geneticin G418-resistant clones

Preparation of polybrene-containing medium (Kawai, and Nishizawa, 1984). Stock solution of polybrene: 10 mg/ml of sterile deionized H_2O filtered with a 0.45-μm filter. Add 3 μl of polybrene (10 mg/ml) to 1.5 ml of DMEM with 10% calf serum (CS) per 6-cm dish. When 10 dishes are used for transfection, prepare polybrene-containing medium at more than 10-fold the volume.

Culture NIH3T3 cells in regular medium such as DMEM with 10% CS (or 5% FBS).

Seed the cells at 1.5×10^5 cells/5 ml DMEM with 10% CS in 6-cm dishes in the afternoon/evening (1 day before the DNA transfection).

Next morning, change the medium to 1.5 ml of polybrene-containing medium.

Add the plasmid vector DNA (5 to 10 μg in 100 μl) into the medium, mix well, and culture for 6 h in a 5% CO_2 incubator at 37 °C.

Prepare 30% DMSO-containing culture medium (DMEM with 10% CS), and keep it at 4 °C in a refrigerator (2.5 ml per 6-cm dish).

Six hours later, change the medium to the preceding DMSO-containing medium. Leave the dishes for 4 min at room temperature.

Remove the DMSO-containing medium by aspiration, and add fresh culture medium (DMEM with 10% CS, 5 ml per dish). Incubate the cells in a 5% CO_2/37 °C incubator.

Next morning, renew the medium. Add Geneticin G418 to a final concentration of 200 μg/ml. Every 3 to 4 days, replace the medium with fresh G418-containing culture medium, and continue the culture for 2 to 3 weeks.

Isolate the G418-resistant cell clones growing as colonies, and transfer them to 24-well dishes.

After expansion of these G418-resistant VEGFR1- or VEGFR2-expressing NIH3T3 cell clones, examine receptor expression levels by Western blotting with an antibodies specific to VEGFR1 or VEGFR2. Isolate the highest and second-highest VEGFR-overexpressing clones from 40 or more clones. Stock aliquots of these clones in a liquid nitrogen tank, and use other aliquots for experiments such as autophosphorylation of receptors and signaling from the receptor (Seetharam *et al.*, 1995; Sawano *et al.*, 1996; Takahashi *et al.*, 2001). Note that the signaling from the VEGFRs in NIH3T3 fibroblasts is slower than that in primary vascular endothelial cells such as SECs and HUVECs (Takahashi *et al.*, 1999).

COMMENTS: in this review, the polybrene-DMSO method (Kawai *et al.*, 1984) is described. Other DNA transfection methods such as the calcium-phosphate-precipitation and lipofection are possible alternate methods. Because the vector plasmid BCMGSneo replicates in an extrachromosomal fashion in cells, without G418 in the culture medium, the transfected cells gradually lose the plasmid vector DNA. So far, regular gene-expression plasmids such as pcDNA have not been suitable for the establishment of permanent cell lines overexpressing VEGFR1 or VEGFR2.

These VEGFR1- or VEGFR2-overexpressing NIH3T3 cells and rat SECs and HUVECs are useful for the analysis of VEGF-dependent cell proliferation with conventional methods such as [³H]thymidine incorporation, BrdU incorporation, and the MTT assay.

2.3. Embryonic stem (ES) cell and embryoid body (EB) culture protocol

2.3.1. Preparation for ES cell culture
2.4.1.1. ES cell lines

A number of ES cell lines of different genetic backgrounds (129 SvJ, C57BL/6 and F1 hybrids of the two) have been confirmed to give comparable patterns of endothelial cell responses in two-dimensional (2D) and 3D angiogenesis assays.

Note that the batch of serum and its content of angiogenic growth factors, in particular VEGF, will dictate the extent of effect of exogenous VEGF. The serum needs to be carefully screened, and preferentially stocks should last for extensive time periods to avoid the trouble of rescreening. Serum-free cultures are an interesting alternative, but, to date, the survival of endothelial cells is very poor in serum-free medium and is, therefore, not an option at present.

2.3.2. Preparation of feeder cells

The feeder cells serve as a coating and a source for production of various substances such as leukemia inhibitory factor (LIF) that maintain ES cells in an undifferentiated state. For a detailed protocol to prepare feeder cells, please see http://www.molgen.mpg.de/~soldatov/protocols/protocol/02/02_03.htm. In short, mouse embryos (E14.5) are decapitated and internal organs (such as heart, liver, lungs) are removed. The remaining tissues are cut into small pieces and trypsinized to single cells. Cells are propagated for 2 to 4 days and then treated with mitomycin C to induce cell cycle arrest. The mitomycin C–treated feeders are then aliquoted and frozen until subsequent use for ES culture.

2.3.3. ES and EB culture medium

ES cells are cultured in a medium prepared as described in Table 13.1. The ES culture medium should be prepared fresh for each time it is used, by mixing the different components. EB culture medium is prepared similar to ES cell medium, but without the ESGRO/LIF, to allow differentiation of the ES cells.

2.3.4. Routine culture of ES cells

Feeder cells are seeded as 200,000 to 250,000 cells/well in a 6-well plate in EB culture medium. Incubate at 37 °C, 5% CO_2.

Table 13.1 ES cell culture medium (500 ml)

Reagents (stocks)	From stock	Final conc.	Catalog number
DMEM*/glutamax	406.5 ml	81.3%	Gibco #61965-026
FBS	75 ml	15%	Gibco #10270-106
1 M HEPES buffer	12.5 ml	25 mM	Gibco # 15360-056
100 mM Sodium pyruvate	6 ml	1.2 mM	Gibco #11360-039
Monothioglycerol 98%	6.2 μl	19 mM	Sigma #M-6145
10^7 U/ml ESGRO/LIF	50 μl	1000 U/ml	Chemicon #ESG1107

* DMEM, Dulbecco's modified Eagles medium; FBS, fetal bovine serum; LIF, leukemia inhibitory factor.

The next day, thaw ES cells in one frozen vial (approximately 1×10^6 cells) quickly and add 3 ml ES medium in one well of a 6-well plate (BD#353046) coated with feeder cells. Change medium later that day or first thing the next day. ES cells grow in colonies and appear "shiny" when they are in good condition.

After these colonies reach subconfluence, they need to be trypsinized and diluted. Remove medium, wash with Versene (PBS/1 mM EDTA).

Add 0.5 to1 ml trypsin-EDTA ($1\times$, Gibco#25300-054) and incubate 5 min at 37 °C.

Resuspend in 4 ml LIF-containing ES growth medium and transfer, preferably 1:10, to a new well.

When ES cells are growing well, splitting can be performed every second day. Otherwise, medium should be changed at least every second day. It is important that the medium has the correct pH; therefore, calibrate the medium in 5% CO_2 if it has turned purple.

2.3.5. Differentiation of ES Cells to EBs

2.3.5.1. Differentiation procedure   During differentiation of EBs, a complete removal of feeder cells is, in most cases, desirable. For this purpose, ES cells are cultured for two extra passages on gelatin-coated dishes (with 1% gelatin/PBS to coat dishes for 15 to 20 min). Alternately, most feeder cells can be removed by letting the trypsinized ES/feeder cell mixture settle on a fresh culture dish for 30 min, which will allow attachment of the feeders but not of the ES cells.

Trypsinize the ES cells as described previously.

Resuspend in 7 ml EB medium. Spin down at 1000 rpm and resuspend cell pellet in 7 ml EB medium. This wash will remove the LIF.

Count the cells.

Dilute the cell suspension to 60,000 cells/ml in EB medium and distribute as 20 μl drops on the inside of a 10-cm bacterial dish lid (BD#351029). Put PBS in the bacterial dish (to humidify the environment) and carefully turn the lid with the drops over and put it back on dish. In these drops, the ES cells will aggregate and differentiate as EBs.

Flush down the EBs in EB medium after 4 days and transfer them either (1) to cell culture dishes, 20 to 40 EBs/6-cm dish, (2) individually, to 8-well chamber glass slide (Falcon, Frankling Lakes, NJ) for 2D culture; or (3) onto a matrix of collagen I gel for 3D cultures.

Change to medium with or without VEGF (or other growth factors) every fourth day.

2.3.5.2. Embryoid body 3D sprouting assay in collagen I Gel   Here, we introduce how to perform the 3D collagen culture with two layers of collagen I gel and how to seed the EBs into the collagen gel.

A sterile 50-ml plastic tube should be placed in a beaker filled with ice. Add the following reagents (sufficient to fill 12 wells in a 12-well plate) in the order shown.

Reagents (stocks)	From stock	Final conc.
0.1 M NaOH	0.5 ml	6.25 mM
10× F12	0.5 ml	~1×
1 M HEPES	100 μl	12.5 mM
7.5% bicarbonate soln	78 μl	0.073%
Glutamax	50 μl	1%
1× F12	2.772 ml	~1×
3 mg/ml collagen I	4 ml	1.5 mg/ml

Gently mix by pipetting up and down. Avoid bubbles.

Add 0.6 ml of the suspension to each well of a 12-well plate (the first collagen layer) and tilt the plate for the collagen to adhere to the walls of the wells. Let the gel polymerize at 37 °C for 3 h. There is some collagen left in the tube. Put the tube at 37 °C for 3 h to check polymerization. Polymerization of the first collagen layer can be done the day before seeding the EBs.

Seed 5 to 15 days 4 EBs on top of the first collagen layer in as small a volume as possible (not more than one drop of medium) with a Pasteur pipette.

Directly add 0.6 ml of fresh collagen solution (the second collagen layer)/ well and tilt the plate for the collagen to adhere to the walls of the wells. Let the gel polymerize for 3 h at 37 °C, then add prewarmed medium with and without growth factor and proceed to culture for the required time period.

2.3.6. Immunohistochemical (IHC)/Immunofluorescent (IF) staining of 2D EBs (Days 8 to 12)
2.3.6.1. First Day

Wash EBs in the 8-well chamber once in Tris-buffered saline (TBS; be careful as the EBs are easily torn by hard flushing from the pipette).

Fix EBs in 140 μl of zinc fix containing 0.2% Triton X-100 overnight at 4 °C or, at room temperature for 4 h. Zinc fix: 100 mM Tris HCl, containing 37 mM zinc chloride, 23 mM zinc acetate, and 3.2 mM calcium acetate.

Wash once in TBS.

ONLY FOR IHC: Block endogenous peroxidase activity with 3% H_2O_2 in methanol for 10 min (1:10 dilution of 30% hydrogen peroxide in methanol).

Wash twice in TBS.

Block unspecific binding with TBT buffer for 30 min–1 h at room temperature. (TBT: 3% BSA, 0.1% Tween-20 in TBS).

For IHC: Incubate with primary Ab (e.g., rat anti-mouse CD31, BD#553370, 1:2000) diluted in TBT overnight at 4 °C. Use one well in the 8-well chamber as a control (i.e., no primary antibody).

For IF: Costain by incubation with primary Ab (e.g., rat anti-mouse CD31, BD#553370, goat anti-mouse Flk1, R&D AF644, final concentration 200 ng/ml) diluted in TBT overnight at 4 °C. Use two wells in the 8-well chamber for the control of the two antibodies, respectively (no primary antibodies).

2.3.6.2. The secondary day

Wash three times in TBS, 0.05%Tween-20.

For IHC: Incubate with biotinylated secondary Ab for 1 h at room temperature.

For IF: Incubate with fluorescent secondary Ab (e.g., Alexa donkey anti-goat 488 and Alexa donkey anti-rat 594) for 0.5 to 1 h at room temperature. From this step on, avoid exposure to light. Remember to add the corresponding secondary antibody in the control well to show specificity of the primary antibody.

Wash three times in TBS, 0.05%Tween-20.

For IHC: Incubate with streptavidin-HRP for 30 min at room temperature.

For IF: Incubate with Hoechst 33342 (to visualize nuclei; final concentration 1 μg/ml) for 10 min at room temperature.

Wash carefully four times with TBS, 0.05%Tween-20. This is an important step.

For IHC: Use AEC kit (2 drops of buffer to 5 ml double distilled H_2O, 3 drops of AEC and 2 drops of peroxide solution). Terminate by adding an excess volume of H_2O. Wash with H_2O a couple of times.

For IF: Have a quick look in the microscope.

Remove the plastic part of the slide with the device that comes with the slides. Let dry on the bench.

For IHC. Add 2 drops of Ultramount aqueous mounting medium (DAKO, Glostrup, Denmark) at the edge of slide and spread over the slide with a cover glass. Put slides in a metal box to protect against dust and put in an oven at 70 °C for 30 min.

For IF: Add several (4) drops of Fluoromount G (Southern Biotechnology, Birmingham, AL), cover the slide with a cover glass, avoiding bubbles. Put at room temperature overnight. For a strong surface add EUKITT mounting medium (O. Kindler GmBH, Freiburg, Germany) and dry for

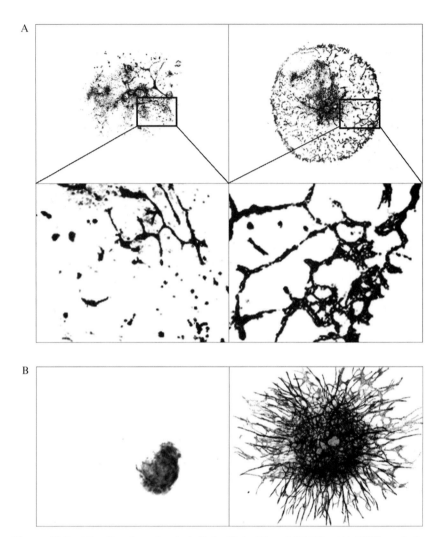

Figure 13.2 Visualization of endothelial cells in 2D and 3D EBs. (A). Differentiation of endothelial cells, visualized by IHC staining, occurs to some extent in the absence of VEGFA165 because of endogenous production of growth factors (left). In the presence of VEGFA165 (right) endothelial cells will arrange in capillary structures in the periphery of the EB (see magnification in lower panels). (B) Visualization of angiogenic sprouts in VEGF-A165-treated 3D EB (right panel) by IF staining. Left panel shows control culture without added growth factor.

a few hours. Wipe the slide with ethanol. Observe in the microscope and store the slides in the dark at 4 °C. See Fig. 13.2A for IHC staining of 2D EBs with and without VEGF-A165 treatment.

2.3.7. Fluorescent staining of 3D EBs

Wash 3D EBs cultured in collagen I in PBS, 2×10 min at room temperature.

Fix in 4% p-formaldehyde/PBS for 30 min at room temperature.

Add PBS until proceeding with next step.

Gently remove the upper layer of the collagen slab with a forceps and place it in an empty well in 24-well plate with PBS, for further processing.

Wash in PBS for 2×15 min or longer at room temperature.

Block and permeabilize in 3% BSA, 0.2% Triton X-100 in PBS for 2 h at room temperature.

Add primary Ab 1:2 000 (e.g., rat anti-mouse CD31 from BD Pharmingen) in PBT (3% BSA, 0.1% Tween 20 in PBS), at 4 °C overnight.

Wash in PBS/0.1% Tween 20, 3×1 h (or overnight at 4 °C).

Add secondary Ab 1:1 000 (Alexa donkey anti-rat 568) in PBT for 1 h or overnight at 4 °C.

Wash in PBS/0.05% Tween 20, $3 \times$ 1 h or longer.

Stain with Hoechst 33342 at a final concentration of 10 μg/ml in PBS for 10 min, followed by extensive washing. Additional washing will decrease the background. See Fig. 13.2B for immunofluorescent staining of 3D EBs after treatment with VEGF-A165.

2.4. Matrigel *in vivo* plug assay, section, and whole mount staining

Procedure is given below for translating the *in vitro* effects (e.g., of angiogenic growth factors to *in vivo* effects) with Matrigel/growth factor mixtures for subcutaneous injection in mice.

2.4.1. Before the assay: Preparation of animals and reagents

Order 5-week-old female mice (nude mice is optimal) from suppliers.

Order Matrigel (BD biosciences, #354248, high concentration, 18 to 22 mg/ml).

Prepare different growth factors in their stock solution in PBS (e.g., 1 μg/μl human VEGF-A165, Peprotech, #100-20).

Prepare (see later) 125 μM sphingosine-1-phosphate (S1P, Avanti Polar Lipids, Cat # 860492P) stock solution in 4 mg/ml fatty acid free BSA (Sigma, cat # A8806-5G).

2.4.2. Prepare S1P methanol solution (maximal storage time, 1 year)

Prepare SIP methanol solution (maximal storage time, 1 year)

Add S1P to methanol (0.5 mg/ml).

Heat mixture (45 to 65 °C; i.e., the boiling point of methanol) and sonicate until S1P is suspended (note, does not get clear but forms a slightly hazy suspension).

Transfer desired amounts of methanol stock solution to glass vessel.

Place glass vessel into a warm (approximately 65 to 70 °C) waterbath.

Rotate glass vessel during evaporation to distribute S1P as a thin film on the tube wall.

Store S1P at −20 °C in closed container.

S1P BSA solution (maximum storage time, 3 months)

Add BSA (4 mg/ml) in water at 37 °C to glass vessel containing S1P to make a 125 μM S1P solution.

Incubate for 30 min at 37 °C and vortex occasionally. A clear solution will form at this step.

2.4.3. The day before injection

Put Matrigel on ice in cold room (4 °C) to thaw.

Cool sterile PBS, 50-ml Falcon tubes, 1.5-ml Eppendorf tubes, 1-ml tips, 200-μl tips, 20-μl tips, 1-ml syringes, 23-gauge needles, at 4 °C.

If nude mice are not used, shave mice to reveal the entire belly; it is easier to inject and take plugs out later on.

2.4.4. Matrigel preparation at the injection day

Note: All the procedures are done on ice, in a tissue culture hood.

Each animal will have two Matrigel implants of 400 μl one each side of the ventral mid line. One plug will contain the angiogenic growth factor; the other one will serve as a control plug.

The overall purpose is to mix VEGF-A165 with Matrigel to obtain final concentration of 5 $\mu g/ml$ (130 nM), containing 1 μM S1P, and then inject 400 μl/plug/right side of each animal. There will be 2 μg VEGF-A165/plug and one growth factor–containing plug/mouse. The Matrigel with 1 μM S1P only serves as control and is used for injection on the left side of the mouse.

Remember to prepare an extra volume of each sample, because there will be losses as a result of viscosity, (i.e., prepare 500 to 550 μl Matrigel mixture for one injection of 400 μl; note that different Matrigel batches have different viscosities).

2.4.4.1. Procedure

When four mice are used in the experiment and, therefore, injection of 8 plugs, prepare a total Matrigel mix volume of 4.4 ml (550 μl/plug × 8), which includes 2.2 ml DMEM/BSA/S1P and 2.2 ml Matrigel stock.

Prepare 0.5 mg/ml fatty acid-free BSA in serum-free DMEM. Transfer 2.2 ml to a prechilled 50-ml Falcon tube (soln 1).

Add 35 μl of 125 μM S1P into "soln 1" to get 1 μM S1P.

Transfer 2.2 ml thawed liquid Matrigel stock into prechilled 50-ml Falcon tube with prechilled 1-ml tip (soln 2).

Add soln 1 to soln 2. Mix by careful inversion. Avoid bubbles!! There is risk of loosing volume in this step.

Aliquot the Matrigel mixture into 1.5-ml Eppendorf tubes; 1 ml/tube. There are 4 tubes in total which will allow injection of 4 mice; 2 plugs/mouse.

Add 5 μl 1 μg/μl VEGF-A165 into two tubes (final concentration 5 μg/ml). There should be no growth factors in the remaining 2 tubes.

Turn gently upside down two to three times to mix. Spin briefly in precooled 4 °C centrifuge.

2.4.4.2. Matrigel injection

Anesthetize mouse number 1.

Fill up two syringes (labeled control and test), one with Matrigel/S1P (control), and another with Matrigel/VEGF-A165/S1P (test sample). Be careful to slowly suck up the Matrigel into syringe up to 400 μl (avoid bubbles). Immediately place syringe flat on ice.

Inject control plug subcutaneously on the left side of the belly, above the leg just slightly toward the middle. Try to control the spreading of the Matrigel by pressing skin around plug. A balloon-shape is recommendable. Also avoid having the plug spread close to mid line to avoid the two plugs from fusing.

After injection, take the needle out carefully without mechanical damage to the surrounding tissue and immediately squeeze the injection point with your fingers to prevent the Matrigel from leaking out. Hold for 1 to 2 min (until the Matrigel starts polymerizing).

Wipe away Matrigel that may have leaked out.

Inject the second plug with growth factors on the right side of the midline and not too close to the leg. Make sure the two plugs do not merge. See Fig. 13.3A for visualization of plugs after subcutaneous injection.

Proceed with the next mouse, starting with anesthesia.

2.4.4.3. Collecting matrigel plugs

The optimal time point for collecting the plugs is on day 7 in our experience. After this time, the Matrigel starts to become reorganized and is replaced by invading tissue.

Euthanize mice one by one.

Stretch mouse on support, cut skin and muscle open over belly to reveal the Matrigel plugs.

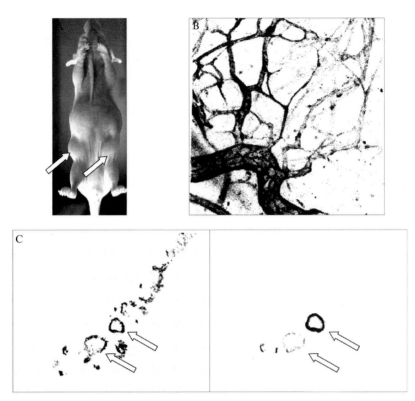

Figure 13.3 Subcutaneous Matrigel plugs in nude mice. (A) Nude mouse with two Matrigel plugs (arrows) on either side of the midline.(B) Immunofluorescent staining to visualize CD31 expression in vessels formed in VEGF-A165–treated Matrigel plugs after whole mount fixation. (C) Visualization of vessels after CD31 immunofluorescent staining of cryo-sectioned Matrigel plugs (left). Right panel (αSMA staining) shows that certain vessels are surrounded by supporting pericytes (indicated by arrows).

For whole-mount fixation, plugs should be collected with the underlying small piece of muscle. This is done to avoid mechanical wear and tear on the plug when handling with a forceps during the subsequent staining process. Do not save the skin, because it becomes very rigid after fixation and will interfere with the microscopy. Indeed, it is important to remove all other tissue on the surface of the plug. Wash the plug in PBS and fix in 4% *p*-formaldehyde in PBS for 1 h at room temperature. Store in PBS at 4 °C until processing for whole-mount fluorescence staining.

For cryo-sectioning, fill the plastic moulds with OCT compound. Cut out plug together with some extra skin and muscle around it to easily identify the edge of the plug later after sectioning and staining. Hold plug in one hand by two fingers and, with dissecting scissors, cut off the tip of the plug so you can see the Matrigel. Be careful to not squeeze out material, because it is quite soft. Hold plug with forceps with bent ends, let cut side down, and place plug in the middle of the mould so it is surrounded by OCT. Pay attention not to mix up the moulds between control/treated. Freeze mould quickly in isopentane/dry ice. Store in plastic bags at −80 °C.

2.4.4.4. Fluorescent whole-mount staining of matrigel plug

Take the plug from 4 °C refrigerator.

Wash in PBS for 2 × 15 min or longer at room temperature.

Block and permeabilize in 3% BSA/0.2% Triton X-100/PBS for 3 h at room temperature.

Add primary Ab (e.g., rat anti-mouse CD31 from BD Pharmingen at a 1:2000 dilution, or various phospho site–specific antibodies against signal transduction proteins) in PBT at 4 °C overnight.

Wash in PBS/0.1% Tween 20, 3×1 h.

Add secondary Ab, 1:1000 Alexa donkey anti-rat 594) and e.g., α smooth muscle actin (αSMA)-FITC (1:500) in PBT overnight at 4 °C (avoid exposure to light from this step).

Wash in PBS/0.05% Tween 20, 3×1 h.

Stain with Hoechst 33342 to visualize nuclei, at a final concentration of 10 μg/ml in PBS for 10 min, followed by extensive washing.

Take Z-stack pictures with confocal microscope with 10× or 20× objective. See Fig. 13.3B for immunostaining of Matrigel plugs after whole-mount fixation.

NOTE: It is tricky to judge Matrigel plug vascularization. It is absolutely critical to compare the effects of VEGF with the negative control and preferably with a positive control such as FGF.

2.4.4.5. Fluorescent staining of matrigel plug sections

Five- to 10-μm frozen sections of plugs are prepared and fixed in cold acetone for 10 min.

Wash in PBS for 5 min (all washes should be done in the plastic containers).

Prepare slide box with two wet sponges to create a humid chamber.

Remove excess of PBS from glass slides by shaking/dabbing with paper. Do not let slides dry, do not touch the specimen area.

Draw a line between two specimens (if needed) on the glass with a special marker. This separates the sections and it will allow with different Abs on the same glass.

Add 100 to 150 μl blocking buffer (PBS/3% BSA) for each specimen and incubate 30 to 60 min at room temperature.

Add primary antibody in blocking solution (e.g., rat anti mouse CD31 1:1 000) overnight at 4 °C in the humidified chamber.

Wash in PBS 3× 5 min, at room temperature.

Add secondary Ab, e.g. donkey anti-rat Alexa 555 diluted 1:1000, together
with αSMA-FITC diluted 1:500, incubate for 1 h at room temperature in
the humidified chamber and avoid light from this step.

Wash 3× 5 min in PBS.

Stain nuclei with Hoechst 33342, at a final concentration of 1 μg/ml, for 10
to 15 min in the dark at room temperature, in a humified chamber.

Wash 3× 5 min in PBS.

Check under the microscope before mounting. For fluorescent secondary
Ab: If there is no background, washing is sufficient.

Remove excess of PBS, place a drop of Fluoromount G per section, avoid
bubbles, put the coverslip carefully on the top, do not press at this may
destroy specimen.

Dry overnight; remember to keep in the dark. See Fig. 13.3C for immunofluorescent staining of sectioned Matrigel plugs.

2.5. Generation of *VEGFR*1 $tk^{-/-}$ Mice

The materials and methods for this experiment are essentially the same as
those for regular gene targeting in mouse ES cells. Thus, here only the basic
strategy for producing *vegfr1* $tk^{-/-}$ mice is described.

2.5.1. Preparation of VEGFR1 (flt-1) genomic DNA from the mouse 129Sv-strain genomic DNA library (Stratagene)

Screen the genomic library with a nylon-membrane filter and label *vegfr1*
cDNA as a hybridization probe (nucleotides 2411-2705) covering exon
17 to 19 of the murine *vegfr1* gene to isolate the genomic DNA clones
that encode for the transmembrane domain and the amino-terminal
portion of the tyrosine kinase domain (Hiratsuka *et al.*, 1998).

Construct a targeting vector with the following elements in order from the
5′-end: The diphtheria toxin A gene (pMC1 DT-A); approximately
1.5 kbp of *HindIII-DraI* genomic DNA derived from intron 16 of the
vegfr1 gene; a neocassette (pPGK-neopA); approximately 7 kbp genomic
vegfr1 DNA, which includes exons 18 and 19.

Transfect the purified recombinant DNA into mouse J1 ES cells by electroporation, and isolate G418-resistant cell clones. From the resistant clones,
identify ES cell clones that have undergone the correct homologous
recombination event by Southern blot analysis.

Inject ES cell clones containing the targeted gene into C57BL/6 blastocysts
to obtain male chimeric mice. Cross the male chimeric mice with
female C57BL/6 mice to yield mice heterozygous for the *flt-1*^{tk} mutation
(*flt1*$^{tk+/-}$ mice). Determine the genotypes of the pups obtained by crosses
between *flt1*$^{tk+/-}$ mice by Southern blot analysis and/or PCR analysis to
obtain *flt1*$^{tk-/-}$ homozygotes.

COMMENTS: Because the mice are missing the amino terminal part of the tyrosine kinase domain, the *flt-1* gene products expected from the gene targeting are two protein products: (1) a truncated VEGFR1 protein that has intact extracellular and transmembrane domains but is missing the entire tyrosine kinase domain, terminated after the transmembrane domain with 6 amino acids derived from the intron 16 sequence, and (2) an intact soluble VEGFR1 protein that is encoded by exons 1 through 13.

The *flt-1 (vegfr1) tk*$^{-/-}$ mice are essentially viable and healthy in specific pathogen-free (SPF) conditions. The viability of *vegfr1 tk*$^{-/-}$ mice indicates that the extracellular and transmembrane regions are essential and sufficient for exhibiting the negative regulatory role of VEGFR1 on angiogenesis early in embryogenesis (Fong *et al.*, 1995; Hiratsuka *et al.*, 1998). These mice are useful in determining the importance of VEGFR TK–derived signaling in various pathologic conditions at adult stages (Shibuya, 2006; Shibuya and Claesson–Welsh 2006).

2.6. Generation of VEGFR2 knockin mice at 1173-tyrosine or 1212-tyrosine site

The tyrosine kinase activity of VEGFR2 is stronger than that of VEGFR1, and VEGFR2 seems to control the major signaling pathways toward DNA synthesis and migration of vascular endothelial cells (Shibuya and Claesson-Welsh, 2006). In primary endothelial cells such as SECs and HUVECs, the major pathway from VEGFR2 to DNA synthesis is initiated from the autophosphorylation of VEGFR2 at tyrosine 1173 (in humans, tyrosine 1175), activation of phospholipase Cγ, activation of protein kinase C (PKC), and activation of the c-Raf/MEK/MAPK pathway. At this time (early 2008) only two knock in mice with the amino acid residues in VEGFR2 have been generated (Sakurai *et al.*, 2005). The strategy for the preparation of these knock in mice follows,

2.6.1. Isolation of VEGFR2 (flk-1) genomic DNA from a mouse genomic DNA library and preparation of the target vector DNA

Screen a mouse 129Sv genomic DNA library (Stratagene) with a nylon-membrane filter and labeled *vegfr2* cDNA as a hybridization probe (corresponding to the carboxyl-terminal half of the VEGFR2 tyrosine kinase domain and carboxyl tail) to isolate the genomic DNA clones which cover *flk-1 (vegfr2)* exons 18 to 28.

Molecularly subclone the 0.65 kbp *NcoI-EcoRI* fragment of genomic *vegfr2* DNA carrying exon 27 to another plasmid DNA and carry out site-directed mutagenesis on this DNA.

Introduce the substitution mutations Y (tyrosine) 1173 to F (phenylalanine) (TAT to TTC) and Y1212F (TAT to TTC). To distinguish the wild-type and mutant alleles by restriction enzyme digestion, introduce restriction sites near the Y1173F and Y1212F mutations without changing the amino acid codons. In the Y1173F targeting vector, a *AccIII* restriction site was introduced at position S1188 (TCT to TCC). In the Y1212F targeting vector, a *PvuII* restriction site was introduced at position A1216 (GCA to GCT).

Construct a targeting vector that contains approximately 8 kbp of genomic DNA carrying *vegfr2* exons 19 to 26 in the 5′ region, a *neomycin (neo)* resistance gene (pMC1-NeopolyA) at the *NcoI* site in intron 26, approximately 1.5 kbp of genomic DNA carrying exon 27 with targeted mutations, and a *diphtheria toxin A subunit* gene (pMC1-DT-A) at the 3′-end.

2.6.2. Generation of mutant mice

Transfect the linearized target vector DNA into mouse 129 SvJ ES cells by electroporation, and select ES cell clones resistant to 400 *μg/ml* G418. By Southern blotting and by DNA sequencing, select proper ES cell clones correctly targeted for Y1173F and Y1212F.

Inject each ES cell clone targeted for Y1173F and Y1212F into C57BL/6 blastocysts to obtain chimeric mice, and mate the chimeric mice with C57BL/6 females. For deletion of the neoresistance gene, mate the *flk1*^{1173Fneo} and *flk1*^{1212Fneo} heterozygous males with *CAG-cre* transgenic females. Confirm deletion of the *neo* gene by Southern blotting.

Intercross the *flk1*^{1173F} and *flk1*^{1212F} heterozygote mice to generate homozygous mutant mice. Check the genotype of mice by Southern blotting and genomic PCR. Confirm the gene targeting by nucleotide sequencing of the full-length mRNAs of *1173F*-mutated and *1212F*-mutated *flk1* genes obtained from E8.5 *flk1*$^{1173F/F}$ or *flk1*$^{1212F/F}$ homozygous embryos.

3. CONCLUSIONS

Herein we provide protocols for established methods. Clearly, these protocols can be modified and the same results can be obtained in more than one way. Moreover, the number of commercially available reagents and kits will facilitate and improve the quality of the laboratory work. The essential message remains, however, that signal transduction analyses have to take the biologic context into account, and this ambition will have to remain for the efforts to be meaningful. Work done this far indicates that VEGF signal transduction is only accurately represented when studied in endothelial cells compared with the conventional transformed tissue culture models. Furthermore, it is evident that molecular mechanisms in a critical step in angiogenesis, formation of three-dimensional lumenized structures, cannot be studied in a 2D assay system. Even though this leads to methodologic challenges, there are strong rewards in the form of immediate clinical applications.

ACKNOWLEDGMENTS

Work in the authors' laboratories was supported by grants from the Swedish Cancer Foundation, the Swedish Science Council and European Commission FP6 Integrated Project Angiotargeting (contract LSHG-CT-2004-504743) and Lymphangiogenomics (contract LSHG-CT-2004-503573), from Grant-in-Aid Special Project Research on Cancer-Bioscience 17014020 and for Scientific Research-B 17390110 from the Ministry of Education, Culture, Sports, Science and Technology of Japan, and from the program "Promotion of Fundamental Research in Health Science" of the Organization for Pharmaceutical Safety and Research (OPSR).

REFERENCES

Ferrara, N., Hillan, K. J., and Novotny, W. (2005). Bevacizumab (Avastin), a humanized anti-VEGF monoclonal antibody for cancer therapy. *Biochem. Biophys. Res. Commun.* **333,** 328–335.

Folkman, J., and Haudenschild, C. (1980). Angiogenesis by capillary endothelial cells in culture. *Trans. Ophthalmol. Soc. UK.* **100,** 346–353.

Fong, G. H., Rossant, J., Gertsentein, M., and Breitman, M. L. (1995). Role of the Flt-1 receptor tyrosine kinase in regulating the assembly of vascular endothelium. *Nature* **376,** 66–70.

Hiratsuka, S., Minowa, O., Kuno, J., Noda, T., and Shibuya, M. (1998). Flt-1 lacking the tyrosine kinase domain is sufficient for normal development and angiogenesis in mice. *Proc. Natl. Acad. Sci. USA* **95,** 9349–9354.

Jakobsson, L., Domogatskaya, A., Tryggvason, K., Edgar, D., and Claesson-Welsh, L. (2007a). Laminin deposition is dispensable for vasculogenesis but regulates blood vessel diameter independent of flow. *FASEB J.* Epub 2007 Dec 11.

Jakobsson, L., Kreuger, J., and Claesson-Welsh, L. (2007b). Building blood vessels–stem cell models in vascular biology. *J. Cell Biol.* **177,** 751–755.

Jakobsson, L., Kreuger, J., Holmborn, K., Lundin, L., Eriksson, I., Kjellen, L., and Claesson-Welsh, L. (2006). Heparan sulfate in trans potentiates VEGFR-mediated angiogenesis. *Dev. Cell.* **10,** 625–634.

Karasuyama, H., Kudo, A., and Melchers, F. (1990). The proteins encoded by the VpreB and lambda 5 pre-B cellspecific genes can associate with each other and with mu heavy chain. *J. Exp. Med.* **172,** 969–972.

Kawai, S., and Nishizawa, M. (1984). New procedure for DNA transfection with polycation and dimethyl sulfoxide. *Mol. Cell Biol.* **4,** 1172–1174.

Kreuger, J., Nilsson, I., Kerjaschki, D., Petrova, T., Alitalo, K., and Claesson-Welsh, L. (2006). Early lymph vessel development from embryonic stem cells. *Arterioscler. Thromb. Vasc. Biol.* **26,** 1073–1078.

Maru, Y., Yamaguchi, S., Takahashi, T., Ueno, H., and Shibuya, M. (1998). v-Ras cooperates with integrin to induce tubulogenesis in sinusoidal endothelial cell line. *J. Cell. Physiol.* **176,** 223–234.

Sakurai, Y., Ohgimoto, K., Kataoka, Y., Yoshida, N., and Shibuya, M. (2005). Essential role of Flk-1 (VEGF receptor 2) tyrosine residue 1173 in vasculogenesis in mice. *Proc. Natl. Acad. Sci. USA* **102,** 1076–1081.

Sawano, A., Takahashi, T., Yamaguchi, S., Aonuma, T., and Shibuya, M. (1996). Flt-1 but not KDR/Flk-1 tyrosine kinase is a receptor for Placenta Growth Factor (PlGF), which is related to vascular endothelial growth factor (VEGF). *Cell Growth Differ.* **7,** 213–221.

Seetharam, L., Gotoh, N., Maru, Y., Neufeld, G., Yamaguchi, S., and Shibuya, M. (1995). A unique signal transduction from FLT tyrosine kinase, a receptor for vascular endothelial growth factor VEGF. *Oncogene* **10,** 135–147.

Shibuya, M. (2000). Development of the liver vascular system. *In* "Morphogenesis of Endothelium." (W. Risau and G. M. Rubanyi, eds.), pp. 175–188. Harwood Academic Publishers, Switzerland.

Shibuya, M., and Claesson-Welsh, L. (2006). Signal transduction by VEGF receptors in regulation of angiogenesis and lymphangiogenesis. *Exp. Cell Res.* **312,** 549–560.

Shibuya, M. (2006). Differential roles of vascular endothelial growth factor receptor-1 and receptor-2 in angiogenesis. *J. Biochem. Mol. Biol.* **39,** 469–478.

Takahashi, T., Ueno, H., and Shibuya, M. (1999). VEGF activates protein kinase C-dependent, but Ras-independent Raf-MEK-MAP kinase pathway for DNA synthesis in primary endothelial cells. *Oncogene* **18,** 2221–2230.

Takahashi, T., and Shibuya, M. (2001). The overexpression of PKCδ is involved in vascular endothelial growth factor–resistant apoptosis in cultured primary sinusoidal endothelial cells. *Biochem. Biophys. Res. Commun.* **280,** 415–420.

Takahashi, T., Yamaguchi, S., Chida, K., and Shibuya, M. (2001). A Single autophosphorylation site on KDR/Flk-1 is essential for VEGF-A–dependent activation of PLC- and DNA synthesis in vascular endothelial cells. *EMBO J.* **20,** 2768–2778.

Yamane, A., Seetharam, L., Yamaguchi, S., Gotoh, N., Takahashi, T., Neufeld, G., and Shibuya, M. (1994). A new communication system between hepatocytes and sinusoidal endothelial cells in liver through vascular endothelial growth factor and Flt tyrosine kinase receptor family (Flt-1 and KDR/Flk-1). *Oncogene* **9,** 2683–2690.

CHAPTER FOURTEEN

ANALYSIS OF LOW MOLECULAR WEIGHT GTPASE ACTIVITY IN ENDOTHELIAL CELL CULTURES

Erika S. Wittchen *and* Keith Burridge

Contents

Abstract

GTPases control a myriad of cellular processes, including cell migration, proliferation, polarity, and cell adhesion, which includes cell–cell junction regulation. Angiogenesis requires many of these diverse cellular events to occur appropriately. In this chapter, we describe the techniques for *in vitro* assessment of GTPase activity in endothelial cells grown in monolayer culture as a model system for studying their role during angiogenesis.

1. INTRODUCTION

Angiogenesis, the process by which new blood vessels are formed, requires a complex set of signaling events to occur in a well-defined spatial and temporal manner. Given their role in signaling pathways involving the regulation of the cytoskeleton, endothelial cell migration, and cell adhesion, perhaps it is not surprising that Rho GTPases have been implicated in many

Department of Cell and Developmental Biology, Lineberger Comprehensive Cancer Center, University of North Carolina at Chapel Hill, Chapel Hill, North Carolina

Methods in Enzymology, Volume 443
ISSN 0076-6879, DOI: 10.1016/S0076-6879(08)02014-4

of the cellular processes associated with angiogenesis. For instance, both Rho (van Nieuw *et al.*, 2003) and Rac (Garrett *et al.*, 2007) are recognized as downstream effectors of VEGF signaling, leading to increased cell migration. Increased vascular permeability induced by VEGF and other inflammatory mediators such as thrombin, lysophosphatidic acid, and histamine (Wojciak-Stothard *et al.*, 2001) has also been attributed to activation of Rho GTPase and subsequent increased endothelial actomyosin contractility leading to permeability (Birukova *et al.*, 2004; Moy *et al.*, 1998). Conversely, activation of the small GTPase Rap1 increases barrier function in endothelial cells, increases cortical actin (Cullere *et al.*, 2005), and inhibits leukocyte transendothelial migration (an effect presumably caused by "tighter" cell–cell junctions) (Wittchen *et al.*, 2005). The final stage of angiogenesis involves morphogenic remodeling (i.e., sprouting, branching, and lumen formation). Again, GTPases seem to be centrally involved in these processes. Rho activity must be temporally regulated during branching and sprouting (Mavria *et al.*, 2006), and Rac/cdc42 activity is important for the formation and fusion of endothelial vacuoles, which drives vascular lumen formation (Connolly *et al.*, 2002; Kamei *et al.*, 2006).

Historically, the cellular activity of Rho family GTPases had to be inferred on the basis of qualitative observations of cytoskeletal morphology (e.g., formation of stress fibers and focal adhesions was interpreted to be indicative of Rho activation [Ridley and Hall, 1992a], whereas formation of membrane ruffles and lamellipodia were taken to mean activation of Rac and/or Cdc42 [Nobes and Hall, 1995; Ridley *et al.*, 1992b]). However, the development of biochemical assays to directly assess activation of GTPases in cells has provided much more detailed insight into the function of and signaling events that GTPases are involved in. GTPases are proteins that act as molecular switches that cycle between an active (GTP bound) and inactive (GDP bound) form. GTP binding, and subsequent activation of Rho proteins, is facilitated by guanine nucleotide exchange factors (GEFs), whereas inactivation of Rho proteins by hydrolysis of GTP $\rightarrow$ GDP is catalyzed by GTPase-activating proteins (GAPs). The Rho protein activity assays, or affinity pull-downs, take advantage of the fact that effector proteins specifically recognize only the GTP bound (active) form of the protein; therefore, one can use GST fusion protein constructs of the binding domain of these effectors as bait in an affinity pull-down assay to detect the amount of active GTPase in any given cell lysate. Furthermore, effector binding typically results in a significant reduction in the intrinsic GTP hydrolysis, thus providing, essentially, a snapshot of the Rho protein activity state at the time of cell lysis. By use of different effector constructs, one can determine the activity level of many different Rho family GTPases (Rho, Rac, and Cdc42). A similar approach can be used with the Ras subfamily member Rap1. Table 14.1 identifies the effector domain constructs used for detecting activity for several of these proteins in endothelial cells. In the

Table 14.1   Effector domain constructs used for GTPase activity assays

For detection of active	Use the effector GTPase binding domain construct	Reference
RhoA, B, C	GST-RBD (Rho binding domain)-Rhotekin	Ren *et al.*, 1999
Rac	GST-PBD (p21 binding domain)-PAK	Sander *et al.*, 1998
Cdc42	GST-PBD-PAK (same as for Rac activity assay)	Sander *et al.*, 1999
Rap1, 2	GST-RBD (Rap binding domain)-RalGDS	Franke *et al.*, 1997

following "Methods and Results" section, we will describe the preparation of the affinity-binding beads and the basic pull–down protocol for Rho, Rac/Cdc42, and then Rap1 in endothelial cells. Examples that use known positive regulators of particular GTPases will illustrate the techniques. Finally, experimental conditions that must be taken into account, such as cell density and a comparison of different serum-starvation procedures, are also discussed.

2. Methods and Results

To illustrate the following protocols, we have used human umbilical vein endothelial cells (HUVECs) as a model endothelial cell line in which to analyze GTPase activity. HUVEC were obtained from Lonza and grown on tissue culture plastic ware in EGM-2 media unless otherwise indicated. Other endothelial cell types (arterial, microvascular, lymphatic), and culture on specific substrates (Matrigel, collagen) may also be analyzed by these methods.

2.1. Critical points for success

To reduce spontaneous hydrolysis of GTP in your samples, the following two key points must be always be observed!

2.1.1. Minimize handling time

- Prepare everything in advance: Organize and label tubes, ensure all solutions are premade with protease inhibitors added, protein assay standards/solutions are already prepared when appropriate, prewash and dispense fusion protein beads into each sample tube.

- Unless experimentally necessary (i.e., seeding unequal number of cells, comparing different cell types, or differentially treating cells with compounds that may cause (for example) a loss of cell adhesion and therefore you expect there will be a difference in protein concentration between samples, do not do a protein assay. Often it is unnecessary, and, in fact, may be detrimental if there is a lot of GTP hydrolysis occurring while the protein concentration is being determined.

2.1.2. Maintain low temperature

- Prechill all tubes and solutions (place on ice), perform all centrifugation and washing steps at 4 °C (preferably in a refrigerated centrifuge or in a cold room equipped with a centrifuge)

2.2. Rho activity assays

2.2.1. Preparation of GST-RBD-Rhotekin beads
2.2.1.1. Solutions
2.2.1.1.1. Lysis buffer

50 mM Tris-HCl, pH 7.4
50 mM NaCl
5 mM MgCl$_2$
1% Triton X-100
1 mM DTT
Protease inhibitor cocktail

2.2.1.2. Bead preparation protocol

1. Prepare a 10-ml overnight culture of pGEX(GST)-RBD-Rhotekin (Ren et al., 1999) in LB media containing 100 μg/ml ampicillin.
2. The next day, dilute 1:50 into 500 ml LB/Amp, grow culture for ~3 h at 37 °C (OD$_{600}$ ≈ 0.2).
3. Induce with 0.3 mM IPTG for ~16 h at *room temperature*.
4. Pellet bacteria by centrifugation at 5000 rpm for 10 min at 4 °C. (Pellets with media removed may be stored at −20 °C for several months).
5. Resuspend pellet in 10 ml lysis buffer.
6. Split suspension in half and sonicate up to 6 times, 20 sec each time, on ice.
7. Clear lysate by centrifugation at 10000 rpm for 10 min at 4 °C.

8. Rotate supernatant for 2 h at 4 °C with 1 ml of a 50% slurry of glutathione Sepharose 4B (prewashed with lysis buffer).
9. Wash beads 3× 10 ml lysis buffer *without* Triton X-100.
10. After final wash, resuspend beads in equal volume of the same buffer.
11. Analyze concentration of fusion protein on the beads by assaying 10 µl of the bead slurry with Pierce CBB protein assay kit. Yield should be approximately 3 to 6 mg/ml. Gel analysis will reveal a 33-kDa product.
12. GST-RBD-Rhotekin beads should be stored at 4 °C and used within 1 week.

2.2.2. Rho activity assay
2.2.2.1. *Solutions and antibodies*
2.2.2.1.1. *Lysis buffer A*

50 mM Tris, pH 7.6
500 mM NaCl
0.1% Sodium dodecyl sulfate (SDS)
0.5% Deoxycholic acid (DOC)
1% Triton X-100
10 mM MgCl$_2$
Protease inhibitors
1 mM Sodium orthovanadate
1 mM NaF

2.2.2.1.2. *Wash buffer B*

50 mM Tris, pH 7.6
150 mM NaCl
1% Triton X-100
10 mM MgCl$_2$
1 mM sodium orthovanadate
1 mM NaF

2.2.2.1.3. *RhoA antibodies*

We blot GST-RBD-Rhotekin pull downs with an anti-Rho monoclonal antibody from Santa Cruz (clone 26C4); this antibody is specific for RhoA.

2.2.2.2. *Pull-down protocol (Rho)*   Endothelial cells typically have relatively low levels of active RhoA, thus the minimum size culture dish we use is 60 mm (This usually approximates 300 to 500 µg total protein). For specific pull-down of active Rho, we use 30 µg of the Rho-binding domain of the effector protein Rhotekin fused to GST (GST-RBD-Rhotekin) per sample.

1. On treatment, immediately place culture dishes on ice (tilted on an angle to aid complete removal of media and rinse buffer).
2. Aspirate media, rinse with ice-cold HEPES-buffered saline (HBS). It is *essential* to *completely* aspirate the rinse buffer to avoid diluting the lysates during the next step.
3. Add the smallest volume possible of the appropriate lysis buffer (200 μl for $\leq$60-mm dishes, 300 μl for $\geq$100-mm dishes), and scrape cells into buffer with a soft-bladed cell scraper.
4. Disperse cell debris by pipetting up and down several times, then immediately pipet into prechilled labeled Eppendorf microfuge tubes.
5a. Clear lysates by centrifugation (14 000 rpm for 5min) at 4 °C.
5b. While centrifuging, prepare fusion protein beads as follows:

   By use of a wide-bore pipet tip and mixing the bead slurry often to ensure an even suspension, aliquot an amount of bead slurry to give 30 to 100 μg of GST fusion protein per tube.

   Wash beads once with 1 ml lysis buffer.

   Collect beads by pulse centrifugation for 30 sec and completely aspirate buffer with a 27-guage (or smaller) syringe needle attached to a vacuum system.

   By holding the needle opening against the side of the tube, and aided by the small diameter of the needle, we attain complete removal of buffer without aspiration of the beads.

6. Collect cleared supernatant into second labeled/chilled tube.
7. For a loading control, save $\sim$50 μl of lysate for gel analysis of total GTPase protein.
8. (Optional) Measure protein concentration with Pierce CBB protein assay reagent. If necessary, adjust samples to equal concentration and volume.
9. Incubate equal volumes of cell lysates with beads for no more than 30 min with rotation for constant mixing at 4 °C.
10. Collect beads and bound proteins by centrifugation for 30 sec, and wash three times with 1 ml of wash buffer, carefully aspirating supernatant after each wash step. To avoid accidental aspiration of beads, leave $\sim$100 μl buffer in each tube.
11. After the last wash step, completely aspirate the remaining buffer with the 27-guage syringe setup.
12. Resuspend beads with 48 μl 2$\times$ SDS-PAGE sample buffer, boil 5 min. At this stage, samples can be stored at $-$20 °C until gels are run.
13. Run samples on 15% SDS-PAGE gel: reserved lysates from step 7 represent "Total" (GTP and GDP bound pool), boiled beads from step 12 represent "Active" (GTP bound), transfer to PVDF membrane for Western blots.

As a positive control for Rho activation in endothelial cells, it can be useful to have a side-by-side treatment with a Rho-activating compound such as calpeptin. Calpeptin inhibits certain phosphatases such as SHP-2 (Schoenwaelder *et al.*, 2000) that act upstream of RhoA signaling pathways. It has been shown (in more than one cell system) to potently induce stress fiber formation through activation of Rho (Schoenwaelder and Burridge, 1999). Figure 14.1 shows an example of an experiment in which HUVECs were treated with calpeptin for 30 min before assaying for Rho activity with a pull-down assay as described here. There is a clear increase in the amount of active Rho on calpeptin treatment. The loading control (cell lysate) confirms that the total amount of Rho is not altered.

2.3. Rac/Cdc42 activity assays

2.3.1. Preparation of GST-PBD-PAK beads

2.3.1.1. Bead preparation protocol

1. Prepare a 10-ml overnight culture of pGEX(GST)-PBD-PAK (Sander *et al.*, 1998, 1999) in LB media containing 100 μg/ml ampicillin.
2. The next day, dilute 1:20 into 200 ml LB/Amp, grow culture for another 1 to 3 h at 37 °C (OD$_{600}$ $\approx$ 0.6).
3. Induce with 0.3 mM IPTG for 3 h at 37 °C.
4. Follow steps 4 though 11 as for the purification of GST-RBD-Rhotekin (as previously).
5. GST-PBD-PAK beads are also stored at 4 °C in 50% glycerol but are much more stable than the GST-RBD beads, so can be used up to a month later without loss of activity.

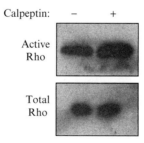

Calpeptin: −   +

Active Rho

Total Rho

Figure 14.1 Detection of Rho activation after stimulation with calpeptin. HUVEC cultures were treated for 30 min with 100 μg/ml calpeptin (+) or DMSO vehicle control (−). A pull-down assay was performed with 50 μg GST-RBD-Rhotekin fusion protein immobilized on glutathione sepharose beads. Samples of lysates (representative of "Total Rho") and boiled beads (representative of "Active Rho") were subjected to SDS-PAGE, followed by Western blotting with an anti-RhoA antibody. Calpeptin treatment induces activation of Rho.

2.3.2. Rac/Cdc42 activity assay protocol
2.3.2.1. Solutions and antibodies
2.3.2.1.1. Lysis and wash buffer B

50 mM Tris, pH 7.6
150 mM NaCl
1% Triton X-100
10 mM MgCl$_2$
Protease inhibitors

2.3.2.1.2. Rac and Cdc42 Antibodies:

For detection of Rac1, we use anti-Rac1 mAb from BD Transduction Labs (clone 102); for detection of Cdc42, we use anti-Cdc42 mAb (clone 44) also from BD Transduction Labs. These antibodies blot best when incubating with TBST instead of 5% milk powder (unpublished observation).

2.3.2.2. Pull-down protocol (Rac/Cdc42)
Endothelial cells have relatively high levels of cellular Rac compared with levels of Rho. Levels of Cdc42 are also higher than for Rho, although not as high as for Rac. Smaller culture dishes (35 mm) may be used for assaying Rac and Cdc42 activities. Furthermore, both Rac and Cdc42 activity can be assessed from the same experiment. To do this, use one third of the sample for the Rac blot and two thirds of the sample for the Cdc42 blot. For specific pull down of active Rac and Cdc42, we use 50 μg of the GST-PBD-PAK fusion protein construct per sample. With lysis and wash buffer B where indicated, follow steps 1 through 13 from *Pull-down Protocol* as described previously for Rho GTPase.

2.4. Rap1 activity assays

2.4.1. Preparation of GST-RBD-RalGDS beads
2.4.1.1. Solutions and antibodies
2.4.1.1.1. Rap1 activity assay lysis and wash buffer

75 mM Tris-HCl, pH 7.4
75 mM NaCl
1% NP-40
0.5% Deoxycholic acid
0.1% SDS
1 mM Sodium orthovanadate
1 mM NaF
Freshly added protease inhibitors

2.4.1.1.2. Antibodies   For detection of Rap1, we use a rabbit polyclonal anti-Rap1 antibody from Santa Cruz Biotechnology (sc-65). This antibody does not distinguish between Rap1A and Rap1B.

2.4.1.2. Bead preparation protocol

1. Prepare a 25-ml overnight culture of pGEX(GST)-RBD-RalGDS (Franke *et al.*, 1997) in LB media containing 100 ug/ml ampicillin.
2. The next day, dilute 1:10 into 250 ml LB/Amp, grow culture for another 1 to 3 h or until OD_{600} is 0.6 to 1.0.
3. Induce with 0.1 mM IPTG for 3 h at 37 °C.
4. Pellet bacteria by centrifugation at 4000 rpm for 10 min at 4 °C.
5. Resuspend pellet with 9 ml PBS containing protease inhibitors and 0.5 mM DTT.
6. Split suspension in half and sonicate three times for 1 min each, on ice.
7. Add Triton X-100 to a 1% final concentration, mix for 30 min with rotation at 4 °C.
8. Clear lysate by centrifugation at 12,000 rpm for 10 min at 4 °C.
9. Rotate supernatant for 2 h at 4C with 5 ml of a 50% slurry of glutathione Sepharose 4B (prewashed with Rap1 activity assay lysis and wash buffer).
10. Wash beads three times with 10 ml lysis and wash buffer.
11. After final wash, resuspend beads in equal volume of the same buffer containing 50% glycerol.
12. Analyze concentration of fusion protein on the beads by assaying 10 μl of the bead slurry with Pierce CBB protein assay kit. Yield should be high, approximately 10 mg/ml.
13. GST-RBD-RalGDS beads may be stored at −70 °C in single-use aliquots indefinitely.

2.4.2. Rap1 activity assay protocol

We typically use at least a 60-mm dish of endothelial cells for the Rap activity assay. Alternately, because Rap1 has a much slower intrinsic rate of hydrolysis than Rho family GTPases (Frech *et al.*, 1990), it is possible to assay Rap activity by reusing lysates after the Rac activity pull down. To perform sequential Rap1 activity assays, simply save the supernatant after the GST-PBD-PAK pull down and apply to 50 to 100 μg prewashed GST-RBD-RalGDS beads. Follow steps 1 through 13 from *Pull-Down Protocol* as described for Rho.

Figure 14.2 demonstrates the usefulness of this assay to identify differences in the Rap1 activity state between endothelial cells stimulated in the following way: confluent HUVECs were treated for 15 min with 50 μM of 8pCPT-2-Me-cAMP (Biolog, Germany), a compound that specifically

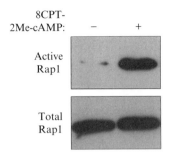

Figure 14.2   Activation of Rap1 by an Epac (GEF)-activating compound. HUVEC cultures were treated for 15 min with 100 μM of the Rap1-activating drug 8CPT-2Me-cAMP (+) or media alone (−). A pull-down assay was performed with 100 μg GST-RBD-RalGDS fusion protein immobilized on glutathione sepharose beads. Samples of lysates (representative of "Total Rap1") and boiled beads (representative of "Active Rap1") were subjected to SDS-PAGE, followed by Western blotting with an anti-Rap1.

activates the Rap1 GEF Epac (Enserink *et al.*, 2002). A Rap activity assay was performed as described in this section, and both total lysates and bead samples were run on a gel and subjected to Western blotting with anti-Rap1 antibodies. The assay shows a robust increase in the level of GTP-bound (active) Rap1 in the samples treated with this compound. 8pCPT-2-Me-cAMP treatment serves as a useful internal control for Rap1 activation when comparing different experimental treatments.

2.5. Experimental considerations

2.5.1. Cell density

One of the most important considerations to take into account when assaying for changes in GTPase activity on stimulation is the basal (unstimulated) activity of the GTPase. Cell density is a key parameter that can impact basal GTPase activity. This is probably due to a combination of factors such as proliferation (cell cycle) status, degree of cell spreading, and perhaps, most importantly for endothelial cells, the impact of signals downstream of cell–cell adhesive junctions (Noren *et al.*, 2001). We analyzed GTPase activity 48 h after plating in HUVEC cultured under three different density conditions (Fig. 14.3A). "Low-density" cell cultures (<50% confluent) are still actively spreading, with a low degree of cell–cell contact between neighbors. "Medium-density" cultures are 100% confluent, just barely covering the entire surface of the dish, and therefore still very spread. "High-density" cultures are densely packed. Although the latter two conditions are both 100% confluent, the key difference is that one is "newly" confluent with cell–cell junctions still actively forming and remodeling, whereas the other is "stably" confluent and quiescent with mature cell junctions.

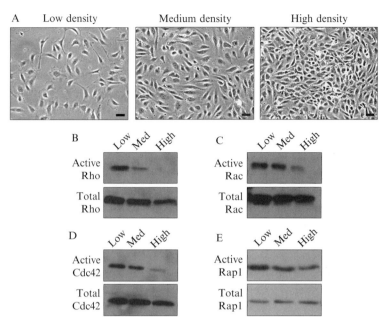

Figure 14.3 Effects of cell density on basal activation state of various GTPases. Equal numbers of HUVEC were seeded onto dishes of varying size (100 mm, 60 mm, and 35 mm in diameter) to obtain cultures that were of "low," "medium," and "high"density after 48 h. This corresponds to ∼50% confluence, early confluence, and stably confluent, respectively. (A) Representative phase images of these three conditions. (B to E) Cultures at each of these densities were subjected to GTPase activity assays. Rho (B) is the most sensitive to cell density, with a significant decrease even comparing low to medium density. Rac (C), Cdc42 (D), and Rap1 (E), show only a small change.

All the GTPases tested are sensitive to some degree to the monolayer density (Fig, 14.3B to E). Rho activity is the most significantly density dependent (Fig. 14.3B), with 100% inhibition when comparing low versus high-density conditions. Rac and Cdc42 activity (Fig. 14.3C, D) is minimally affected between low and medium density; however, at high density, there is approximately an 80% decrease in basal GTPase activity. Rap1 activity (Fig. 14.3E) was slightly inhibited with increasing density. At very low cell densities (∼10% confluence), Rap activity was extremely high (data not shown), most likely because of its known role in cell spreading events (Arthur *et al.*, 2004). In summary, it is critical to take cell density into account when planning experiments.

2.5.2. Serum starvation conditions

The other major point to consider when performing GTPase activity assays also relates to reducing basal levels of activation, this time because of the presence of soluble factors present in serum (growth factors and other

bioactive compounds). Serum-induced GTPase activation might potentially mask small, but physiologically relevant, differences in activity on
treatment; therefore, it is important to bear this in mind when setting up
the experimental conditions. Endothelial cell types from different sources
may respond differently; therefore, it may be necessary to optimize starvation conditions empirically. It is important to strike a balance between
reducing serum stimulation without altering endothelial cell monolayer
morphology (i.e., causing cell retraction, rounding up, or junctional disso-
lution). We tested three different starvation conditions that used HUVEC as
a model endothelial cell type. Recently confluent cultures were either
switched to serum-reduced media (0.1%) overnight or treated for 30 min
and 3 h in completely serum-free media (SFM) the day of the experiment.
Rac (Fig. 14.4A) and Cdc42 activity (data not shown) did not change
dramatically with any of the starvation conditions compared with nonstarved (NS) cells. Likewise, Rap1 activity does not seem to be influenced
to any great degree by the presence of serum under these culture conditions.
As expected (Ridley and Hall, 1992), Rho activity was sensitive to the
presence of serum factors (Fig. 14.4C). Although there was a significant
reduction in basal Rho activation comparing 30 min with 3 h of complete
serum deprivation, the greatest lowering of Rho activity occurred on
overnight incubation in low (0.1%) serum-containing media. For some
endothelial types such as HUVEC, it is not advisable to culture in

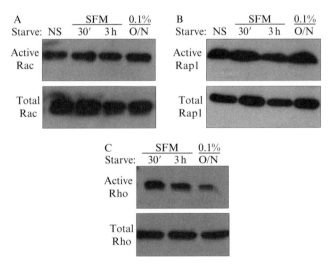

Figure 14.4 Effects of serum-starvation conditions on basal GTPase activity levels.
HUVEC grown to confluence were either nonstarved (NS), grown in low serum over-
night (O/N), or in serum-free media (SFM) for 0.5 to 3 h before being subjected to
GTPase activity assays. Starvation does not significantly affect basal activation of Rac
(A) or Rap1 (B). However, basal Rho activity can be significantly attenuated on both
short-term serum starvation and longer-term serum deprivation (C).

completely serum-free conditions for much longer than 3 h because of undesirable morphologic changes (data not shown); specific conditions must be determined for each endothelial type.

ACKNOWLEDGMENTS

This work was supported by Grants GM-029860, HL-045100, and HL-080166 from the National Institutes of Health.

REFERENCES

Arthur, W. T., Quilliam, L. A., and Cooper, J. A. (2004). Rap1 promotes cell spreading by localizing Rac guanine nucleotide exchange factors. *J. Cell Biol.* **167,** 111–122.

Birukova, A. A., Smurova, K., Birukov, K. G., Kaibuchi, K., Garcia, J. G., and Verin, A. D. (2004). Role of Rho GTPases in thrombin-induced lung vascular endothelial cells barrier dysfunction. *Microvasc. Res.* **67,** 64–77.

Connolly, J. O., Simpson, N., Hewlett, L., and Hall, A. (2002). Rac regulates endothelial morphogenesis and capillary assembly. *Mol. Biol. Cell.* **13,** 2474–2485.

Cullere, X., Shaw, S. K., Andersson, L., Hirahashi, J., Luscinskas, F. W., and Mayadas, T. N. (2005). Regulation of vascular endothelial barrier function by Epac, a cAMP-activated exchange factor for Rap GTPase. *Blood* **105,** 1950–1955.

Enserink, J. M., Christensen, A. E., de Rooij, J., van Triest, M., Schwede, F., Genieser, H. G., Doskeland, S. O., Blank, J. L., and Bos, J. L. (2002). A novel Epac-specific cAMP analogue demonstrates independent regulation of Rap1 and ERK. *Nat. Cell Biol.* **4,** 901–906.

Franke, B., Akkerman, J. W., and Bos, J. L. (1997). Rapid Ca^{2+}-mediated activation of Rap1 in human platelets. *EMBO J.* **16,** 252–259.

Frech, M., John, J., Pizon, V., Chardin, P., Tavitian, A., Clark, R., McCormick, F., and Wittinghofer, A. (1990). *Science* **249,** 169–171.

Garrett, T. A., Van Buul, J. D., and Burridge, K. (2007). VEGF-induced Rac1 activation in endothelial cells is regulated by the guanine nucleotide exchange factor Vav2. *Exp. Cell Res.* **313,** 3285–3297.

Kamei, M., Saunders, W. B., Bayless, K. J., Dye, L., Davis, G. E., and Weinstein, B. M. (2006). Endothelial tubes assemble from intracellular vacuoles *in vivo. Nature* **442,** 453–456.

Mavria, G., Vercoulen, Y., Yeo, M., Paterson, H., Karasarides, M., Marais, R., Bird, D., and Marshall, C. J. (2006). ERK-MAPK signaling opposes Rho-kinase to promote endothelial cell survival and sprouting during angiogenesis. *Cancer Cell.* **9,** 33–44.

Moy, A. B., Bodmer, J. E., Blackwell, K., Shasby, S., and Shasby, D. M. (1998). cAMP protects endothelial barrier function independent of inhibiting MLC20-dependent tension development. *Am. J. Physiol.* **274,** L1024–L1029.

Nobes, C. D., and Hall, A. (1995). Rho, rac, and cdc42 GTPases regulate the assembly of multimolecular focal complexes associated with actin stress fibers, lamellipodia, and filopodia. *Cell* **81,** 53–62.

Noren, N. K., Niessen, C. M., Gumbiner, B. M., and Burridge, K. (2001). Cadherin engagement regulates Rho family GTPases. *J. Biol. Chem.* **276,** 33305–33308.

Ren, X. D., Kiosses, W. B., and Schwartz, M. A. (1999). Regulation of the small GTP-binding protein Rho by cell adhesion and the cytoskeleton. *EMBO J.* **18,** 578–585.

Ridley, A. J., and Hall, A. (1992a). The small GTP-binding protein Rho regulates the assembly of focal adhesions and actin stress fibers in response to growth factors. *Cell* **70**, 389–399.

Ridley, A. J., Paterson, H. F., Johnston, C. L., Diekmann, D., and Hall, A. (1992b). The small GTP-binding protein Rac regulates growth factor-induced membrane ruffling. *Cell7* **70**, 401–410.

Sander, E. E., van Delft, S., ten Klooster, J. P., Reid, T., van der Kammen, R. A., Michiels, F., and Collard, J. G. (1998). Matrix-dependent Tiam1/Rac signaling in epithelial cells promotes either cellcell adhesion or cell migration and is regulated by phosphatidylinositol 3-kinase. *J. Cell Biol.* **143**, 1385–1398.

Sander, E. E., ten Klooster, J. P., van Delft, S., van der Kammen, R. A., and Collard, J. G. (1999). Rac downregulates Rho activity: Reciprocal balance between both GTPases determines cellular morphology and migratory behavior. *J. Cell Biol.* **147**, 1009–1022.

van Nieuw Amerongen, G. P., Koolwijk, P., Versteilen, A., and van Hinsbergh, V. W. (2003). Involvement of RhoA/Rho kinase signaling in VEGF-induced endothelial cell migration and angiogenesis *in vitro*. *Arterioscler. Thromb. Vasc. Biol.* **23**, 211–217.

Wittchen, E. S., Worthylake, R. A., Kelly, P., Casey, P. J., Quilliam, L. A., and Burridge, K. (2005). Rap1 GTPase inhibits leukocyte transmigration by promoting endothelial barrier function. *J. Biol. Chem.* **280**, 11675–11682.

Wojciak-Stothard, B., Potempa, S., Eichholtz, T., and Ridley, A. J. (2001). Rho and Rac but not Cdc42 regulate endothelial cell permeability. *J. Cell Sci.* **114**, 1343–1355.

CHAPTER FIFTEEN

SEMAPHORIN-INDUCED CYTOSKELETAL COLLAPSE AND REPULSION OF ENDOTHELIAL CELLS

Diane R. Bielenberg,*,1 Akio Shimizu,*,1 *and* Michael Klagsbrun*,†

Contents

Abstract

Class 3 semaphorins (SEMA3) are mediators of neuronal guidance first shown to repel axons and collapse axonal growth cones by depolymerization of cytoskeletal F-actin. Subsequently, it was found that SEMA3 could also mediate angiogenesis. SEMA3F binds to its receptor, neuropilin 2 (NRP2), a transmembrane protein expressed on neurons, EC (EC), and tumor cells. *In vitro*, SEMA3F collapses the F-actin cytoskeleton, repels EC, and inhibits EC and tumor cell adhesion and migration in a manner similar to what occurs with axons. In a mouse tumor model,

* Vascular Biology Program, Department of Surgery, Children's Hospital, Harvard Medical School, Boston, Massachusetts
† Department of Pathology, Children's Hospital, Harvard Medical School, Boston, Massachusetts
1 Both authors contributed equally.

Methods in Enzymology, Volume 443
ISSN 0076-6879, DOI: 10.1016/S0076-6879(08)02015-6

SEMA3F is a potent inhibitor of tumor angiogenesis, tumor progression, and metastasis. SEMA3F is encoded in a region of chromosome 3p21.3 that is commonly deleted in small cell lung cancers, suggesting that SEMA3F is a tumor suppressor. SEMA3F may have therapeutic potential. Therefore, this chapter is focused primarily on the detailed methods to purify SEMA3F and to assay its biologic activity, including cytoskeleton collapse and repulsion.

1. INTRODUCTION

The nervous and vascular systems share biologic and molecular properties. For example, both processes are characterized by branching that leads to network formation. Furthermore, the structures of growth cones at the tips of axons and the endothelial tip cells of sprouting capillaries are designed to sense the environment and to react to environmental cues that can be either attractive or repulsive (Gerhardt *et al.*, 2003; 2004). These cues can either promote or inhibit cellular processes such as spreading and migration. In addition, nerves and blood vessels can exist in close proximity and interact with each other. A good example is that arteries, but not veins, are specifically aligned with peripheral nerves in the embryonic mouse limb skin (Mukouyama *et al.*, 2002; 2005). The nerves provide the template for blood vessel branching and arterial differentiation. On a molecular level, nerves provide vascular endothelial growth factor (VEGF), a potent angiogenesis factor needed for arteriogenesis. Another example is the dorsal root ganglia (DRG), where neurons and capillaries are in direct apposition. In DRG, neuronal-derived VEGF is necessary for endothelial cell survival (Kutcher *et al.*, 2004). It has become apparent that factors first shown to mediate axon guidance also mediate angiogenesis. Several ligand/receptor pairs have been shown to have dual activity in mediating axon guidance and angiogenesis including: (1) semaphorins and neuropilins, (2) Ephrins and Ephs, (3) Slits and Robos, and (4) netrins and UNC5 or DCC (Klagsbrun and Eichmann, 2005).

The semaphorins (SEMA) are a large family of axon guidance molecules consisting of eight classes of SEMA proteins and more than 20 *sema* genes (Bielenberg *et al.*, 2007; Neufeld *et al.*, 2008; Raper *et al.*, 2000; Taniguchi *et al.*, 2005; Yazdani *et al.*, 2006). Class 3 semaphorins (SEMA3) consist of seven (A to G) soluble proteins of approximately 100 kDa and are the only class that binds neuropilins (NRP), receptors that mediate axon guidance and angiogenesis. There are two NRP genes (NRP1, NRP2). NRP1 is the receptor for SEMA3A, whereas NRP2 is the receptor for SEMA3F (Chen *et al.*, 1997). NRPs are also receptors of vascular endothelial cell growth factor (VEGF) (Soker *et al.*, 1998). To convey a signal, SEMAs and NRPs also need to interact with plexins, transmembrane proteins whose cytoplasmic domains are substrates for nonreceptor kinases (Fazzari *et al*, 2007; Takahashi *et al.*, 2001; Tamagnone *et al.*, 2000; Winberg *et al.*, 1998; Yaron *et al.*, 2005). Nine plexins have been reported, of which plexin A1 and plexin A2 form complexes with SEMA3F and NRP2 *in vitro*. *In vivo*, NRP2 signaling is mediated by plexin A3 in the mouse embryonic nervous system (Yaron *et al.*, 2005). An exception is SEMA3E, which is not dependent on NRPs but acts directly by means of plexin D1 to repel blood vessels (Gu *et al.*, 2005).

Functionally, class 3 SEMA proteins were first shown to repel axons and collapse growth cones. The earliest indication that class 3 SEMA acted on nonneuronal cells was the demonstration that SEMA3A binds to EC by means of NRP1 *in vitro*, depolymerizes F-actin, inhibits EC migration, retracts EC lamellipodia, and inhibits capillary sprouting in an aortic ring assay (Miao *et al.*, 1999). Subsequently, it was shown that SEMA3A inhibits integrin activation in EC and mediates adhesion of EC to fibronectin and vitronectin (Serini *et al.*, 2003). SEMA3F is a ligand of NRP2 (Chen *et al.*, 1997; Kolodkin *et al.*, 1997) and repels vascular EC and lymphatic EC (Bielenberg *et al.*, 2004).

SEMA3F plays a role in cancer. SEMA3F was first isolated from a region of chromosome 3p21.3 that is commonly deleted in small cell lung cancers (Roche *et al.*, 1996; Sekido *et al.*, 1996; Xiang *et al.*, 1996), suggesting that SEMA3F is a tumor suppressor. Interestingly, p53 induces SEMA3F expression (Futamura *et al.*, 2007). SEMA3F, *in vivo*, is a potent inhibitor of tumor angiogenesis, tumor progression, and metastasis in human melanoma mouse tumor models (Bielenberg *et al.*, 2004). Inhibition of metastasis is consistent with the downregulation of endogenous SEMA3F observed in highly metastatic tumor cell lines (Bielenberg *et al.*, 2004). Overexpression of SEMA3F in human metastatic melanoma cells completely inhibits spontaneous metastasis to lymph nodes and lungs. Tumors overexpressing SEMA3F are poorly vascularized. Histologic analysis reveals that in melanoma xenografts SEMA3F induces the formation of a barrier around the tumors that prevents blood vessel invasion from the surrounding dermis by repulsing the EC (Bielenberg *et al.*, 2004). As a therapeutic, SEMA3F has the advantage that it can target both EC and tumor cells expressing NRP2.

This chapter is focused primarily on semaphorin 3F (SEMA3F), how it can be purified, and how its biologic activity can be assayed.

2. PURIFICATION OF HUMAN RECOMBINANT SEMA3F

SEMA3F, in an NRP2-dependent manner, has a striking effect on tumor cell and EC morphology, causing cytoskeleton collapse, loss of stress fibers, loss of adhesion, loss of contractility, and depolymerization of F-actin rapidly within min. A similar mechanism holds for SEMA3A interactions with cells expressing NRP1. To carry out *in vitro* and *in vivo* SEMA3F analysis, it is important to be able to purify this protein and to assay its biologic activity. This section reports in-depth the method for the purification of SEMA3F.

2.1. Plasmid information

Human *semaphorin 3F* (*SEMA3F*) exists as two alternatively spliced isoforms (Xiang *et al.*, 1996). The shorter isoform (accession # HSU38276) is inserted into pSecTag2A (Invitrogen) at *EcoRI* and *XbaI* restriction enzyme sites. The signal sequence of *SEMA3F* is replaced with the human Igκ signal sequence. The original stop codon of *SEMA3F* is removed and connected to Myc (EQKLISEEDL) and His (HHHHHH) tag sequences at the 3' end. *E. coli* (DH5α, Invitrogen) are transformed with the tagged-*SEMA3F* construct (kindly provided by Dr. Marc Tessier-Lavigne, Genentech Inc., San Francisco, CA) for amplification. Using the Quantum Maxi Kit (BioRad) according to the manufacturer's instructions, 300 ml of Luria-Bertani (LB) culture medium (1% tryptone [Boston Bioproducts], 0.5% Bacto-yeast extract [Amresco], 1% NaCl [Sigma], autoclaved 15 min at 121 °C) yields 3 mg of purified *SEMA3F* plasmid.

2.2. Culture of HEK293 cells

Human embryonic kidney (HEK) 293 cells are obtained from the American Type Culture Collection (ATCC). HEK293 cells are cultured in Dulbecco's modified Eagle medium (DMEM, Invitrogen) (1000 mg/L D-glucose, L-glutamine, 25 mM HEPES, and 110 mg/L sodium pyruvate) supplemented with 10% fetal bovine serum (FBS, Invitrogen), and 1% glutamine-penicillin-streptomycin (GPS, Invitrogen). Cells are incubated at 37 °C in a humidified chamber with 95% air/5% CO_2. Cell density is maintained within 0.5 to 3 × 10^6 cells/10-cm tissue culture dish (BD Biosciences).

2.2.1. Transfection
HEK293 cells (2 × 10^6 cells/dish) are plated into 15 separate 15-cm tissue culture dishes (BD Biosciences) with 20 ml of DMEM growth media (10% FBS, 1% GPS) per dish. After 16 h, cells are transfected with *SEMA3F*. The transfection solution is prepared as a stock cocktail and then added to each dish. The stock transfection solution is prepared in two steps: (1) add 675 μl Fugene6 (Roche) to 22.5 ml DMEM in a 50-ml sterile conical tube (BD Biosciences) and (2) after 5 min, add 225 μg *SEMA3F* plasmid. After 15 min, the transfection solution (1.5 ml) is gently added directly to each

15-cm dish containing HEK293 cells in 20 ml DMEM growth media. (This procedure can be scaled up or down; each 15-cm dish requires 45 μl Fugene6, 15 μg plasmid, 1.5 ml DMEM.) After 16 h, the DMEM growth media is changed to serum-free CD293 medium (Invitrogen, 20 ml/dish). Conditioned media ($\sim$300 ml for 15 dishes) containing SEMA3F protein is collected after 48 h and immediately purified as described below (NOTE: Do not store the conditioned media at 4 °C or at -20 °C). Fresh, serum-free CD293 Media (20 ml/dish) is added to each dish containing transfected-HEK293 cells. After 48 h, a second round of conditioned media is collected ($\sim$600 ml total) and immediately purified according to the protocol for "second round protein purification."

2.3. Preparation of the conditioned medium (before applying to FPLC)

The conditioned media ($\sim$300 ml from the first collection) containing SEMA3F protein is placed in a clean 500-ml polypropylene centrifuge tube (VWR). (The bottle must be used for only this purpose to avoid contamination of bacterial endotoxin). The pH of the conditioned media is adjusted to 7.4 with 6 N NaOH. The conditioned medium is divided equally (by weight) into two 500-ml plastic centrifuge tubes and centrifuged at 4 °C for 15 min at 16,900g with an SLA-3000 angle rotor (Serval) in an RC-5B centrifuge (Serval). After centrifugation, the supernatant is filtered through a Stericup 500-ml unit with a 0.22-μm Durapore® Polyvinylidene Fluoride (PVDF) membrane (Millipore).

2.4. Preparation of the column

A HiTrap high-performance (HP) chelating column with a 5-ml bed volume (GE Healthcare Amersham Biosciences) is washed with 50 ml of deionized water (deH$_2$O) to remove any ethanol. A 0.1 M NiSO$_4$ (Sigma) solution is prepared with deH$_2$O, and 3 ml of the solution is slowly (1 to 2 ml/min) injected into the column through the injection attachment (GE Healthcare) with a 3-ml syringe (BD Biosciences). The column resin changes from a white color to a homogeneous green color due to the NiSO$_4$. Fast protein liquid chromatography (FPLC) pumps (GE Healthcare) and lines are kept at 4 °C in a hinged glass door refrigerator (Fisher) and controlled by FPLC director software (GE Healthcare). All lines of the FPLC are filled with deH$_2$O. The nickel column is connected to the FPLC machine and washed with 50 ml of deH$_2$O at a flow rate of 5 ml/min. Next, the column is detached, and the lines are filled with sterile phosphate-buffed saline (PBS, pH 7.4) with the FPLC director wash function. Finally, the column is reattached and equilibrated with 50 ml of PBS (pH 7.4) at a flow rate of 5 ml/min.

2.5. Recombinant SEMA3F protein purification

Because the SEMA3F protein is tagged with histidine, it can easily be purified with a nickel affinity column. The filtered conditioned media containing SEMA3F ($\sim$300 ml still in the lower collection portion of the plastic Stericup) is placed in a 4 °C hinged glass door refrigerator (Fisher) and connected to the FPLC machine through an input line (tubing). The "pump and flow" command is chosen from the function menu, and the conditioned media is aspirated into the FPLC machine at 5 ml/min. The column flowthrough is collected into a sterilized bottle. To ensure that bubbles are not inadvertently introduced into the system and to purify all of the conditioned media, the flow is stopped with the "pause" function on the FPLC director when the volume of the conditioned media reaches $\sim$50 ml (after $\sim$50 min). The remaining conditioned media is transferred to a 50-ml conical tube (BD Biosciences), the input line is inserted to the bottom of the tube, and the flow is resumed with the "continue" function. Once the column is fully loaded, it is washed with 50 ml of wash buffer (20 mM Na$_2$HPO$_4$, 300 mM NaCl, 0.05 mM imidazole, pH 7.4) at 5 ml/min. The SEMA3F protein is then eluted by washing the column with elution buffer (20 mM Na$_2$HPO$_4$, 500 mM NaCl, 500 mM imidazole, pH 7.4) at 5 ml/min. The protein is collected into 1-ml fractions in low-adhesive tubes (PGC Sciences). The absorbance (280 nm) of each fraction is determined with a spectrophotometer (Beckman), and fractions with an absorbance >0.2 are pooled (usually 2 to 4 fractions). Salt and imidazole are removed from the pooled protein with a PD-10 gel filtration column (GE Healthcare). The PD-10 column is washed with PBS; then the protein (2 to 4 ml) is applied to the column and eluted with 5 ml PBS. (Isotonic PBS buffer is preferred for *in vivo* use of the protein, but to prevent precipitation of the protein during storage or for strictly *in vitro* use, sodium chloride phosphate buffer can be used [500 mM NaCl, 20 mM Na$_2$HPO$_4$].) Ten fractions of 0.5 ml each are collected. The protein concentration of each sample is determined with a DC protein assay (BioRad) according to the manufacturer's protocol. The expected final concentration is 200 to 1000 ng/μl, and the final yield is 2.5 mg. After the column is replenished (as described later), the second batch of conditioned media (another 300 ml) is purified similarly. The total yield from both purifications is 5 mg. The purified SEMA3F protein is stored at −80 °C. Figure 15.1 shows aliquots from a representative purification analyzed by SDS-PAGE and immunoblotted with anti-myc antibodies (Santa Cruz). The SEMA3F protein is enriched more than 50 fold.

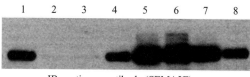

IB: anti-myc antibody (SEMA3F)
1: Unpurified conditioned medium (50 μl)
2: Flow through column fraction (50 μl)
3: Washing column fraction (50 μl)
4: Imidazole elution fraction No.4 (1 μl)
5: Imidazole elution fraction No.5 (1 μl)
6: Imidazole elution fraction No.6 (1 μl)
7: Imidazole elution fraction No.7 (1 μl)
8: Imidazole elution fraction No.8 (1 μl)

Figure 15.1 Purification of myc-tagged-SEMA3F with FPLC. HEK293 cells were transfected with the SEMA3F construct, and conditioned media containing the tagged SEMA3F protein were collected and applied to a HiTrap HP Nickel Chelating Column. Purified proteins were eluted with imidazole. Aliquots from various steps in the procedure were analyzed by immunoblotting (IB) with anti-myc antibody to detect purified SEMA3F and to estimate percent recovery and enrichment.

2.6. Reuse of the column

The HiTrap HP chelating column is reusable. The used column is washed with 50 ml of metal dissociation buffer (20 mM NaH$_2$PO$_4$, 500 mM NaCl, 50 mM EDTA) to remove the nickel divalent cation. The metal dissociation buffer is slowly (2 to 4 ml/min) injected into the column with the injection attachment through a 60-ml syringe (BD Biosciences). All the FPLC tubing lines are filled with deH$_2$O. The column is connected to the FPLC machine and washed with 75 ml deH$_2$O at 5 ml/min. After washing, 3 ml 0.1 M NiSO$_4$ is reapplied to the column manually followed by 50 ml deH$_2$O at 5 ml/min and 50 ml PBS at 5 ml/min as described previously.

3. SEMA3F-INDUCED ENDOTHELIAL CELL COLLAPSE

3.1. Cell culture of HUVEC and HMVEC

Human umbilical vein EC (HUVEC, Lonza) are cultured in endothelial cell growth medium (EGM-2, Lonza) with full supplements (EGM-2 bullet kit: 2% FBS, 0.4% hFGF-2, 0.1% VEGF, 0.1% R^3-IGF-1, 0.1% hEGF, 0.04% hydrocortisone, 0.1% ascorbic acid, 0.1% heparin, 0.1% GA-100). Human microvascular EC (HMVEC, Lonza) are cultured in EGM-2 MV media (Lonza) with full supplements (EGM-2 MV bullet kit: 5% FBS, 0.4% hFGF-2, 0.1% VEGF, 0.1% R^3-IGF-1, 0.1% hEGF, 0.04% hydrocortisone, 0.1% ascorbic acid, 0.1% GA-100). Both cell lines are maintained in 95%

air/5% CO_2 at 37 °C in 10-cm tissue culture dishes. Cell density is maintained at 0.75 to 3×10^6 cells (HUVEC) and 0.4 to 1.6×10^6 cells (HMVEC) per 10-cm dish.

3.2. SEMA3F-induced collapse assay: (Time-lapse) imaging

A glass petri dish (35 mm, Corning) is cleaned and sterilized by adding 3 ml of 100% ethanol for 30 min. The ethanol is aspirated and primary EC (1 to 2×10^4 cells) are plated into the glass dish in 1.5 ml of their appropriate growth medium. After 6 h, the media is changed to 1.5 ml endothelial cell basal medium (EBM-2, Lonza) supplemented with 10% FBS but no additional growth factors. After 16 h, the dish is placed on the microscope stage (Zeiss Axiovert 200) inside an Incubator XL-3 incubation chamber (Zeiss) maintained at 37 °C with constant 5% CO_2 infusion. The cells are examined with a 63× objective (NA 1.4) and a 1.6× Optivar (Zeiss). Differential interference contrast (DIC) images of the EC are captured every 10 sec (image capture time of 50 to 100 msec) with an Orca IIER cooled charged coupled device (CCD) camera (Hamamatsu) for 5 min. The magnified images are continuously monitored visually during the videography and manually adjusted to maintain the focal point. Metamorph software (Universal Imaging of Molecular Devices) controls the electronic shutters and image acquisitions. After 5 min of filming, 500 μl of purified SEMA3F protein is added to the 1.5 ml of media in the culture dish (final concentration is 320 ng/ml), and images are acquired every 10 sec for 30 to 60 additional min. Time-lapse movies depicting "endothelial cell collapse" are generated with the Metamorph image analysis program. Representative DIC images of HUVEC at time 0 and 40 min after SEMA3F addition are shown in Fig. 15.2 C and D.

3.2.1. Pretreatment of cover glasses (for confocal imaging of SEMA3F-induced collapse)

Different cell types require different pretreatments of the cover glasses. For example, primary human EC such as HMVEC and HUVEC prefer glasses pretreated according to the basic method, whereas porcine aortic endothelial cells (PAEC) require silane coating (alternative method). Basic method: Square cover glasses (18 × 18 mm, Baxter Scientific Products) are soaked in 10% HNO_3 solution (Sigma) for 24 h. The glasses are washed with deH_2O three times and autoclaved for 20 min in deH_2O. Alternative method: In a sterile laminar-flow tissue culture hood, sterile cover glasses are rinsed in 0.1 N NaOH solution for 1 min and dried for 30 min by leaning against the wall of a sterile tissue culture dish lid. The lid of a sterile 6-well plate (Costar) is then inverted and covered in clean Parafilm (Alcan Inc). The dry cover glasses are placed on the Parafilm with sterile forceps and covered over the entire surface with 150 μl of 3-aminopropyltrimethoxysilane

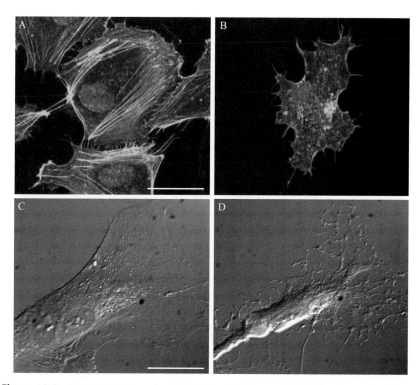

Figure 15.2 SEMA3F-induced endothelial cell collapse. HUVEC undergo significant and rapid morphologic changes in response to SEMA3F. (A, B) Confocal microscopy after Alexafluor 488 Phalloidin staining (white color) to show F-actin stress fibers. Nuclei are stained with bis-benzamide (Hoescht). HUVEC treated with SEMA3F (320 ng/ml) for 30 min (B) show greatly diminished stress fibers compared with control (A). Panels A, B have been previously shown in Shimizu, *et al.*, *J. Biol. Chem*, 2008 (in press). (C, D) Frames selected from time-lapse differential interference contrast (DIC) videography. HUVEC control (C) and the same cell after SEMA3F treatment (320 ng/ml) at 40 min (D). SEMA3F-treated cells shrink by up to 30 to 40% and are less adherent. Scale bars = 20 μm.

(APTS, Sigma) for 10 min. The 6-well dish is inverted and placed on top of the lid to avoid evaporation of the APTS. After 10 min, the APTS is removed by suction, and the cover glasses are washed three times with deH$_2$O. Each glass is then covered with 600 μl of 0.5% glutaraldehyde solution for 30 min. The cover glasses are washed three times with deH$_2$O and dried in the laminar flow hood under ultraviolet light for 30 min.

SEMA3F-induced collapse assay: confocal microscopy. One pretreated cover glass is placed into the bottom of each well in a 6-well tissue culture plate. Primary EC (1 to 2 $\times$ 10^4 cells/well) are plated into each well of the 6-well dish in 1.5 ml of their appropriate growth medium and incubated at 37 °C in 95% air/5% CO$_2$. After 6 h, the media is changed to 1.5 ml

EBM-2 supplemented with 10% FBS but no additional growth factors and
again incubated at 37 °C. After 16 h, purified SEMA3F protein (320 ng/ml)
is added to half of the wells and incubated at 37 °C. After 30 to 60 min, the
plates are placed in the sterile hood, and the cover glasses are loosened from
the bottom of the wells with a bent 23-gauge needle. The cover glasses are
transferred to Parafilm with small, serrated forceps (RS-5135, Roboz) and
immediately covered with PBS to avoid drying. The cells on the cover
glasses are fixed in 4% paraformaldehyde (diluted with PBS) for 12 min at
room temperature, washed twice with PBS, permeabilized with 0.2%
Triton X-100 (Sigma) in PBS for 5 min, and washed again twice
with PBS. Each cover glass is covered with 160 μl of Alexafluor 488
phalloidin (Invitrogen, 1:250 dilution) and incubated at room temperature
for 1 h in the dark. Cover glasses are washed twice with PBS and incubated
for 1 min in bis-benzamide/Hoescht 33258 (Sigma, 1:10,000 dilution in
deH$_2$O). Cover glasses are rinsed with deH$_2$O, inverted, and mounted with
Gel/Mount (Biomeda) onto a slide glass (25 × 75 × 1 mm, Fisher). The
edges of the cover glass are sealed with nail polish. The actin cytoskeleton of
the EC is visualized and imaged with a Leica TCS SP2 AOBS confocal laser
scanning system attached to a Leica DMIRE2 inverted microscope and
equipped with a 63× oil objective (NA 1.4), a 488-nm argon ion laser (F-
actin), and a 405-nm diode (nuclei). Leica Confocal Software and NIH
ImageJ are used to scale recorded images. A line average of 2 and a frame
average of 4 are used. Representative confocal images of phalloidin staining
of F-actin–containing stress fibers in control HUVEC and depolymerized
stress fibers in SEMA3F-treated HUVEC (collapsed) is shown in Fig. 15.2 A
and B.

4. ENDOTHELIAL CELL REPULSION ASSAY

This assay is designed to test the repulsive activity of Class 3 semaphorins on EC. This assay requires a point source of SEMA3F protein and
uses tumor cells transfected with SEMA3F as the source. Another requirement is that the cells secreting the SEMA3F should be capable of detaching
and reattaching throughout the dish, usually a property of tumor cells. The
SEMA3F-expressing cells repel the EC if they are expressing the appropriate neuropilin receptor. For example, we have found that SEMA3A can
repel EC expressing NRP1 but not cells expressing NRP2 (Bielenberg,
unpublished data), and SEMA3F can repel EC expressing NRP2 but not
cells expressing NRP1 (Bielenberg *et al.*, 2004). The EC are plated in a
confluent monolayer. After a cell expressing SEMA3A/F adheres to
the confluent monolayer of EC, it secretes SEMA3A/F and the surrounding
EC begin to repel from the source of the ligand in a NRP1/2-dependent

manner. The higher the concentration of SEMA3 and the longer the assay goes, the larger the zone of repulsion.

4.1. Preparation of the repulsion assay

The setup of the EC repulsion assay is illustrated in Fig. 15. 3 (steps 1 through 3). Before use, glass Superfrost™ Plus microscope slides (Fisher Scientific) and 5-mm diameter glass cloning cylinders (VWR) are sterilized by autoclave. In a laminar flow tissue culture hood, the lids are removed from 10-cm square plastic tissue culture dishes (Fisher) and set aside. Sterile forceps are used to align three sterile glass microscope slides in each square dish. The slides should not touch one another. Three glass cloning cylinders (5 mm) are aligned on each slide with approximately 5- to 7-mm space between each cloning tube. A 2% agarose solution is prepared by adding 2 g agarose to 100 ml of tap water and boiling for 2 to 4 min in a microwave oven. The agarose is used to affix the cylinders in one position as well as to seal the bottoms of the cylinders so that the cell suspensions do not leak out. Ten ml of the 2% agarose solution is pipetted over the slides and between the cloning tubes in each dish, avoiding the center of the cloning tubes. The agarose is allowed to solidify in the laminar flow hood. Once the agarose is solidified, the inside of each cylinder is checked for any agarose that may have leaked into the cylinder. In the case of leakage, a sterile 18-gauge needle is used to remove the agarose from inside the cylinder.

SEMA3F–induced repulsion assay. The EC repulsion assay is illustrated in Fig. 15.3 (steps 4 through 10). EC are placed in the upper and lower cylinder on each slide. Primary EC (HUVEC or HMVEC) are trypsinized (0.5 M trypsin/EDTA, Lonza) for 2 to 3 min, neutralized with trypsin neutralizing solution (TNS, Lonza), and pelleted in a 15-ml conical tube (BD Biosciences) with a desktop centrifuge (Beckman) at 1000 rpm for 5 min at 4 °C. The TNS is removed, and the primary EC are resuspended to a final concentration of 100,000 cells/ml in EGM-2 media with full supplements (described previously). Alternately, spontaneously immortalized EC such as PAEC stably expressing NRP2 (PAEC NRP2) are resuspended to a final concentration of 100,000 cells/ml in F12 nutrient mixture HAM media (Invitrogen) containing 10% FBS and 1% GPS. For SEMA3F–expressing tumor cells, such as A375SM human melanoma cells (kindly provided by Dr. Isaiah J. Fidler, M. D. Anderson Cancer Center, Houston, TX) transfected with human SEMA3F (Bielenberg et al., 2004), cells are resuspended to a final concentration of 100,000 cells/ml in minimal essential media (Invitrogen Corp.) containing 10% FBS and 1% GPS. The cell suspensions (10,000 cells/100 μl) are pipetted into the cloning cylinders with a P200 pipet and sterile pipet tips. The pipet tip should be inserted to the bottom of the cylinder before dispensing the cell suspension to avoid bubbles. One ml of serum-free media is added on top of the agarose away

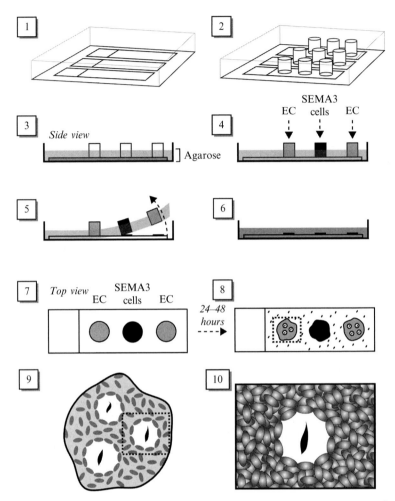

Figure 15.3 Illustration of SEMA3F-induced repulsion assay. Various steps are shown that demonstrate how the repulsion assay is carried out. (1) Slides are aligned in dishes. (2) Glass cylinders are placed on each slide. (3) Slides are covered with agarose (2%). (4) Cells are placed in the cylinders and incubated 12 to 15 h. (5) Agarose and cylinders are removed. (6) Serum-free media is added. (7) Top view of the circular cell areas before incubation. (8) After incubation, SEMA3F-expressing cells adhere throughout the slide and EC repulsion occurs. (9) Higher magnification view of cells in step 8. (10) Higher magnification view of cells in step 9.

from the cylinders to keep the environment moist and to avoid the media in the cylinder from drying out overnight. The lids are replaced on top of the 10-cm plastic tissue culture dishes, and the cells are allowed to adhere to the glass slides overnight (12 to 15 h) in an incubator at 37 °C and 95% air/5% CO_2. The following day, the cells are observed under the microscope to ensure that the cells in each cylinder have formed a confluent monolayer. Any dishes in which the cells have not formed a monolayer can be placed back in the incubator to continue to grow and fill out the area of the cylinder; in this case, it may be necessary to add any media that may have evaporated to the top of the cylinder. Dishes in which the cells have formed a confluent monolayer are placed in the laminar flow sterile tissue culture hood. The lids are removed from the dishes one at a time, and the agarose layer is cut with a No. 22 scalpel (or spatula) between each slide. With a sterile forceps, the agarose pieces are carefully removed from each slide. The media within the cylinders are not removed before the agarose removal. The agarose containing the cylinders is set aside. The slides are quickly covered with 10 ml of serum-free endothelial media (Ham's F12 for PAEC or EBM for HUVEC or HMVEC). The dish should now contain three slides with three confluent circular areas of cells on each slide. A top view of one slide is shown in Fig. 15.4 and depicted in Fig. 15.3, step 7. The lids are replaced on the dishes, and the dishes are incubated at 37 °C for an additional 24 h. The following day, tumor cells expressing SEMA3F will have detached and reattached all throughout the slide with some cells adhering to the confluent monolayer of EC at either end of the slide (depicted in Fig. 15.3, step 8). A "zone of clearance" or an area void of cells will be present around the tumor cells expressing SEMA3F if the EC express NRP2. Longer incubation times (48 to 72 h) will result in larger zones of clearance, but eventually elevated SEMA3F levels in the media may result in adverse effects on EC survival. After repulsion is confirmed microscopically, the media are removed from the dish, and slides are fixed in 10% neutral buffered formalin for 10 min at room temperature. Slides are washed with phosphate-buffered saline (PBS, pH 7.2) and stored at 4 °C for immunocytochemistry (ICC) to detect SEMA3F-expressing cells. This assay may be modified to analyze the functional activity of other class 3 SEMAs, including the repulsive activity of SEMA3A on EC expressing NRP1.

4.2. Immunocytochemistry (ICC)

To differentiate the tumor cells expressing SEMA3F from the EC in the repulsion assay, the slides can be stained with antibodies to tumor-specific antigens (such as $S100\beta$ for melanoma cells or EpCam for carcinoma cells) or with antibodies to myc if the SEMA3A/F-transfected protein is myc-tagged. Slides are removed from PBS, quickly wiped dry with WypAll X-60 lint-free towels (Kimberly-Clark Professional), and the areas containing cells are encircled with a Pap Pen (ECM Biosciences). PBS is dripped onto the cells (inside the hydrophobic circles) to keep the cells hydrated, and the slides are placed horizontally in a slide chamber (Market Lab). PBS is removed and replaced with 3% H_2O_2 in methanol for 12 min at room temperature to block endogenous peroxidases. Slides are washed in PBS

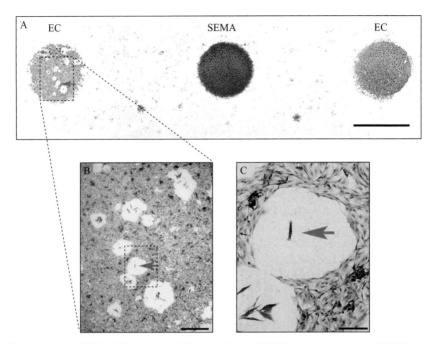

Figure 15.4 SEMA3F-induced EC repulsion. PAEC overexpressing NRP2 are repelled by melanoma cells overexpressing SEMA3F. (A) Scanned image of an entire slide after the repulsion assay and immunostaining (with melanoma-specific antibody) procedures. This scan is an actual image of that depicted in the illustration of Figure 15.3, step 8. PAEC NRP2 cells (EC) were plated on the left and right side of the slide. SEMA3F-expressing cells (SEMA) were plated in the center of the slide and have migrated throughout the slide. All cells were counterstained with hematoxylin. Scale bar = 5 mm. (B) Magnification from outlined area in A. Numerous zones of clearance are apparent because of the repulsion of the EC by very few tumor cells overexpressing SEMA3F. Scale bar = 0.5 mm. (C) Magnification from outlined area in B. Scale bar = 0.1 mm. Arrows point to the same tumor cell overexpressing SEMA3F in each image.

three times for 3 min each. Nonspecific proteins are blocked with protein blocking solution (3% sheep serum, 2% goat serum in PBS) for 15 min at room temperature. Slides are incubated overnight at 4 °C in primary rabbit anti-human S100β (Dako, diluted 1:200 in protein blocking solution) to detect human melanoma cells expressing SEMA3F or in primary mouse monoclonal anti-myc IgG1 (Santa Cruz, diluted 1:200 in protein blocking solution) to detect tumor cells expressing myc-tagged SEMA3F. The following day, the slides are washed in PBS three times for 3 min each and blocked in protein blocking solution for 10 min at room temperature. Slides are incubated in secondary horseradish peroxidase (HRP)–conjugated goat anti-rabbit IgG F(ab')$_2$ (Jackson ImmunoResearch, diluted 1:200 in protein blocking solution) or HRP-conjugated rat anti-mouse

IgG1 (Pharmingen, diluted 1:200 in protein blocking solution) at room temperature for 1 h. Slides are washed in PBS three times for 3 min each to remove excess antibody and rinsed briefly in H_2O to remove salts. Slides are developed in chromogen (3,3′-diaminobenzidine (DAB), Vector Laboratories) for 2 to 10 min. The reaction is carefully monitored under a light microscope, and positive cells appear brown in color. Rinsing in H_2O terminates the reaction. Cells are counterstained with hematoxylin solution Gill 3 (Sigma). The hematoxylin is differentiated (from purple color to a blue color) with Tacha's bluing solution (Biocare Medical). The slides are rinsed with water, dried, and mounted with Permount (Fisher) and Fisherbrand microscope cover glasses (22 × 40 mm).

REFERENCES

Bielenberg, D. R., Hida, Y., Shimizu, A., Kaipainen, A., Kreuter, M., Kim, C. C., and Klagsbrun, M. (2004). Semaphorin 3F, a chemorepulsant for EC, induces a poorly vascularized, encapsulated, nonmetastatic tumor phenotype. *J. Clin. Invest.* **114,** 1260–1271.

Bielenberg, D. R., and Klagsbrun, M. (2007). Targeting endothelial and tumor cells with semaphorins. *Cancer Metastasis Rev.* **26,** 421–431.

Chen, H., Chedotal, A., He, Z., Goodman, C. S., and Tessier-Lavigne, M. (1997). Neuropilin-2, a novel member of the neuropilin family, is a high affinity receptor for the semaphorins Sema E and Sema IV but not Sema III. *Neuron* **19,** 547–559.

Fazzari, P., Penachioni, J., Gianola, S., Rossi, F., Eickholt, B. J., Maina, F., Alexopoulou, L., Sottile, A., Comoglio, P. M., Flavell, R. A., and Tamagnone, L. (2007). Plexin-B1 plays a redundant role during mouse development and in tumour angiogenesis. *BMC Dev. Biol.* **7,** 55.

Futamura, M., Kamino, H., Miyamoto, Y., Kitamura, N., Nakamura, Y., Ohnishi, S., Masuda, Y., and Arakawa, H. (2007). Possible role of semaphorin 3F, a candidate tumor suppressor gene at 3p21.3, in p53-regulated tumor angiogenesis suppression. *Cancer Res.* **67,** 1451–1460.

Gerhardt, H., Golding, M., Fruttiger, M., Ruhrberg, C., Lundkvist, A., Abramsson, A., Jeltsch, M., Mitchell, C., Alitalo, K., Shima, D., and Betsholtz, C. (2003). VEGF guides angiogenic sprouting utilizing endothelial tip cell filopodia. *J. Cell Biol.* **161,** 1163–1177.

Gerhardt, H., Ruhrberg, C., Abramsson, A., Fujisawa, H., Shima, D., and Betsholtz, C. (2004). Neuropilin-1 is required for endothelial tip cell guidance in the developing central nervous system. *Dev. Dyn.* **231,** 503–509.

Gu, C., Yoshida, Y., Livet, J., Reimert, D. V., Mann, F., Merte, J., Henderson, C. E., Jessell, T. M., Kolodkin, A. L., and Ginty, D. D. (2005). Semaphorin 3E and plexin-D1 control vascular pattern independently of neuropilins. *Science* **307,** 265–268.

Klagsbrun, M., and Eichmann, A. (2005). A role for axon guidance receptors and ligands in blood vessel development and tumor angiogenesis. *Cytokine Growth Factor Rev.* **16,** 535–548.

Kolodkin, A. L., Levengood, D. V., Rowe, E. G., Tai, Y. T., Giger, R. J., and Ginty, D. D. (1997). Neuropilin is a semaphorin III receptor. *Cell* **90,** 753–762.

Kutcher, M. E., Klagsbrun, M., and Mamluk, R. (2004). VEGF is required for the maintenance of dorsal root ganglia blood vessels but not neurons during development. *FASEB J.* **15,** 1952–1954.

Miao, H. Q., Soker, S., Feiner, L., Alonso, J. L., Raper, J. A., and Klagsbrun, M. (1999). Neuropilin-1 mediates collapsin-1/semaphorin III inhibition of endothelial cell motility: Functional competition of collapsin-1 and vascular endothelial growth factor-165. *J. Cell Biol.* **146**, 233–242.

Mukouyama, Y. S., Shin, D., Britsch, S., Taniguchi, M., and Anderson, D. J. (2002). Sensory nerves determine the pattern of arterial differentiation and blood vessel branching in the skin. *Cell* **109**, 693–705.

Mukouyama, Y. S., Gerber, H. P., Ferrara, N., Gu, C., and Anderson, D. J. (2005). Peripheral nerve-derived VEGF promotes arterial differentiation via neuropilin 1-mediated positive feedback. *Development* **132**, 941–952.

Neufeld, G., and Kessler, O. (2008). The semaphorins: Versatile regulators of tumour progression and tumour angiogenesis. *Nat. Rev. Cancer,* in press.

Raper, J. A. (2000). Semaphorins and their receptors in vertebrates and invertebrates. *Curr. Opin. Neurobiol.* **10**, 88–94.

Roche, J., Boldog, F., Robinson, M., Robinson, L., Varella-Garcia, M., Swanton, M., Waggoner, B., Fishel, R., Franklin, W., Gemmill, R., and Drabkin, H. (1996). Distinct 3p21.3 deletions in lung cancer and identification of a new human semaphorin. *Oncogene* **12**, 1289–1297.

Sekido, Y., Bader, S., Latif, F., Chen, J. Y., Duh, F. M., Wei, M. H., Albanesi, J. P., Lee, C. C., Lerman, M. I., and Minna, J. D. (1996). Human semaphorins A(V) and IV reside in the 3p21.3 small cell lung cancer deletion region and demonstrate distinct expression patterns. *Proc. Natl. Acad. Sci. USA* **93**, 4120–4125.

Serini, G., Valdembri, D., Zanivan, S., Morterra, G., Burkhardt, C., Caccavari, F., Zammataro, L., Primo, L., Tamagnone, L., Logan, M., Tessier-Lavigne, M., Taniguchi, M., Puschel, A. W., and Bussolino, F. (2003). Class 3 semaphorins control vascular morphogenesis by inhibiting integrin function. *Nature* **424**, 391–397.

Soker, S., Takashima, S., Miao, H. Q., Neufeld, G., and Klagsbrun, M. (1998). Neuropilin-1 is expressed by endothelial and tumor cells as an isoform-specific receptor for vascular endothelial growth factor. *Cell* **92**, 735–745.

Takahashi, T., and Strittmatter, S. M. (2001). Plexina1 autoinhibition by the plexin sema domain. *Neuron* **29**, 429–439.

Tamagnone, L., and Comoglio, P. M. (2000). Signalling by semaphorin receptors: Cell guidance and beyond. *Trends Cell Biol.* **10**, 377–383.

Taniguchi, M., Masuda, T., Fukaya, M., Kataoka, H., Mishina, M., Yaginuma, H., Watanabe, M., and Shimizu, T. (2005). Identification and characterization of a novel member of murine semaphorin family. *Genes Cells* **10**, 785–792.

Winberg, M. L., Noordermeer, J. N., Tamagnone, L., Comoglio, P. M., Spriggs, M. K., Tessier-Lavigne, M., and Goodman, C. S. (1998). Plexin A is a neuronal semaphorin receptor that controls axon guidance. *Cell* **95**, 903–916.

Xiang, R. H., Hensel, C. H., Garcia, D. K., Carlson, H. C., Kok, K., Daly, M. C., Kerbacher, K., van den Berg, A., Veldhuis, P., Buys, C. H., and Naylor, S. L. (1996). Isolation of the human semaphorin III/F gene (SEMA3F) at chromosome 3p21, a region deleted in lung cancer. *Genomics* **32**, 39–48.

Yaron, A., Huang, P. H., Cheng, H. J., and Tessier-Lavigne, M. (2005). Differential requirement for Plexin-A3 and -A4 in mediating responses of sensory and sympathetic neurons to distinct class 3 Semaphorins. *Neuron* **45**, 513–523.

Yazdani, U., and Terman, J. R. (2006). The semaphorins. *Genome Biol.* **7**, 211.

CHAPTER SIXTEEN

PERICYTE ISOLATION AND USE IN ENDOTHELIAL/PERICYTE COCULTURE MODELS

Brad A. Bryan *and* Patricia A. D'Amore

Contents

Abstract

Vascular assembly, patterning, and maintenance is a complex and highly regulated process that begins with the formation of a primary capillary plexus by means of angiogenesis or vasculogenesis and ends when the primitive vessels have been remodeled into quiescent, differentiated vessels. Differentiated or "mature" microvessels are characterized in large part by their association with pericytes, and failure of these interactions results in severe, and often lethal, defects that have been implicated in a number of human pathologic conditions,

Schepens Eye Research Institute, and Departments of Ophthalmology and Pathology, Harvard Medical School, Boston, Massachusetts

Methods in Enzymology, Volume 443
ISSN 0076-6879, DOI: 10.1016/S0076-6879(08)02016-8

including tumor angiogenesis, diabetic microangiopathy, ectopic tissue calcification, stroke, and dementia. This chapter describes methods that can be used to isolate and culture primary pericytes, as well as to study pericyte–endothelial cell interactions with *in vitro* cell culture systems.

 1. INTRODUCTION

Pericytes are vascular mural cells that make specific cell-to-cell contacts with the endothelium of capillaries, precapillary arterioles, postcapillary venues, and collecting venules (Allt and Lawrenson, 2001). These cells are obligatory constituents of blood microvessels and serve as essential regulators of vascular development, stabilization, maturation, and remodeling. Although related in function and assumed to belong to the same lineage as vascular smooth muscle cells (VSMC) that surround larger blood vessels, pericytes can be distinguished from VSMC by their unique location, morphology, and marker expression; however, a continuum of cellular phenotypes exists, ranging from classical VSMC to typical pericyte when distributed along intermediate to small-sized vessels. In contrast to VSMC that compose a separate layer in the vascular wall termed the media, pericytes are physically embedded within the endothelial basement membrane of microvessels (Fig. 16.1) Moreover, although VSMC mediate vascular tone and contraction of large vessels, pericytes facilitate cell communication to the underlying endothelium of microvessels.

The intimate nature and critical function of pericyte–endothelial cell (EC) interactions is reflected by the numerous signaling pathways regulating diverse cellular processes from the initial endothelial sprout formation through the formation of fully differentiated quiescent vasculature. Newly sprouting ECs secrete PDGF-B ligands that bind to their cognate receptors located on pericytes, resulting in robust pericyte proliferation and migration to the sight of the newly formed endothelial sprout (Betsholtz, 2004; Hoch and Soriano, 2003). Indeed, knockout of PDGF-B or the PDGF receptor leads to perinatal death caused by vascular dysfunction because of lack of pericyte coverage on endothelium (Hellstrom *et al.*, 1999; Leeven *et al.*, 1994; Lindahl *et al.*, 1997; Soriano, 1994). After pericyte binding to EC, the platelet-derived secreted sphingolipid, sphingosine-1-phosphate (SIP), is capable of activating pericyte G-protein–coupled receptors and triggers cytoskeletal, adhesive, and junctional changes, resulting in strengthened contact of pericytes to ECs through N-cadherin adherence junctions (Balabnov and Dore-Duffy, 1998; Kono *et al.*, 2004; Liu *et al.*, 2000). Angiopoietin/Tie2 signaling is responsible for modulating vessel maturation and stabilization, whereby the pericyte-derived Ang1 binds to the endothelial-specific Tie2 receptor, leading to stable and quiescent vasculature (Davis *et al.*, 1996;

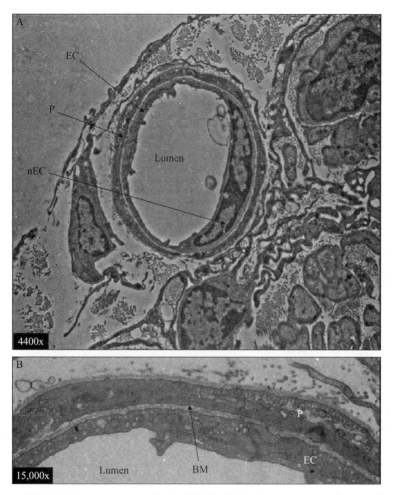

Figure 16.1   Pericyte coverage of endothelial cells. Pericytes are vascular mural cells physically embedded within the endothelial basement membrane of capillaries, precapillary arterioles, postcapillary venues, and collecting venules and coordinate cell communication to the underlying endothelium. Transmission electron micrographs ([A] original magnification, 4400×, [B] original magnification, 15,000×) of mouse retina cross-sections showing a pericyte surrounding and integrating with the endothelial basement membrane of a capillary. EC, Endothelial cell; P, pericyte; nEC, nucleus of endothelial cell; BM, basement membrane.

Sato *et al.*, 1995; Sundberg *et al.*, 2002; Suri *et al.*, 1996; 1998; Thurston *et al.*, 1999; Uemura *et al.*, 2002). In contrast, during new vessel sprouting from the mature vasculature, ECs express the antagonistic Ang2 ligand that competitively inhibits Ang1 binding to Tie2 and, therefore, destabilizes the endothelium, allowing sprouting ECs to migrate and proliferate to form new vessels

(Maisonpierre *et al.*, 1997; Witzenbichler *et al.* 1998). EC and pericyte-derived TGFβ ligands primarily modulate context-dependent effects on ECs, whereby proliferation and differentiation are mediated through Alk1/Smad1/5 and Alk5/Smad2/3, respectively (Goumans *et al.*, 2002). Aberrant TGFβ signaling reportedly disrupts pericyte function by eliciting defects in the EC basement membrane because of an incorrect balance of proliferation and differentiation of ECs (Lebrin *et al.*, 2004; Torsney *et al.*, 2003); however, it is possible that TGFβ may directly act on pericytes given that EC-derived TGFβ leads to differentiation of VSMC (Carvalho *et al.*, 2004; Hirshi *et al.*, 1998). Moreover, mouse genetic knockouts resulting in altered TGFβ signaling lead to severe pericyte and EC defects (Arthur *et al.*, 2000; Chang *et al.*, 1999; Dickson *et al.*, 1995; Larsson *et al.*, 2001; Li *et al.*, 1999; Oh *et al.*, 2000; Oshima *et al.*, 1996; Urness *et al.*, 2000; Yang *et al.*, 1999).

A single pericyte is capable of contacting multiple neighboring ECs, whereby it assimilates complex information and, subsequently, coordinates the intracellular signaling cascades mentioned previously. Pericyte coverage of the endothelium is highly variable, depending on the particular vascular bed of the organ. For example, pericyte/endothelial coverage can range from very low as reported for striated muscle (1:100) or extremely abundant as reported in the retina and central nervous system (1:1) (Shepro and Morel, 1993). Pericyte morphology can be visually distinguished from ECs (cobblestones), fibroblasts (long spindle shaped with extended filopodia), and VSMC (fusiform and compact); however, pericytes do exhibit morphologic variability, depending on the particular vascular beds they reside in. For example, CNS pericytes are flattened or elongated stellate-shaped solitary cells with multiple cytoplasmic processes encircling the capillary endothelium and contacting a large abluminal vessel area, whereas mesangial pericytes of the kidney glomerulus are rounded, compact, and contact only a minimal area in the underlying endothelium (Armulik *et al.*, 2005).

Moreover, the wide morphologic diversity of pericytes mirrors dynamic variability at the molecular level given that pericyte markers such as smooth muscle α-actin, desmin, NG-2, PDGF-R, aminopeptidase A and N, and RGS5 (Fig. 16.2) are also expressed in cells other than pericytes, and pericytes from differing vascular beds and developmental stages exhibit striking flexibility in their marker expression (Chan-Ling *et al.*, 2004; Gerhardt *et al.*, 2000; Hughes and Chan-Ling, 2004). Since pericytes exhibit diverse morphology and marker expression, careful consideration regarding this matter should be used when determining what organ and development stage from which to isolate pericytes. Successfully isolating pure cultures of pericytes depends on demonstrating the presence of a variety of characteristics along with the absence of the most likely contaminant cells such as ECs or pigmented epithelium. Also, unlike ECs, which express nonmuscle actin, or VSMC, which predominantly express muscle actin, pericytes express both forms of actin. Exclusion of likely cellular

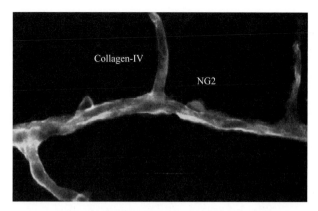

Figure 16.2   Pericyte-marker detection. Numerous markers have been used to identify pericytes, including smooth muscle α-actin, desmin, NG-2, PDGF-R, aminopeptidase A and N, and RGS5. Perhaps the most commonly referenced pericyte marker is NG2, which effectively differentiates between endothelial cells shown here stained for collagen IV.

contaminants can be determined by detecting the presence of von Willebrand factor expression and the ability to uptake acetylated low-density lipoproteins that are specific for ECs (Blann *et al.*, 1997; Voyta *et al.*, 1984), or the presence of black granules characterizing the pigmented epithelium of the retina.

Our laboratory has investigated the various roles of pericyte–EC interactions during the process of angiogenesis (Antonelli-Orlidge *et al.*, 1989; Darland and D'Amore, 2001; Darland *et al.*, 2003; Ding *et al.*, 2004; Herman and D'Amore, 1985; Hirshi *et al.*, 1998; Orlidge and D'Amore, 1987), and we describe in this chapter a number of techniques that can be used to examine the physiologic relationship between these cell types during vessel formation, maturation, and regression. We explain in detail a method to isolate and culture retinal pericytes, and, furthermore, illustrate how to successfully study pericyte–EC interactions with 2-dimensional and 3-dimensional coculture assays.

2. PROTOCOL FOR ISOLATING PERICYTES

Because highest density of pericytes to ECs is observed in the retina, we have routinely isolated primary pericytes from bovine retina because of its ease of dissection; however, the described techniques may be applied to the isolation of pericytes from the microvasculature of any tissue or organism. After the first week of isolation, irregularly shaped pericyte cells will be easily observed migrating from the capillary fragments and doubling approximately every 3 to 4 days; however, pericytes will never form a confluent monolayer (Fig. 16.3). Contaminant colonies of ECs will be noted during the first week

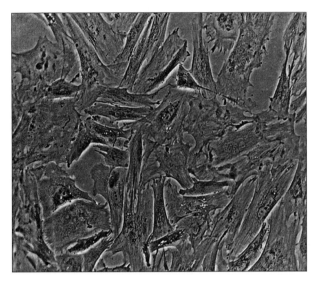

Figure 16.3 Bovine retinal microvessel pericytes. Pericytes isolated as described and grown for 3 weeks. The pericytes will never form a confluent monolayer, rather, after reaching this density, the cells will begin to form nodules that represent multiple layers of cells. (Figure reprinted courtesy of Springer-Verlag from D'Amore [1990], Cell Culture techniques in Heart and Vessel Research).

of primary culture; however, these colonies are characterized by classic endothelial morphology of polygonal, contacted cells that tend to grow as colonies. These endothelial colonies will not grow beyond approximately 100 cells and will lift from the plate within the first 2 weeks of culture, leaving a pure population of pericytes.

Because the plating efficiency of trypsinized pericytes is poor (approximately 50%), and pericytes require an extended trypsinization time to be detached from the flask (5 to 10 min), we generally only use the first passage of pericytes per experiment. Moreover, we are concerned that similar to VSMC whose phenotype has been shown to significantly change with increasing passage number (Orlidge and D'Amore, 1986), pericytes may dedifferentiate with passage.

2.1. Pericyte isolation and culture

1. Sterilely remove two to three entire eyes from bovine donors and incubate the tissue in a 20% Betadine-PBS solution for 10 min.
2. Wash two times in PBS.
3. Puncture the eye approximately 5 mm posterior to the limbus (the line of contact between the cornea and the sclera of the eye) and cut around the globe at this level. Remove the vitreous body. With a blunt probe,

gently separate the retina from the eyecup, adding a small amount of PBS if the tissue is dry and adherent. When the entire retina is free, cut it at its connection to the optic nerve and transfer to a 100-mm dish containing 10 ml of PBS.

4. Cut away the adherent pigment epithelium (black layers) and wash the retina thoroughly with multiple washes of PBS. Remove any residual pigmented epithelial cells (black areas) by teasing them away from the retinal tissue.

5. Mince the tissue with scalpel blades until the final tissue pieces should be small enough to fit into the bore of a 5-ml pipette.

6. Wash the minced tissue with PBS and gentle centrifugation (800g).

7. Suspend the tissue in 0.2% collagenase-BSA solution and shake at 40 rpm on a rotary shaker at room temperature for 1 h.

8. Pipette the solution against the side of the tube three times to free the capillary fragments that have been loosened by the enzyme treatment.

9. Pipette the solution onto Nitex mesh that is loosely stretched over the top of a funnel and secured to the sides with tape. Wash the tissue through the mesh with 20 ml of DMEM supplemented with 5% calf serum. The single cells and smaller vessel fragments (2 to 6 cells) will pass through the mesh, and larger vessel fragments and pieces of parenchyma and connective tissue will be retained in the mesh.

10. The tissue that has passed through is then pelleted by gentle centrifugation (800g)

11. Resuspend the pelleted cells in DMEM supplemented with 5% calf serum and 80 U/ml penicillin/streptomycin and wash twice in the same medium.

12. Plate the cells into one T-25 flask with DMEM supplemented with 10% calf serum and 80 U/ml penicillin/streptomycin.

13. The following day, wash the cells gently and refeed with the same medium.

14. Refeed every 3 days thereafter.

The pericytes do not achieve confluence, but will grow to a density that represents approximately 80% coverage of the flask within 3 to 4 weeks.

3. PROTOCOL FOR MULTIPOTENT MESENCHYMAL CELLS AS PERICYTE PRECURSORS

As with the use of most primary cell culture lines, there exists significant limitations in working with primary pericytes, such as difficult isolation procedures, limited cell survival over numerous passages, and low to nonexistent transfection efficiency. To overcome these limitations, our laboratory has demonstrated that C3H/10T$_{1/2}$ (10T$_{1/2}$) mesenchymal cells (ATCC #CCL-226) and primary ECs form capillary-like structures in both collagen and matrigel. In 3D cultures of endothelial cells, tubelike structures efficiently form within 18 to 24 h and subsequently regress thereafter (Darland and D'Amore, 2001); however, the heterotypic cell–cell interactions that are present when pericytes are cocultured with ECs leads to a measurable stabilization of vessels compared with monocultures of endothelial cells. *In vivo*, pericyte processes are seen to wrap around the length of the EC tube (Fujiwara and Uehara, 1984), and similar cell morphology is seen in $10T_{1/2}$ cells in 3D coculture assays. Moreover, the $10T_{1/2}$ cells in the 3D cocultures adopt a pericyte/VSMC phenotype on the basis of expression of differentiation-associated proteins such as smooth muscle α-actin and NG2 proteoglycan (Darland and D'Amore, 2001). Described in the following is a protocol for culturing and differentiating $10T_{1/2}$ cells as an alternative to the use of primary pericytes.

3.1. Culture and differentiation of $10T_{1/2}$ cells

1. Culture $10T_{1/2}$ cells in DMEM supplemented with 10% fetal calf serum, 233.6 μg/ml glutamine, 25 mM glucose, and 80 U/ml penicillin/streptomycin. Always maintain at subconfluent levels when passaging $10T_{1/2}$ cells.
2. Immediately before use in $10T_{1/2}$/EC coculture assays, allow $10T_{1/2}$ cells to reach 100% confluence and maintain at this level of confluence for 3 to 5 days.
3. Trypsinize $10T_{1/2}$ cells with 0.05% trypsin, 0.02% EDTA for 5 min and use in coculture assays as you would primary pericytes.

4. PROTOCOL FOR TWO-DIMENSIONAL PERICYTE-EC COCULTURE

Because pericytes exist in intimate association with ECs by forming a single layer that covers varying amounts of the abluminal EC surface, it is important to determine the contribution that pericytes make to EC function and whether this is due to direct contact between these two cells types or secreted paracrine factors. Although lacking the complex environment occurring *in vivo*, coculturing pericytes with ECs in a tissue culture dish allows for a rapid, though some may argue oversimplified, means to study the interactions between these two cell types with regard to growth factor–mediated signaling, migration, and proliferation. Indeed, the development of *in vitro* pericyte–endothelial coculture systems has been indispensable in dissecting and analyzing the cellular interactions and potential mediators involved in pericyte recruitment and differentiation. As described in detail in the following, we have routinely used two methods of 2D coculture: one in which contact is allowed between the cells and a second in which contact is prevented. These two methods allow differentiation between contact-dependent and contact-independent effects of pericytes on EC function and visa versa.

4.1. Coculture with direct contact

The most rapid, simple, and relatively inexpensive method to study pericyte-EC interactions with respect to basic biologic processes such as proliferation and migration is to use direct coculture of these cell types in the same tissue culture dish. For instance, with this system we have demonstrated that ECs can recruit undifferentiated mesenchymal cells and direct their differentiation into mural cell lineage (Hirshi *et al.*, 1998). Although advantageous in terms of simplicity, care must be taken to not overinterpret the results of 2D coculture. For example, pericytes and ECs grown in 2D coculture do not result in the formation of lumen-bearing vessels likely because the multifaceted physiologic environment that exists *in vivo* (i.e., presence of basement membrane) is lacking in this system, although complex cell–cell interactions can still be observed (Fig. 16.4).

1. Arrest pericyte cell division by treatment with mitomycin C (10 μg/ml) for 2 h.
2. Rinse the cells with PBS and remove them from the dish by trypsinization (0.05% trypsin, 0.02% EDTA).

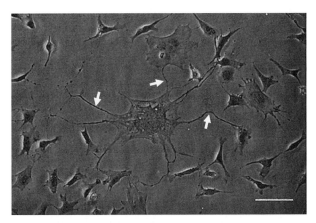

Figure 16.4 Pericyte interactions with ECs in culture. ECs and pericytes plated at a 10:1 ratio, illustrating a centrally located pericyte extending many processes (arrows) that contact multiple ECs. (White bar = 10 μm) (Figure reprinted courtesy of Rockefeller University Press from Orlidge and D'Amore [1987]. *J. Cell. Biol.*)

3. Plate pericytes at a density of 20,000 cells per well into 24-well tissue culture dishes in their respective growth media.
4. Allow cells to attach overnight.
5. Determine the number of pericytes that attached to the well and, into each well, add an equal number of ECs suspended in DMEM supplemented with 10% FCS and 80 U/ml penicillin/streptomycin.
6. Incubate cultures for 48 h (or as necessary) and assay as desired.

Controls for this assay consist of cell division–arrested pericytes and ECs cultured alone under identical conditions.

4.2. Coculture without direct cell contact by use of transwells

To determine the role of cell contact or proximity on pericyte–EC interactions, cocultures can also be grown without direct contact with transwells. In this system, cells are cocultured in the same well, but prevented from physical interaction by growing ECs in the transwell inserts 1 to 2 mm above the layer of growth arrested pericytes.

1. Arrest pericyte cell division by treatment with mitomycin C (10 μg/ml) for 2 h.
2. Rinse the cells with PBS and remove them from the dish by trypsinization (0.05% trypsin, 0.02% EDTA).
3. Plate pericytes at a density of 20,000 cells per well into 24-well tissue culture dishes in their respective growth media.
4. Determine the number of pericytes that attached to the well and add an equal number of ECs into the transwells (Corning, Inc.).
5. After EC attachment, place transwell inserts containing ECs into the wells containing the cell division-arrested pericytes.
6. Add 1 ml of DMEM supplemented with 10% fetal calf serum and 80 U/ml penicillin/streptomycin.
7. Incubate cultures for 48 h (or as necessary) and assay as desired.

Controls for this assay consist of cell division–arrested pericytes and ECs cultured alone under identical conditions.

4.3. Coculture without direct cell contact by use of the under-agarose assay

One additional method for pericyte–EC coculture is the under-agarose assay (Fig. 16.5), which has been used in our laboratory to examine the effects of ECs on both pericyte chemotaxis and spontaneous migration (Hirshi et al., 1998). This procedure makes use of a tissue culture dish filled with an agarose

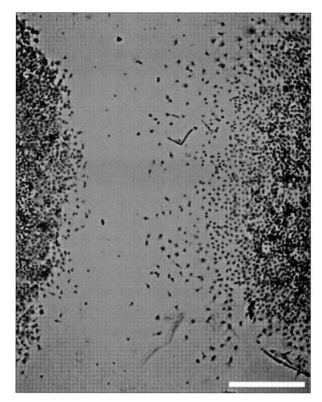

Figure 16.5 Under-agarose migration assay. Bovine aortic endothelial cell (BAEs) and pericytes were cocultured in an under-agarose assay for 48 h, and then fixed and stained with Coomassie blue. Pericytes (right) migrated directionally toward the BAEs (left). (White bar = 800 μm) (Figure reprinted courtesy of Rockefeller University Press from Hirschi *et al.* [1998]. *J. Cell Biol.*)

mixture containing two wells physically spaced 2.4 mm apart—one containing pericytes and the other ECs. Chemoattractants diffuse from the wells in the agarose to form a gradient. Cells in nearby wells can be monitored as they migrate in the direction of the chemoattractant source.

1. With a heating block, dissolve a 2% agarose solution in sterile distilled water.
2. Cool to 48 °C in H2O bath.
3. Mix the agarose solution with an equal volume of prewarmed 2 × DMEM supplemented with 2% calf serum and 80 U/ml penicillin/streptomycin.
4. Pipette 5 ml of the agarose medium into a 60- × 15-mm tissue culture dish and allow hardening until firm at 4 °C for 30 min.

5. With a plexiglas template and stainless steel punch, cut two agarose wells 2.4 mm in diameter and spaced 2.4 mm apart. Take care not to scratch the plastic underneath, because this will inhibit cell migration.

6. Remove the agarose plugs from the wells with a needle. Do not use vacuum suction, because this will result in distortion of the well.

7. Trypsinize pericytes and ECs with 0.05% trypsin, 0.02% EDTA for 5 min.

8. Centrifuge trypsinized pericytes and ECs in separate centrifuge tubes at low speed (800g) for 3 min.

9. Remove the supernatant from each tube and resuspend the cells at 8 × 10^5 cells/ml in DMEM supplemented with 2% FCS and 80 U/ml penicillin/streptomycin.

10. Plate 25 μl of ECs (2 × 10^4 cells/well) into one well, and plate 25 μl of pericytes (2 × 10^4 cells/well) into the adjacent well.

11. Incubate for 48 h.

12. Fix with 2% paraformaldehyde in PBS for 30 min.

13. Stain with 0.1% Coomassie blue for 30 min.

14. View with inverted microscope.

As a control, the distance migrated of ECs cultured alone or pericytes cultured alone is subtracted from the distances migrated by each cell type under the experimental conditions.

5. PROTOCOL FOR 3-DIMENSIONAL PERICYTE-EC COCULTURE

Although the 2D coculture system is a highly simplified model to study pericyte–EC interactions with respect to proliferation and migration, great improvements in our understanding of EC–pericyte interactions have been achieved through the use of 3D coculture systems. Because ECs and pericytes normally reside and interact within a 3D environment, the added presence of a basement membrane matrix, such as collagen or Matrigel, more closely resembles the *in vivo* physiologic environment in that vessels are formed such that EC–pericyte cocultures performed in a 3D matrix results in the formation of lumen-bearing vessels (Fig. 16.6). In the following, we describe the 3D collagen sandwich coculture method used by our laboratory that we have found to be highly reproducible and effective for the study of pericyte–EC interactions during tube formation *in vitro*. Moreover, solely collagen use as a basement membrane allows for precisely defined conditions, whereas the components of Matrigel (an extracellular matrix secreted by the Engelbrecht–Holm–Swarm mouse sarcoma cell line) are largely unknown. However, the assay works efficiently with either basement membrane.

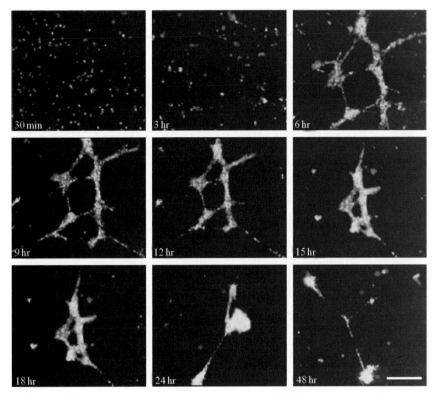

Figure 16.6   Cord formation in 3D cocultures. Pericytes and ECs were labeled with fluorescent vital dyes and cocultured in a 3D sandwich for 48 h. Photographs were taken every 3 h over the course of 48 h. The cells migrated toward one another, forming the primary cord lattice by 6 h. Cords were present until 18 to 20 h after plating and then condensed to form large aggregates, connected by thin filamentous projections. (White bar = 300 μm) (Figure reprinted courtesy of Springer-Verlag Publishers from Darland and D'Amore [2001]. Angiogenesis.)

5.1. 3D Coculture in a collagen sandwich

1. On ice, prepare collagen gels by mixing eight parts PureCol (Inamed Biomaterials) with one part 10× DMEM and one part 0.1 M NaOH.
2. Add 175 μl-aliquots of the collagen mixture to each well of a 48-well plate and allow to polymerize for 45 min at 37 °C (5% CO_2).
3. Add 1× 10^6 cells per ml of endothelial cells and 1× 10^6 cells per ml of pericytes per well containing the polymerized collagen matrix. Allow the cells to adhere to the collagen matrix.
4. On ice, prepare fresh collagen as described above and add 75 μl of the collagen mixture into each well. Allow to solidify for 45 min.

5. Add 250 μl endothelial cell basal medium (EBM) supplemented with 10% horse serum (HS) and 80 U/ml penicillin/streptomycin to the top of the collagen sandwich.
6. Incubate at 37 °C. Tube formation will occur within 24 to 48 h.

6. PROTOCOL FOR DIFFERENTIAL LABELING OF PERICYTES AND ENDOTHELIAL CELLS

One major obstruction in the use of direct pericyte–EC cocultures is identification of each cell type in a mixed culture. Overexpression of GFP and its color variants is a classical method to distinguish between two cell types in coculture; however, most primary ECs and pericytes fail to achieve any practical level of transfection efficiency. To overcome this limitation in working with primary endothelial and pericyte cultures, our laboratory uses several commercially available fluorescent cell tracking dyes whose half-life in nondividing cells reaches approximately 10 to 100 days, depending on the particular dye. PKH67 and PKH26 (Sigma), which fluoresce green and red, respectively, are cell membrane–labeling dyes with almost undetectable cell-to-cell transfer and provide a robust differential labeling of endothelial cells and pericytes (Fig. 16.7).

6.1. PKH67/PKH26 fluorescent labeling of endothelial cells and pericytes

1. Trypsinize pericytes and endothelial cells with 0.05% trypsin, 0.02% EDTA for 5 min.
2. Resuspend cells at a density of 2×10^6 cells/100 μl.
3. Add an equal volume of 40 mM PKH dye (PKH67 for ECs; PKH26 for pericytes) and mix with gentle swirling.
4. Incubate cell mixtures for 3 to 5 min at room temperature.
5. Terminate dye uptake with the addition of 7 ml PBS supplemented with 200 μl FCS.
6. Centrifuge cells at 800g over a 3-ml bed of FCS and further wash with 10 ml of DMEM supplemented with 2% FCS and 80 U/ml penicillin/streptomycin.
7. To allow for full recovery from the labeling procedure, plate cells for 24 h before use in coculture.
8. After completion of the coculture experiment, view cellular labeling with fluorescent microscopy. The excitation and emission spectra for PKH26 are 551 nm and 567 nm, and for PKH67 are 490 nm and 502 nm, respectively.

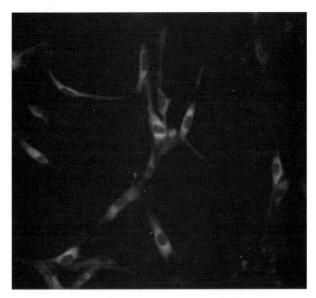

Figure 16.7 Differential labeling of pericytes and ECs. PKH67 and PKH26 are cell membrane labeling dyes with almost undetectable cell-to-cell transfer and provide a robust differential labeling of endothelial cells and pericytes.

ACKNOWLEDGMENTS

Support for this study was provided by the National Institutes of Health grants EY05318, EY015435, and CA45548. P. A. D. is a Research to Prevent Blindness Senior Scientific Investigator. We would like to thank Dr. Tony Walshe and Dr. Magali Saint-Geniez of the Schepens Eye Research Institute, Boston, MA, for their contributions of images to this manuscript.

REFERENCES

Allt, G., and Lawrenson, J. G. (2001). *Cells Tissues Organs* **169,** 1–11.

Antonelli-Orlidge, A., Saunders, K. B., Smith, S. R., and D'Amore, P. A. (1989). *Proc. Natl. Acad. Sci. USA* **86,** 4544–4548.

Antonelli–Orlidge, A., Smith, S. R., and D'Amore, P. A. (1989). *Am. Rev. Respir. Dis.* **140,** 1129–1131.

Armulik, A., Abramsson, A., and Betsholtz, C. (2005). *Circ. Res.* **97,** 512–523.

Arthur, H. M., Ure, J., Smith, A. J., Renforth, G., Wilson, D. I., Torsney, E., Charlton, R., Parums, D. V., Jowett, T., Marchuk, D. A., Burn, J., and Diamond, A. G. (2000). *Dev. Biol.* **217,** 42–53.

Balabanov, R., and Dore-Duffy, P. (1998). *J. Neurosci. Res.* **53,** 637–644.

Betsholtz, C. (1994). *Genes Dev.* **8,** 1875–1887.

Betsholtz, C. (2004). *Cytokine Growth Factor Rev.* **15,** 215–228.

Blann, A. D., Seigneur, M., Steiner, M., Boisseau, M. R., and McCollum, C. N. (1997). *Eur. J. Clin. Invest.* **27**, 916–921.

Carvalho, R. L., Jonker, L., Goumans, M. J., Larsson, J., Bouwman, P., Karlsson, S., Dijke, P. T., Arthur, H. M., and Mummery, C. L. (2004). *Development* **131**, 6237–6247.

Chang, H., Huylebroeck, D., Verschueren, K., Guo, Q., Matzuk, M. M., and Zwijsen, A. (1999). *Development* **126**, 1631–1642.

Chan-Ling, T., Page, M. P., Gardiner, T., Baxter, L., Rosinova, E., and Hughes, S. (2004). *Am. J. Pathol.* **165**, 1301–1313.

Darland, D. C., and D'Amore, P. A. (2001). *Angiogenesis* **4**, 11–20.

Darland, D. C., Massingham, L. J., Smith, S. R., Piek, E., Saint-Geniez, M., and D'Amore, P. A. (2003). *Dev. Biol.* **264**, 275–288.

Davis, S., Aldrich, T. H., Jones, P. F., Acheson, A., Compton, D. L., Jain, V., Ryan, T. E., Bruno, J., Radziejewski, C., Maisonpierre, P. C., and Yancopoulos, G. D. (1996). *Cell* **87**, 1161–1169.

Dickson, M. C., Martin, J. S., Cousins, F. M., Kulkarni, A. B., Karlsson, S., and Akhurst, R. J. (1995). *Development* **121**, 1845–1854.

Ding, R., Darland, D. C., Parmacek, M. S., and D'Amore, P. A. (2004). *Stem Cells Dev.* **13**, 509–520.

Fujiwara, T., and Uehara, Y. (1984). *Am. J. Anat.* **170**, 39–54.

Gerhardt, H., Wolburg, H., and Redies, C. (2000). *Dev. Dyn.* **218**, 472–479.

Goumans, M. J., Valdimarsdottir, G., Itoh, S., Rosendahl, A., Sideras, P., and ten Dijke, P. (2002). *EMBO J.* **21**, 1743–1753.

Hellstrom, M., Kalen, M., Lindahl, P., Abramsson, A., and Betsholtz, C. (1999). *Development* **126**, 3047–3055.

Hirschi, K. K., Rohovsky, S. A., and D'Amore, P. A. (1998). *J. Cell Biol.* **141**, 805–814.

Hoch, R. V., and Soriano, P. (2003). *Development* **130**, 4769–4784.

Hughes, S., and Chan-Ling, T. (2004). *Invest. Ophthalmol. Vis. Sci.* **45**, 2795–2806.

Herman, I. M., and D'Amore, P. A. (1985). *J. Cell Biol.* **101**, 43–52.

Kono, M., Mi, Y., Liu, Y., Sasaki, T., Allende, M. L., Wu, Y. P., Yamashita, T., and Proia, R. L. (2004). *J. Biol. Chem.* **279**, 29367–29373.

Larsson, J., Goumans, M. J., Sjostrand, L. J., van Rooijen, M. A., Ward, D., Leveen, P., Xu, X., ten Dijke, P., Mummery, C. L., and Karlsson, S. (2001). *EMBO J.* **20**, 1663–1673.

Lebrin, F., Goumans, M. J., Jonker, L., Carvalho, R. L., Valdimarsdottir, G., Thorikay, M., Mummery, C., Arthur, H. M., and ten Dijke, P. (2004). *EMBO J.* **23**, 4018–4028.

Leveen, P., Pekny, M., Gebre-Medhin, S., Swolin, B., Larsson, E., Lindahl, P., Johansson, B. R., Leveen, P., and Betsholtz, C. (1997). *Science* **277**, 242–245.

Li, D. Y., Sorensen, L. K., Brooke, B. S., Urness, L. D., Davis, E. C., Taylor, D. G., Boak, B. B., and Wendel, D. P. (1999). *Science* **284**, 1534–1537.

Liu, Y., Wada, R., Yamashita, T., Mi, Y., Deng, C. X., Hobson, J. P., Rosenfeldt, H. M., Nava, V. E., Chae, S. S., Lee, M. J., Liu, C. H., Hla, T., Spiegel, S., and Proia, R. L. (2000). *J. Clin. Invest.* **106**, 951–961.

Maisonpierre, P. C., Suri, C., Jones, P. F., Bartunkova, S., Wiegand, S. J., Radziejewski, C., Compton, D., McClain, J., Aldrich, T. H., Papadopoulos, N., Daly, T. J., Davis, S., Sato, T. N., and Yancopoulos, G. D. (1997). *Science* **277**, 55–60.

Oh, S. P., Seki, T., Goss, K. A., Imamura, T., Yi, Y., Donahoe, P. K., Li, L., Miyazono, K., ten Dijke, P., Kim, S., and Li, E. (2000). *Proc. Natl. Acad. Sci. USA* **97**, 2626–2631.

Orlidge, A., and D'Amore, P. A. (1987). *J. Cell Biol.* **105**, 1455–1462.

Orlidge, A., and D'Amore, P. A. (1986). *Microvasc. Res.* **31**, 41–53.

Oshima, M., Oshima, H., and Taketo, M. M. (1996). *Dev. Biol.* **179**, 297–302.

Sato, T. N., Tozawa, Y., Deutsch, U., Wolburg-Buchholz, K., Fujiwara, Y., Gendron-Maguire, M., Gridley, T., Wolburg, H., Risau, W., and Qin, Y. (1995). *Nature* **376,** 70–74.

Shepro, D., and Morel, N. M. (1993). *FASEB J.* **7,** 1031–1038.

Soriano, P. (1994). *Genes Dev.* **8,** 1888–1896.

Sundberg, C., Kowanetz, M., Brown, L. F., Detmar, M., and Dvorak, H. F. (2002). *Lab. Invest.* **82,** 387–401.

Suri, C., Jones, P. F., Patan, S., Bartunkova, S., Maisonpierre, P. C., Davis, S., Sato, T. N., and Yancopoulos, G. D. (1996). *Cell* **87,** 1171–1180.

Suri, C., McClain, J., Thurston, G., McDonald, D. M., Zhou, H., Oldmixon, E. H., Sato, T. N., and Yancopoulos, G. D. (1998). *Science* **282,** 468–471.

Thurston, G., Suri, C., Smith, K., McClain, J., Sato, T. N., Yancopoulos, G. D., and McDonald, D. M. (1999). *Science* **286,** 2511–2514.

Torsney, E., Charlton, R., Diamond, A. G., Burn, J., Soames, J. V., and Arthur, H. M. (2003). *Circulation* **107,** 1653–1657.

Uemura, A., Ogawa, M., Hirashima, M., Fujiwara, T., Koyama, S., Takagi, H., Honda, Y., Wiegand, S. J., Yancopoulos, G. D., and Nishikawa, S. (2002). *J. Clin. Invest.6* **110,** 1619–1628.

Urness, L. D., Sorensen, L. K., and Li, D. Y. (2000). *Nat. Genet.* **26,** 328–331.

Voyta, J. C., Via, D. P., Butterfield, C. E., and Zetter, B. R. (1984). *J. Cell Biol.* **99,** 2034–2040.

Witzenbichler, B., Maisonpierre, P. C., Jones, P., Yancopoulos, G. D., and Isner, J. M. (1998). *J. Biol. Chem.* **273,** 18514–18521.

Yang, X., Castilla, L. H., Xu, X., Li, C., Gotay, J., Weinstein, M., Liu, P. P., and Deng, C. X. (1999). *Development* **126,** 1571–1580.

Author Index

Subject Index

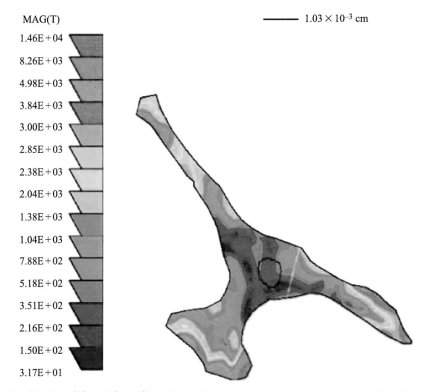

MAG(T)

1.46E+04	
8.26E+03	
4.98E+03	
3.84E+03	
3.00E+03	
2.85E+03	
2.38E+03	
2.04E+03	
1.38E+03	
1.04E+03	
7.88E+02	
5.18E+02	
3.51E+02	
2.16E+02	
1.50E+02	
3.17E+01	

1.03×10^{-3} cm

Cynthia A. Reinhart-King, Figure 3.1 Color contour map depicting the traction forces exerted by a bovine aortic endothelial cell on a deformable polyacrylamide substrate derivatized with an RGD-containing peptide. The image was obtained by use of the LITRC traction algorithm written by Micah Dembo at Boston University; he is also the inventor of the basic theory that underlies traction force microscopy.

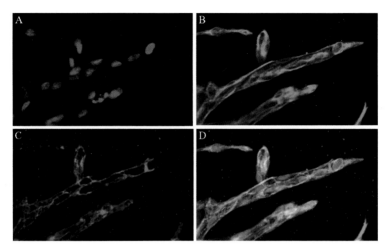

Martin N. Nakatsu and Christopher C. W. Hughes, Figure 4.2 Adherens junctions connect the EC in a sprout. Fibrin gel cultures were established on chamber slides for 10 days. Fibroblasts were then removed and the gels fixed in paraformaldehyde overnight at 4 °C before staining. (A) Cell nuclei are revealed by DAPI staining. (B) Intermediate filaments containing vimentin are apparent in both tip and trunk cells. (C) Staining for β-catenin reveals adherens junctions between individual EC, including between the tip cell and the first trailing trunk cell. The presence of adherens junctions in the tip cell suggests that an epithelial–mesenchymal transition (EMT) is not occurring. (D) Merged images show cell nuclei (blue), vimentin (green), and adherens junctions (red).

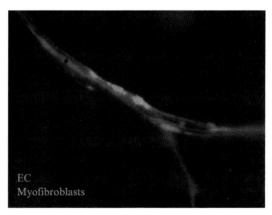

Martin N. Nakatsu and Christopher C. W. Hughes, Figure 4.3 Myofibroblast-endothelial interactions can be modeled. EC were transduced with a pBN-mCherry retroviral vector, and myofibroblasts were transduced with a pBN-GFP retroviral vector. Transduced EC were coated on beads and embedded in fibrin gels through which were distributed transduced myofibroblasts. Over several days, myofibroblasts (green) migrate toward and wrap around EC sprouts (red), appearing to behave as pericytes and smooth muscle cells do *in vivo*. Photograph courtesy of Xiaofang Chen and Steve George, UC Irvine.